Introductory Mathematics fo

Introductory Mathematics for Economics and Business

Second Edition

K. Holden and A. W. Pearson

M

First published 1975 by
David & Charles (Holdings) Ltd

Published as *Introductory Mathematics for Economists*
1983 by The Macmillan Press Ltd
Reprinted 1986, 1988, 1990

This edition published 1992 by
THE MACMILLAN PRESS LTD
Houndmills, Basingstoke, Hampshire RG21 2XS
and London
Companies and representatives
throughout the world

ISBN 0–333–57649–7 hardcover
ISBN 0–333–57650–0 paperback

A catalogue record for this book is available
from the British Library

Printed in Hong Kong

Contents

Preface

This book introduces to students with a limited mathematics background the essential mathematics needed for economics and business courses. It is not intended to replace more formal mathematics texts and, by design, does not include proofs and derivations of all the formulae used; these are included only where they aid understanding. We hope that this approach will be suitable for those students who, as a result of their earlier experiences in this subject area, do not regard themselves as having any mathematical ability. While reading the book they are advised to check the algebra by jotting the working down on paper and to use a calculator to check the numbers. Many examples are included in the text and the exercises (with worked answers given in Appendix C) are intended to help such students. At the end of each chapter there are also some revision exercises (without answers).

The material is selected so as to increase in difficulty as the book progresses. Chapter 1 revises many basic concepts in algebra and introduces linear equations, with immediate applications in simple economic models of markets and the national economy. Chapters 2 and 3 are the natural generalisations of elementary matrix algebra and non-linear equations. Chapter 4, on series, covers many applications in finance as well as providing the groundwork for calculus in Chapters 5–7. Chapter 5, on differential calculus, includes profit maximisation for a firm, simple inventory models and other applications of marginal concepts. Integration, the subject of Chapter 6, covering both standard analytical techniques and numerical methods, is also applied to simple marginal concepts, including consumer's and producer's surplus. Relationships with many variables are examined in Chapter 7, on partial differentiation, which ends with maximisation subject to a constraint. A variation on this problem is treated in Chapter 8, on linear programming. Chapters 9–10 consider dynamic relationships – in continuous terms in Chapter 9 and in discrete terms in Chapter 10. There is an extensive treatment of trigonometric functions in Appendix A, and an introduction to set theory in Appendix B. Each of these can be treated as optional, with little loss of the main text, except for parts of Chapters 9 and 10 which need trigonometric functions. Detailed numerical answers to the exercises are provided in Appendix C.

The order in which the material is covered can be varied. For example, integral calculus may be followed by differential equations, or matrix algebra may be deferred to precede linear programming.

We have intentionally excluded any coverage of statistics, since we believe that any brief treatment would be inadequate for most courses. We have also ignored computer programming because the wide variety of languages and packages now available mean that any general discussion is unhelpful to students and teachers. The omission of these topics should not be taken as evidence that we believe they are unimportant.

We are grateful to Roger Latham (Queen's University, Ontario, Canada) and David Peel (University College, Aberystwyth), formerly of the University of Liverpool, for reading through an early draft of the manuscript. Our thanks are particularly due to John Thompson (Liverpool Polytechnic) who made detailed comments on the whole manuscript which helped us to remove many errors. Peter Stoney, Tim Worrall and Ken Cleaver, of the University of Liverpool, kindly read particular sections and made helpful comments which improved the manuscript. Since none of these saw the final manuscript then, as always, the responsibility for any errors lies with the authors. Thanks are also due to Simon Blackman, for help with the graphs and Jenny Holden, for valuable assistance with preparation of the manuscript and the index.

This book is a much revised and expanded version of *Introductory Mathematics for Economists* (Macmillan, 1983) which was originally published by David and Charles (Holdings) Ltd in 1973.

Finally, a word of warning. This book is introductory and intentionally uses simple models. These are not meant to represent the real world and many of the numbers chosen – for example, in supply and demand analysis – and the assumptions imposed are arbitrary and for illustrative purposes only. The world is complicated and it is only by simplifying it that we can hope to gain an understanding of it. But once we do master simple models, more complex and realistic models can be considered.

K.H.
A.W.P

Chapter 1
Linear Equations

1.1 Some preliminaries

Since the purpose of this book is to help those students who have a weak background in mathematics to understand applications in economics and business studies, it is useful to start by reviewing some basic concepts. One fundamental idea is that of a *number*. The type of number that is familiar in everyday life is known as a *real number*. That is, it can be used to represent something 'real', such as 2 and 3 representing the numbers of people in two cars. In this case the numbers are positive whole numbers or *positive integers*, and are part of the familiar sequence 1, 2, 3, 4, . . . where the three dots indicate 'and so on'. The positive integers increase without limit and are said to be *unbounded*. They are used in counting. If the sequence is continued in the negative direction the values are 0, −1, −2, −3, . . . Below we will see that the value zero is rather special in that it does not satisfy some of the standard rules of algebra, and can be interpreted as being positive or negative or neither. The numbers −1, −2, −3, . . . are the *negative integers* and while their interpretation is less obvious than for positive integers, they are familiar for measuring sub-zero temperatures and in economics they can represent debts (negative assets) or losses (negative profits).

While the integers are a useful starting point in discussing real numbers, another way of looking at real numbers is to classify them as rational and irrational numbers. *Rational numbers* are defined as being any numbers which can be expressed as the ratio of two integers, where it is assumed that the second integer is not zero. Since any integer n, say, can be expressed as $n/1$ then all the integers are rational numbers. Also, fractions, such as 2/5 and −5/6 are rational numbers. Some fractions can be represented precisely as decimals, such as 1/2 = 0.5000, while others are non-terminating, repeating decimals, such as 1/3 = 0.333333 . . . and 2/11 = 0.181818 . . . *Irrational numbers* are

1

numbers which cannot be expressed as the ratios of two integers and are non-terminating, non-repeating decimals such as $\sqrt{2} = 1.414213\ldots$ and the value $\pi = 3.141592\ldots$

To sum up, real numbers can be sub-divided into rational and irrational numbers. Rational numbers can also be sub-divided into fractions and integers. In elementary applications of mathematics in economics and business studies only real numbers are needed. However, in section 3.3 below a new type of number, known as an *imaginary number*, is introduced and, when combined with a real number, gives a *complex number*.

The values of the real numbers are fixed or constant. This contrasts with a *variable* which is anything that can take on different values. An example of a variable is the price of apples since it can change as time passes or take different values at any particular time in different places. Other examples of variables in economics are profits, quantities demanded, the level of inventories, consumption, income, imports and exports. Variables can be *discrete*, that is they can only take on a limited number of values between any given two values, or *continuous*. Examples of discrete variables are the number of students in a class and the amount of cash (measured in pounds or dollars) that is in one's hand. Notice that, as in this latter example, discrete variables need not be integers. Continuous variables can, in principle, take on any value between any two given values. For example, if the weights of two bags of apples are 2.10kg and 2.11kg, it is possible to find another bag of apples with a weight anywhere between these two, such as 2.10419345kg.

In applying mathematics in economic and business studies it is frequently convenient to represent variables by symbols, so that price might be represented by p, quantity sold by q, revenue by r and total cost by TC. This makes it easier to see how the different variables are related.

For example, it is frequently the case that the total cost (TC) of producing an output, q, of goods is related to the level of q. That is, TC depends on q or

$$TC = f(q)$$

which is read 'TC is a function of q'. The symbol f, together with the brackets, is a shorthand way of saying that the two variables TC and q are related in some way.

The statement that one variable depends on another – in this example that TC is related to q – may not appear to be very useful, but in fact it can be the starting point in many applications of economic theory and management science to practical problems. The equation is a statement of the *behavioural relationship* linking the two variables. Another form of equation is an *identity*, which is a definition or statement that two variables or expressions are always equivalent, such as total revenue, *TR*, is price, *p*, times the quantity sold, *q*. The third form of equation which is common in economics is the *equilibrium condition* in which two or more variables are set equal to one another, as in supply and demand analysis where, in equilibrium, the quantity demanded is set equal to the quantity supplied.

Behavioural equations can arise either from introspection (as do many ideas in economics and management science) or from observing what happens in the real world. In the relationship between total cost and the level of output, *TC* is called the *dependent variable* and *q* is the single *independent variable*, and *TC* depends on, or is determined by, *q*. In stating this a concept of *causality* is being introduced: a change in *q* causes a change in *TC*. A further distinction which is important in economic modelling is that between *endogenous variables*, whose value is determined or explained by the model being considered, and *exogenous variables*, whose value is determined outside the model. Whether a variable is endogenous or exogenous is determined by the circumstances of the model. For example, a particular fruit-grower may believe that the price of apples is determined by the world market and so is exogenous for each individual producer, while for a fruit-growers' federation, their total production may affect the world price, making the price endogenous.

The precise form of the relationship between total cost and output has not been specified but it might be assumed that for each value of *q* there is only one value of *TC*, so that the relationship is said to be *single valued*. Also, it might be expected that as the level of output, *q*, increases, *TC* increases.

Continuing with the example, to make progress some assumptions have to be made about the form of the relationship (and they would need to be checked in any particular practical situation). Here we might assume:

(a) Some parts of the total cost of production are necessarily incurred if there is to be any production and they do not increase as production is increased. These are known as *fixed costs* and include such items as rent, rates, and wages of the labour force which within certain limits are not affected by changes in the level of output.

(b) There are other parts of the total cost of production which increase as the level of output increases. These are known as *variable costs* and include such items as raw materials, power, and parts of the labour force which can be employed on the process if and when required.

It is clear that in many practical situations the split between fixed and variable costs cannot be made very clearly and that in the long run all costs tend to be variable. However, a simple breakdown into these two categories can prove to be useful in establishing such points as the breakeven level of production, as we shall see later. But first, let us consider an example in which the available information about the production process indicates that the fixed costs amount to £100 and that the variable costs associated with manufacturing 100 units of the product amount to £300. We will assume that the variable costs are directly proportional to the number of units of output and hence that the total cost varies linearly with the level of output.

The relationship between costs and output can then be written

Total cost = fixed cost + (variable cost per unit) × (level of output)

or $$TC = a + b \times q$$

where a and b are two constants which represent fixed cost and variable cost per unit respectively.

1.2 Graphical representation

For a two-variable relationship a common method of presentation of information is by means of a *graph*. This is a two-dimensional diagram with two *axes*, one of which represents TC (in our example the total cost of production) and the other q (the level of output). These axes are generally drawn at right angles to each other and their point of intersection, the origin, O, is where both total costs and the level of output are zero. By choosing suitable scales to represent different values of the variables we can construct a graph from any given set of data.

In our example in Fig. 1.1, when the level of output is 0, the fixed costs and hence the total costs are equal to £100, i.e. when $q = 0$, $TC = 100$. Also, when the level of output is 100, the total costs are £400, i.e. when $q = 100$, $TC = 400$.

Since we are assuming that total cost increases uniformly and continuously with output we can join the two points by a straight line. Hence $TC = a + bq$ is an equation of a straight line, and in this case

a = fixed costs = £100 and

b = variable cost per unit of output = £3

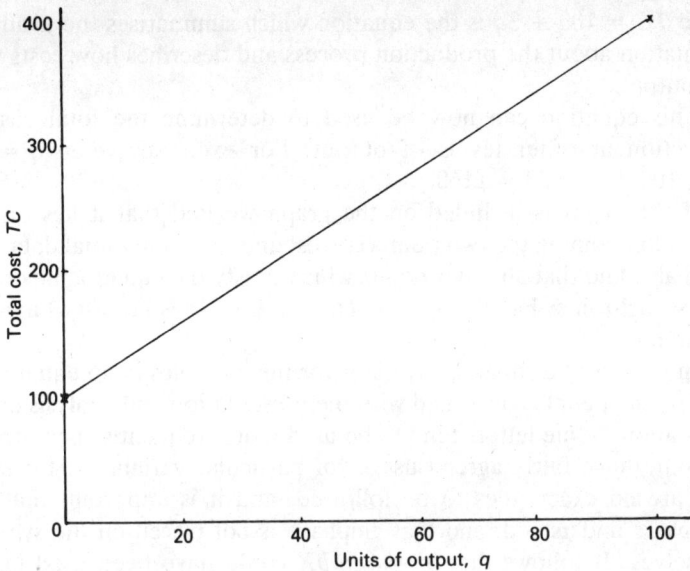

Fig. 1.1

Effectively we are determining the values of a and b by using the information about the two points to solve the equation $TC = a + bq$.

This information enables us to locate two points on the graph with the *coordinates* (0, 100) and (100, 400), where the two numbers in brackets refer to the values of the variables measured along the horizontal and the vertical axes respectively. In general, any point on this graph (Fig. 1.1) is represented by the coordinates (q, TC).

For the first point: $q = 0$, $TC = 100$ and so

$$100 = a + b \times 0$$

$$a = 100$$

and for the second point $q = 100$, $TC = 400$ and so

$$400 = a + 100b$$

Substituting the value $a = 100$ in this equation gives

$$400 = 100 + 100b$$

Subtracting 100 from each side of this equation leaves us with

$$300 = 100b \quad \text{and so} \quad b = 3$$

Hence $TC = 100 + 3q$ is the equation which summarises the available information about the production process and describes how costs vary with output.

This equation can now be used to determine the total cost of production at other levels of output. For example, when $q = 20$, $TC = 100 + 3 \times 20 = £160$.

If this point is included on the graph we find that it lies on the straight line joining the two points corresponding to the initial data. We would also find that all other points which satisfy the equation lie on the same straight line. For this reason $TC = 100 + 3q$ is known as a *linear equation*.

In general the choice of symbols for the variables in an equation is made by the people concerned with the presentation and analysis of the information. Some letters tend to be used more frequently than others, and some have fairly agreed usage for particular variables. However, there are no exact rules to be followed, and it is important that the principle is understood, and that emphasis is not placed on the symbols themselves. It follows that $Y = a + bX$ could have been used in the example we have just considered, with Y replacing TC and X replacing q. The form of the equation is unaltered. In the following sections a number of different letters will be used to represent the variables in an equation. This is not intended to confuse, but to assist in understanding that the form of an equation does not depend upon the symbols which represent the variables. However, in all practical cases a clear indication should be given of what the symbols do represent.

It is possible to determine the form of a linear equation from two pairs of values of the variables. This follows graphically because only one straight line can be drawn through any two given points and algebraically because there are two constants to be determined and two pairs of values provide two equations in a and b which can, in general, be solved.

1.3   Intercept and slope

In the equation $Y = a + bX$, a is often referred to as the *intercept term* because it is the value at which the straight line intercepts the vertical axis, i.e. it is the value of the function when $X = 0$, and b is often referred to as the *slope* or gradient because it is a measure of the inclination of the line to the horizontal axis.

We can now make use of such an equation in a number of ways. For example we can examine the effect which would be produced by an alteration in a and b, e.g. in fixed and variable costs, as follows:

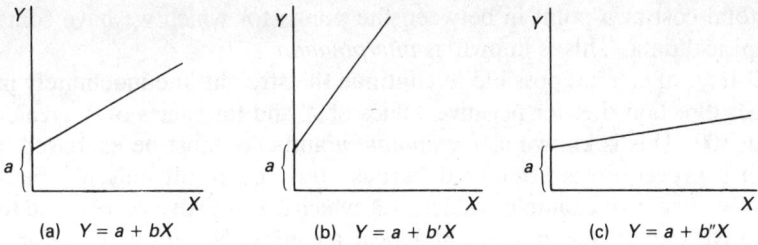

(a) $Y = a + bX$ (b) $Y = a + b'X$ (c) $Y = a + b''X$

Fig. 1.2

1. Fixed costs remain the same but variable cost changes. In Figure 1.2 a is the same in all three cases, but in graph (b) the variable cost is greater than in graph (a) and in graph (c) variable cost is less than graph (a).

That is, b' is greater than b (written $b' > b$)

and b'' is less than b (written $b'' < b$)

The line will then rotate about the point at which it crosses the cost axis.

2. Variable cost remains the same but the fixed costs change. The line will then move up or down parallel to itself, i.e. with the same slope. In Fig. 1.3 b is constant in all three cases, but graph (b) shows higher fixed costs than graph (a) and graph (c) shows lower fixed costs than graph (a). That is,

$$a' > a \quad \text{and} \quad a'' < a.$$

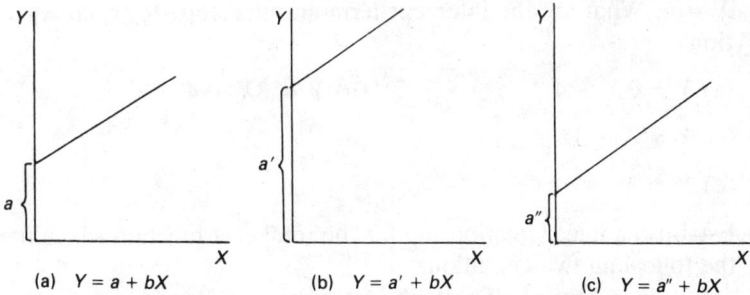

(a) $Y = a + bX$ (b) $Y = a' + bX$ (c) $Y = a'' + bX$

Fig. 1.3

1.4 Interpolation and extrapolation

We have seen that a relationship between two variables can be shown graphically, or alternatively it can be represented by a mathematical expression. This equation can then be used to determine the value of total cost at a point in between the points for which we have some empirical data. This is known as *interpolation*.

It is, of course, possible to continue the straight line indefinitely in either direction, i.e. for negative values of X and for values of X greater than 100. This is known as *extrapolation* and care must be exercised if such a procedure is attempted because the end result may not be a sensible one. For example, values of X which are negative correspond to negative output and have no practical meaning. We must, therefore, eliminate such values by specifying that X must be greater than or equal to zero.

But what about values of X greater than 100? These may or may not be possible, and we have to be careful when we are considering extrapolation that we are certain the fixed and variable costs will remain the same when more than 100 units are produced. If they do not, then we must add the condition that X cannot be greater than 100, or alternatively, X must be less than or equal to 100.

The cost function can then be written

$$Y = 100 + 3X \qquad (0 \leqslant X \leqslant 100)$$

where the constraint, written inside the brackets, implies that the cost function is only a valid representation of the process for outputs from 0 to 100 units.

1.5 Exercises

1. Draw the graphs of the following equations for the range $X = 0$ to $X = 6$. What are the intercept term and the slope for each equation?

   (a) $Y = 3X + 4$ (b) $Y = 3X - 4$

   (c) $Y = 4 - 3X$ (d) $Y = 3X$

   (e) $Y = 4$

2. Establish a linear relationship for the total cost function which fits the following two conditions:
   (a) at an output of 10 units the total cost is £70, and
   (b) at an output of 20 units the total cost is £120.
   What do you infer about the fixed and variable cost? Sketch the graph of the total cost function for outputs less than 30.

3. Establish a linear relationship for the total cost function from the following data:

(a) the variable cost is £6 per unit, and

(b) the fixed cost is £15.

What output would make the total cost £75? Sketch the graph of the total cost function for outputs less than 20.

4. The total cost of production (Y) of a product is related to the number of units of output (X) by the equation

$$Y = 3 + 2X$$

(a) Sketch the graph of this line for $0 \leqslant X \leqslant 25$

(b) What are the fixed cost and variable cost?

(c) What wil be the total cost of producing 15 units of output?

(d) If the total cost is 45, how many units of output will be produced?

1.6 Simultaneous linear equations

We have seen how to obtain the equation of a straight line from two pieces of information. These allowed two equations to be obtained which had to be satisfied simultaneously and, by using algebra, the values of the unknowns, a and b, were determined.

Before turning to applications in economics and business it is useful to review some of the basic rules of algebra. While these may seem obvious, it is still worthwhile considering them because they do not always transfer to later applications (such as with determinants, matrices and sets). Let x, y and z represent any three numbers or be particular values of three variables. Then, if they are non-zero, x, y and z satisfy the following:

1. Commutative rule of addition and multiplication:

$$x + y = y + x \text{ and } xy = yx$$

Thus the order in which two numbers are added or multiplied does not matter. Notice that this does not apply to subtraction and division, since $5 - 3 = 2$ while $3 - 5 = -2$, and $8/2 = 4$ while $2/8 = 0.25$.

2. Associative rule of addition and multiplication:

$$(x + y) + z = x + y + z = x + (y + z)$$

$$(xy)z = xyz = x(yz)$$

3. Distributive rule: $x(y + z) = xy + xz$

or
$$(x + y)z = xz + yz$$

4. Operations involving zero: If x is non-zero,

$$0/x = 0,\ x/0 = \infty,\ 0/0 \text{ is indeterminate.}$$

5. Any valid equation remains valid if we add the same expression to or substract the same expression from both sides of the equation.

6. Any valid equation remains valid if we multiply or divide both sides of the equation by the same non-zero expression.

These are useful for simplifying equations. For example, the expression $(x + 2)(y - 1)$ can be expanded as

$$(x + 2)\ (y - 1) = x(y - 1) + 2(y - 1) = xy - x + 2y - 2$$

Also, in the equation

$$x - 2y = 2x - y + 3$$

the terms in x and y can be collected together on the left hand side by subtracting $2x$ from each side, and adding y. This results in

$$x - 2y - 2x + y = 2x - y + 3 - 2x + y$$

or, $$-x - y = 3$$

and multiplying by -1, $x + y = -3$.

In economics and business studies many situations arise where there are two or more equations which have to be satisfied simultaneously. For illustration let us return to our earlier example in which the cost of production could be represented by a linear equation

$$Y = a + bX$$

This relationship is obviously important but does not tell us anything about the profitability of the overall operation. This can only be assessed if the price is known at which the resulting product might be sold along with the quantity that can be sold at this price.

Under conditions of perfect competition it is usual to assume that the price of the article is fixed and that at this price it is possible to sell all that can be produced. This would be the situation confronting an individual producer whose output formed only a very small part of the total demand. In this case his demand curve would be horizontal (i.e. price $= p$) (Fig. 1.4).

The revenue function can then be represented by a linear equation which can be written as

Total revenue = price per unit × number of units sold

i.e. $R = pX$

Thus when $X = 0$, $R = 0$.

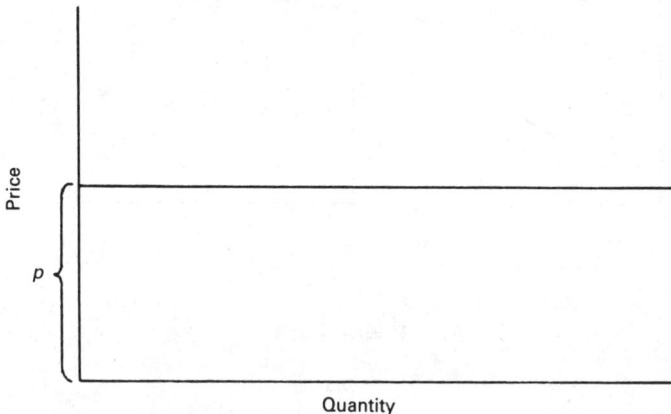

Fig. 1.4

Since in this case the revenue function starts at zero and the cost function at the value a, it is obvious that the two lines will intersect at a positive level of output only if the constant p is greater than the constant b.

Assuming that this is so we would have the situation in Fig. 1.5.

Notice that we are using one set of axes for two graphs. We can do this because the horizontal axis represents the number of units of output and this scale is the same for both equations. The vertical axis represents cost and revenue respectively for the two equations but these are both measured in £ and, therefore, the same axis with the same scale can be used.

The point K where the lines intersect is often referred to as the *breakeven point* because at this point the total revenue equals the total cost and the net revenue is zero. The volume of output corresponding to this point is equal to X_0, which is obtained by dropping a perpendicular from the point of intersection to the horizontal axis. If sales are below X_0 a loss will be made and if sales are above X_0 a profit will be made.

The graphs which have been drawn show the cost and revenue equations and from them we can determine the breakeven point.

The same result can also be obtained from the following set of equations:

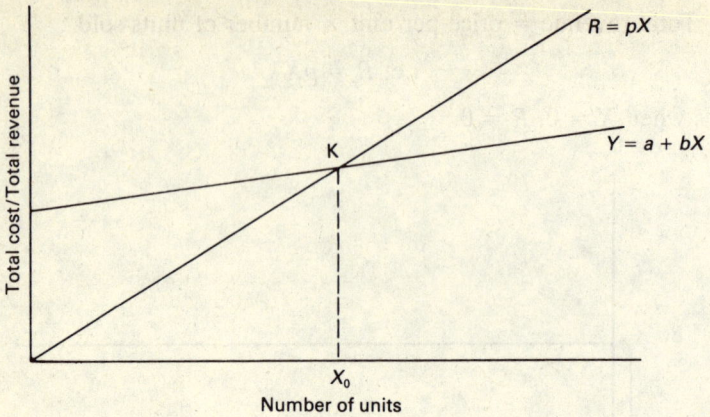

Fig. 1.5

$$Y = a + bX \qquad (1)$$

$$R = pX \qquad (2)$$

with the added condition that at the breakeven point

$$R = Y \qquad (3)$$

These are three equations in three unknowns (R, Y and X) and are then a complete description of the system. By using (3) we can equate (1) and (2) to determine X_0, the breakeven point:

$$R = Y$$

$$\therefore pX = a + bX$$

This equation can be solved for X by grouping terms involving X on one side of the equality. Subtracting bX from each side gives

$$pX - bX = a + bX - bX$$

or $\qquad pX - bX = a$

We can now take out the factor X from the terms on the left hand side because it is common to both and obtain

$$X(p - b) = a$$

We then complete the algebraic manipulation by dividing both sides of the equation by the term $(p - b)$ to give

$$X = \frac{a}{(p-b)}$$

None of the changes we have made has in any way altered the relationship between the variables, although we have ended up with an expression which enables us to calculate the breakeven point for the process for any value of p, given that we know the values of a and b. In descriptive terms,

$$\text{Breakeven point} = \frac{\text{fixed cost}}{\text{price} - \text{variable cost}}$$

The difference between the price and the variable cost at a particular level of output is often termed the *contribution* per unit of product, because it is the net revenue which is produced by each extra unit sold. The breakeven point X_0 is equal to the ratio of the fixed cost to the contribution per unit of product. Since the producer's net revenue (represented by N) is the difference between his revenue and his costs, it is given by

$$N = R - Y = pX - a - bX$$
$$= (p - b)X - a$$

Example

Total cost of production is given by $Y = 25 + 6X$ and price is fixed at £11. What is the value of output at the breakeven point, and what is the net revenue if 20 units of output are produced?
At the breakeven point

$$\text{total cost} = \text{total revenue}$$

that is,
$$Y = pX = 11X$$
$$\therefore \qquad 25 + 6X = 11X$$
$$25 = 5X$$
$$X = 5$$

At the breakeven point the output is 5 units. The net revenue is given by

$$N = pX - Y = 11X - 6X - 25 = 5X - 25$$

When $X = 20$, $N = 5 \times 20 - 25 = 100 - 25 = 75$. Hence, a net revenue of £75 is obtained when the output is 20 units.

1.7 Demand and supply

For any particular product there are, in general, a number of potential consumers and the quantity which each consumer buys depends, at least in part, upon the price at which the product is offered on the market. It is safe to assume, therefore, that the quantity demanded of any good is affected by its price and that for most goods the relationship is an inverse one. That is, the quantity demanded decreases as the price is increased and vice versa. This can be expressed mathematically, and in the simplest case it is possible to think of it as being represented by a straight line (Fig. 1.6).

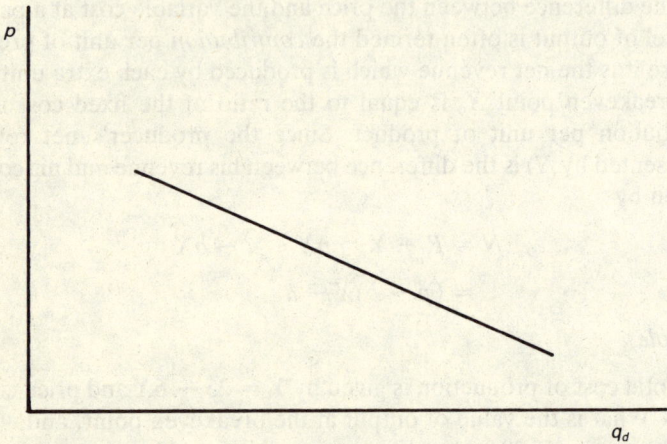

Fig. 1.6

Although the quantity demanded is, by convention, represented along the horizontal axis the equation is usually written algebraically in the form

$$q_d = f(p) = a + bp$$

In this example the constant b would be negative because the line is downward sloping and the relationship is such that the quantity demanded q_d decreases as the price is increased.

In a similar way the market, in a competitive situation, can be considered to be supplied by a large number of independent producers. The number of such producers and the level of their individual outputs for any product is determined by the cost of production and the price which the market is prepared to pay for the product. It can be assumed that the supply of any product is affected by the price which can be obtained for it, and, in this case, the quantity supplied increases as the price increases (Fig. 1.7).

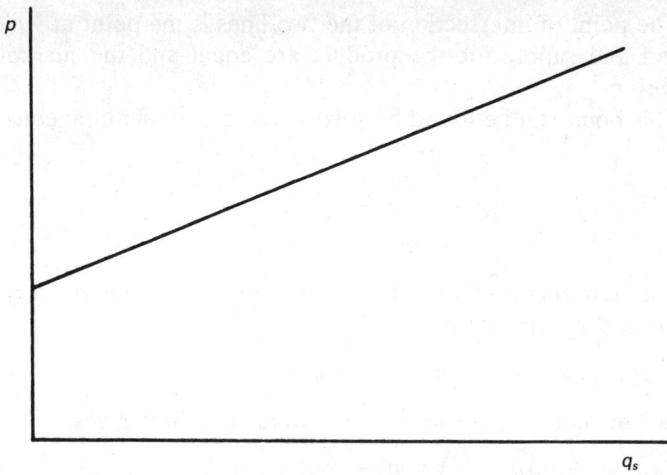

Fig. 1.7

Algebraically this can be written

$$q_s = a' + b'p$$

The two equations can be drawn on the one set of axes (Fig. 1.8).

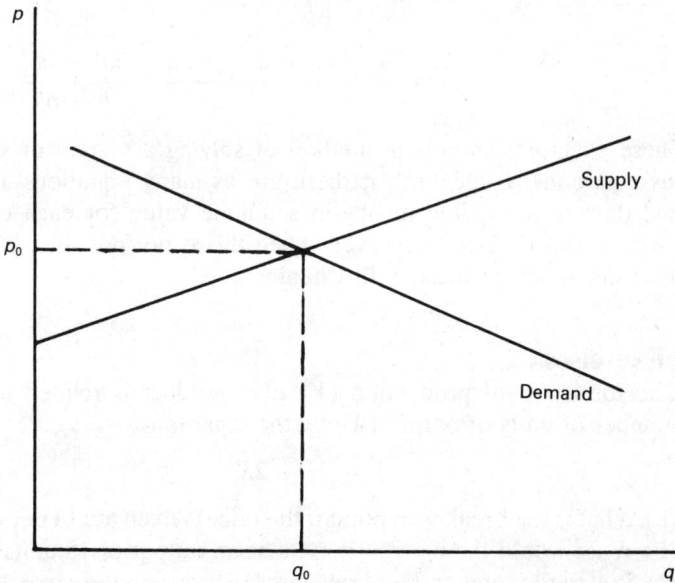

Fig. 1.8

The point of intersection of the two lines is the point at which the demand and supply for the product are equal and the market is in equilibrium.

This point can be found by solving the pair of simultaneous equations:

$$q_d = a + bp$$

$$q_s = a' + b'p$$

using the information that at the equilibrium point $q_d = q_s = q_0$, say.

From these three equations

$$a + bp_0 = a' + b'p_0$$

Subtracting a and $b'p_0$ from each side and cancelling gives

$$p_0(b - b') = a' - a$$

$$p_0 = \frac{a' - a}{b - b'}$$

and

$$q_0 = a + b(p_0)$$

$$= a + b \left[\frac{a' - a}{b - b'} \right]$$

$$= \frac{ab - ab' + ba' - ba}{b - b'} = \frac{ba' - ab'}{b - b'}$$

These examples show one method of solving a system of simultaneous equations. In general, if there are as many equations as unknowns, then it is possible to obtain a unique value for each of the unknowns. Cases do arise, however, when this is not possible, but we will defer discussion of these until Chapter 2.

1.8   Exercises

1. The total cost of production (Y) of a product is related to the number of units of output (X) by the equation

$$Y = 33 + 2X$$

   (a) What is the breakeven point if the price is fixed at £13 per unit?
   (b) What would the net revenue be at an output of 15 units?
   (c) Sketch the total cost and revenue functions on the same graph.

(d) Why would the producer not fix the price at £1 per unit?
2. A producer has a fixed cost of £50 and a variable cost of £5 per unit of output if his output is less than 200 units.
   (a) What is the equation of his total cost function (which is linear)?
   (b) What is the breakeven point if the producer fixed his price at £10 per unit?
   (c) What is the producer's net revenue if the output is 12 units and the price is (i) £5, (ii) £10, (iii) £15?
3. A producer is willing to supply a market with an output q_s according to the relationship

$$q_s = 25p - 10$$

where p is the price.
The demand function for the product is

$$q_d = 200 - 5p$$

where q_d is the quantity demanded at price p.
   (a) What is the equilibrium price and quantity?
   (b) If the producers supply function changes to

$$q_s = 20p - 25$$

   what is the new equilibrium position?
4. Draw the graphs of the following equations and where possible find the values of p and q which satisfy the equations simultaneously.

(a) $2p + 3q = 17$ (b) $6p - q = 3$
    $5p - 4q = 8$ $4p + 7q = 2$

(c) $2p + 4q = 7$ (d) $2p + q = 1$
    $4p + 8q = 14$ $4p + 2q = 1$

1.9 The effects of taxes on demand

In section 1.7 we discussed the determination of the equilibrium price and quantity when the demand and supply equations are linear. We now continue with this model and examine the effects of taxes on the equilibrium values. Initially we will consider a flat-rate tax of t per unit, which is independent of the selling price and then discuss a tax which is a percentage of the price. The former corresponds to an excise tax such as might be imposed on a bottle of spirits, while the latter is more like a sales or value-added tax.

Let the original demand and supply curves be

$$q_d = 100 - 5p$$

and $$q_s = -20 + 10p$$

where p is the market price, q_d the quantity demanded and q_s the quantity supplied. With no taxes the supplier receives p from each sale and the purchaser pays p. The equilibrium is when $q_d = q_s$ and the equilibrium values are $q_0 = 60$, $p_0 = 8$.

Suppose a flat-rate tax of 3 per unit is imposed. The price received by the supplier p^t is now

$$p^t = p - 3$$

since 3 is paid to the government in tax. The new supply function is therefore

$$q_s = -20 + 10p^t$$
$$= -20 + 10(p - 3)$$
$$= -50 + 10p$$

The effect of the tax is to change the intercept term of the supply function from -20 to -50. The demand curve is unaffected. The new equilibrium is when $q_s = q_d$ or

$$-50 + 10p_1 = 100 - 5p_1$$

or $$15p_1 = 150$$

giving $p_1 = 10$ and $q_1 = 50$

The equilibrium has changed from $p_0 = 8$, $q_0 = 60$ to $p_1 = 10$, $q_1 = 50$ and so the new price is higher and the quantity is lower. This is illustrated in Fig. 1.9 where the original supply curve is labelled S and the after-tax supply curve is S'. As p is on the vertical axis the supply curve has shifted to the left. Also, notice that while the tax is 3 per unit, the increase in price to the consumer is only 2 per unit and so the tax has not been fully passed on to the customers. The tax revenue is $3q_1$ or 150 while the supplier's income is $q_1(p_1 - 3)$ or 350 compared with $p_0 q_0$ or 480 originally.

The above process can also be carried out for general linear supply and demand functions

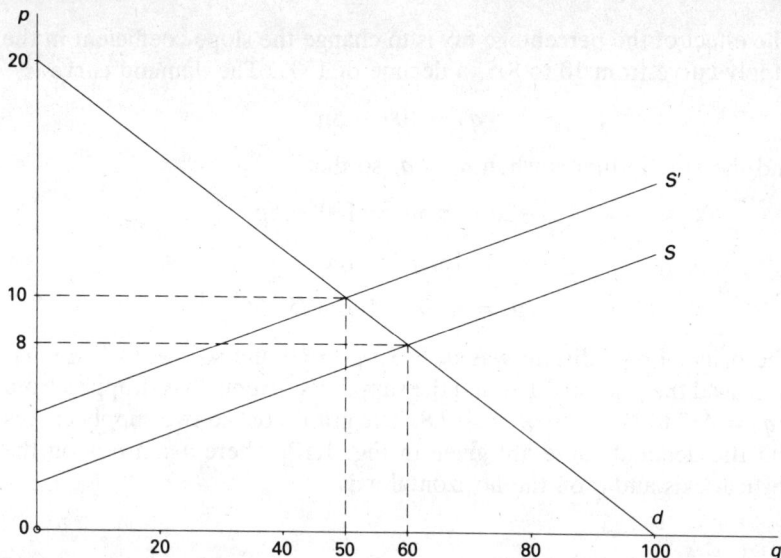

Fig. 1.9

$$q_s = a' + b'p^t \quad (b' > 0)$$

$$q_d = a + bp \quad (b < 0)$$

and $$p^t = p - t$$

The new equilibrium is now

$$p_1 = (a' - a - b't)/(b - b')$$

and $$q_1 = (a'b - ab' - bb't)/(b - b')$$

If the tax t is zero these equations reduce to those at the end of section 1.7 while if $t > 0$, since $b' > 0$, it can be seen that $p_1 > p_0$ and $q_1 < q_0$.

We now turn to the case of a percentage tax which is at a rate of $100r\%$ (so that $r = 0.15$ for a 15% tax). Again the demand curve is unaffected and there is a difference between the market price, p, and the price received by the supplier, p^r. Here $p^r = p - rp$. Taking the example used above and $r = 0.15$, the original supply curve is

$$q_s = -20 + 10p$$

and the tax changes this to

$$q_s = -20 + 10p^r = -20 + 10(1 - r)p$$

$$= -20 + 8.5p$$

The effect of the percentage tax is to change the slope coefficient in the supply curve from 10 to 8.5, a decline of 15%. The demand curve is

$$q_d = 100 - 5p$$

and the equilibrium is when $q_s = q_d$ so that

$$-20 + 8.5p_1 = 100 - 5p_1$$

$$13.5p_1 = 120$$

$$\text{or } p_1 = 8.89 \text{ and } q_1 = 55.56$$

The original equilibrium was $p_0 = 8$, $q_0 = 60$ and so the 15% tax has increased the price by 11% and the supplier's revenue has dropped from $p_0q_0 = 480$ to $(1 - r)p_1q_1 = 419.8$. The graphs of the two supply curves and the demand curve are given in Fig. 1.10, where again p is on the vertical axis and q on the horizontal axis.

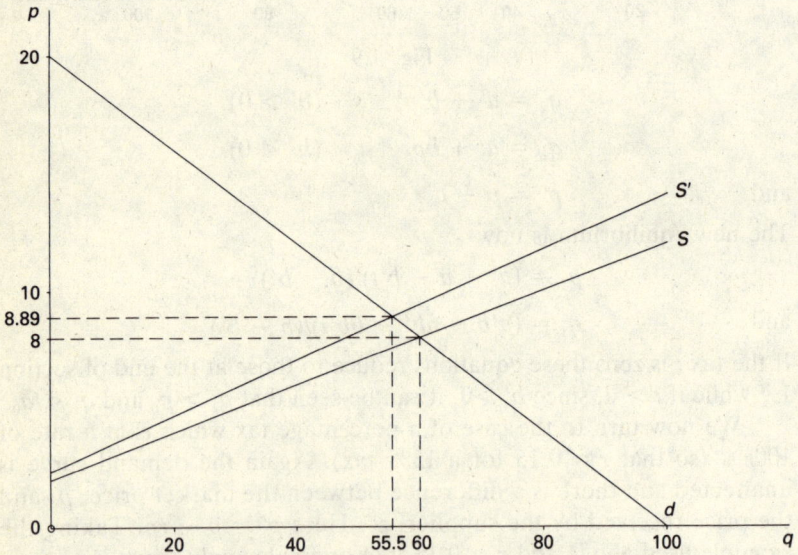

Fig. 1.10

Finally, we repeat the case of a percentage tax for the general linear supply and demand curves. The new supply curve is

$$q_s = a' + b'p^r \qquad (b' > 0)$$

the demand curve is

$$q_d = a + bp \qquad (b < 0)$$

and the supplier's price is

$$p^r = (1 - r)p$$

By substitution, the equilibrium price (p_1) and quantity (q_1) are

$$p_1 = (a - a')/\{(1 - r)b' - b\}$$

and
$$q_1 = \{ab' (1 - r) - a'b\}/\{(1 - r)b' - b\}$$

As previously, if the tax rate is zero these reduce to the results given at the end of section 1.7. It is easily seen that the effect of a percentage tax is to increase the equilibrium price, and less easily seen that the equilibrium quantity will be reduced.

1.10 Simple national income models

As a further illustration of the use of simultaneous equations in economics and business studies we now consider some small macroeconomic models. For simplicity we will start with a two-equation consumption–national income model, in which there is no government sector and no external sector. The first equation is a linear consumption function which relates aggregate consumption, C, and national income, Y, while the second equation, an identity, defines national income as being the sum of aggregate consumption and what we will call nonconsumption, N. Here the two equations explain the endogenous variables C and Y, and there is one exogenous variable, N, which is determined outside the system. Initially we will also assume that N is fixed at some value, N_0. The equations are

$$C = 100 + 0.6Y$$
$$Y = C + N_0$$

Here the *marginal propensity to consume*, the proportion of an extra unit of income which is consumed, is 0.6, and the level of consumption when income is zero is 100. These two equations can be solved to give the equilibrium values of C and Y in terms of N_0. Substituting for C in the second equation gives

$$Y = 100 + 0.6Y + N_0$$

and collecting the Y terms on the left hand side results in

$$0.4Y = 100 + N_0$$

and dividing through by 0.4 gives

$$Y = 250 + 2.5N_0$$

Substituting this into the consumption equation,

$$C = 100 + 0.6(250 + 2.5N_0)$$

and simplifying,

$$C = 250 + 1.5N_0$$

In these solutions the values of Y and C depend on the value of N_0. If N_0 increases by one unit then Y will increase by 2.5. This value, 2.5, is called the *multiplier*, since it gives the effect on aggregate income of a unit increase in non-consumption expenditure.

More generally, if the two equations are

$$C = a + bY$$

$$Y = C + N_0,$$

the solutions are easily shown to be

$$C = (a + bN_0)/(1 - b)$$

$$Y = (a + N_0)/(1 - b)$$

Here the multiplier, from the income equation, is $1/(1 - b)$. Since b is the marginal propensity to consume, $0 < b < 1$, and the multiplier is positive. If this condition is violated strange results can occur. For example if $b = 1$, the multiplier is infinite as are the equilibrium values of C and Y, while if b is negative, both C and Y are negative.

The above two-equation model is obviously very unrealistic and we now expand it by introducing a simple investment function. We will assume that

$$I = c + dY$$

so that investment is a linear function of national income, with $c > 0$ and $0 < d < 1$, making it endogenous. (A model in which investment is affected by the rate of interest is discussed in section 1.11 below.) The national income identity needs to be modified to take account of investment:

$$Y = C + I$$

and the consumption function is as before:

$$C = a + bY$$

To solve these three equations for the equilibrium levels of income, consumption and investment, substitute for C and I in the national income equation to give

$$Y = a + bY + c + dY$$

and rearranging,

$$Y = (a + c)/(1 - b - d)$$

We notice that for Y to be positive, which is to be expected, since both a and c are positive, $(1 - b - d)$ must also be positive and so

$$1 > b + d$$

The solutions for C and I are easily given by substitution as

$$C = (a - ad + bc)/(1 - b - d)$$

$$I = (c - bc + ad)/(1 - b - d)$$

Another obvious modification is the introduction of a government sector which has an income from taxation (T) and expenditure (G). We will assume that G and T are determined by the government and so are exogenous. The national income identity is now

$$Y = C + I + G$$

The consumption function is modified to relate consumption to disposable income $(Y - T)$,

$$C = a + b(Y - T)$$

and investment is determined as before

$$I = c + dY$$

The model consists of three equations in three endogenous variables, C, Y and I. The solution follows the previous method of starting with the national income identity to give

$$Y = a + b(Y-T) + c + dY + G$$

which reduces to

$$Y = (a - bT + c + G)/(1 - b - d)$$

and the solutions for the other variables are

$$C = \{a - ad + bc + bG - b(1-d)T\}/(1 - b - d)$$

$$I = (c - bc + ad - bdT + dG)/(1 - b - d)$$

The *tax multiplier* gives the effect on Y of a unit increase in T. From the solution for Y it can be seen that if T increases by 1, Y changes by $-b/(1 - b - d)$. Since b and $(1 - b - d)$ are both positive, this multiplier is negative. That is, in this model, increasing the tax rate reduces national income. Similarly, the *government expenditure multiplier* is the coefficient on G in the solution for Y and so is $1/(1 - b - d)$, which is positive. This implies that an increase in government expenditure will result in an increase in national income, if the model is correct.

We will discuss more complicated national income models in sections 1.11 and 1.15 below. However, before leaving these simple models it is important to realise that they are not intended to represent any particular economy, the numerical values are purely for illustrative purposes and any policy implications are as valid only as the assumptions underlying the models.

1.11   IS–LM analysis

In section 1.10 a simple model of national income was developed in which the relationships between national income, consumption, investment and taxation were discussed. Here we consider a different approach, known as IS–LM analysis, in which equilibrium occurs in two separate markets. These are the market for goods and the market for money, where interest rates now have an important role. Again, we initially ignore the government and foreign trade sectors.

At the aggregate level the goods market is in equilibrium when income equals expenditure or $Y = C + I$, where Y is national income, C is consumption and I is investment. The *IS-schedule* shows the different combinations of interest rates and national income at which the goods market is in equilibrium. Here it is assumed that investment is affected by the interest rate, which is endogenous. For example, the investment function might be $I = 90 - 60i$ where i is the rate of interest (so for a 5% rate $i = 0.05$). The constant term in this investment function can be interpreted as being autonomous investment – that is, the part of investment which is unrelated to interest rates and income. We will assume that the consumption function is a simple linear one and that

$C = 10 + 0.8Y$. The goods market therefore has the three equations to determine Y, C, I and i:

$$Y = C + I$$

$$I = 70 - 60i$$

$$C = 10 + 0.8Y$$

For equilibrium, the three equations are satisfied simultaneously and so

$$Y = C + I = 10 + 0.8Y + 70 - 60i$$

and simplifying gives $Y = 400 - 300i$ which is the IS-schedule. Notice that there is a negative relationship between Y and i.

The IS-schedule is concerned with the goods market. We now consider the money market where, for equilibrium, money supply, M_s, equals money demand, M_d. In this simple model we will assume that money supply is exogenous and is in fact fixed at $M_s = 250$ and that money demand is given by

$$M_d = L_1 + L_2$$

where $L_1 = 0.2Y$ is the transaction-precautionary demand for money, which is determined by the level of national income, and $L_2 = 190 - 100i$ is the speculative demand for money which is negatively related to the rate of interest. Here, equilibrium occurs in the money market when

$$M_s = M_d$$

$$250 = 0.2Y + 190 - 100i$$

or $$Y = 300 + 500i$$

This is the *LM-schedule* which shows the different combinations of national income and interest rates at which the money market is in equilibrium.

For there to be equilibrium in both the goods and the money markets, the values of Y and i must be the same in the IS- and LM-schedules so that we need to solve the equations simultaneously. They are

$$IS \qquad Y = 400 - 300i$$

$$LM \qquad Y = 300 + 500i$$

which solve to give $Y = 362.5$ and $i = 0.125$ or 12.5%.

This model can be used to examine the effects of changes in the assumptions about the different components. For example, suppose there is an increase in autonomous investment and the investment function changes from $I = 70 - 60i$ to $I = 80 - 60i$, this moves the IS-schedule to

$$Y = C + I = 10 + 0.8Y + 80 - 60i$$

or $\qquad Y = 450 - 300i$

With the LM-schedule from above the new equilibrium values are $i = 0.1875$ or 18.75% and $Y = 393.75$, so that, in this particular model, an increase in autonomous investment results in a higher interest rate and a higher national income. Investment was originally 62.5 and increases to 68.75, so that the increase in autonomous investment is partly off-set by the effects of the increase in the interest rate.

The model can also be extended to make it more realistic. For example, government expenditure (G) could be included by adding G to the national income identity and then treating G as exogenous or endogenous. Further possible extensions include the effects of taxation, the overseas sector and the labour market. In section 1.15 more complex models will be discussed.

1.12   Exercises

1. The demand and supply equations for a particular product are

$$q_d = 200 - 4p$$
$$q_s = -10 + 26p$$

(a) Determine the equilibrium values of p and q and the producer's revenue they imply.

(b) A flat-rate tax of 5 per unit is imposed on each unit sold. Determine the new equilibrium position, the tax revenue at the equilibrium and the producer's revenue.

(c) Instead of the flat-rate tax of (b) above, a tax of 20% of the price is impoosed on each item sold. Determine the new equilibrium position, the tax revenue at the equilibrium and the producer's revenue.

2. The demand and supply equations for a new product are

$$q_d = 300 - 6p$$
$$q_s = -40 + 15p$$

The government are undecided as to whether to impose a tax of 10% on each item sold or to impose a flat-rate tax. Show that the price to the consumer would be the same in either case if the flat-rate tax is approximately 1.7 per unit sold.

3. In a simple national income model, consumption (C) is given by $C = 20 + 0.7\,Y_d$ where Y_d is disposable national income. The national income identity is $Y = C + N$ where N is non-consumption and is exogenous. Determine the equilibrium values of Y and C in the following cases if $N = 4$:

(a) There are no taxes and $Y_d = Y$
(b) A lump-sum tax of $T = 10$ is imposed on income so that disposable income is now $Y_d = Y - T$
(c) The lump-sum tax is replaced by a tax on income at a rate of 25% so that $Y_d = Y - (0.25\,Y)$
(d) The lump-sum and the percentage tax on income are combined so that $Y_d = Y - 10 - 0.25Y$.

4. In a simple IS–LM model of an economy with two sectors the goods market is described by

$$C = 15 + 0.8Y$$

$$I = 75 - 100i$$

$$Y = C + I$$

where Y is the national income, C is consumption, I is investment and i is the interest rate. The money market has the transaction-precautionary demand for money given by

$$L_1 = 0.1Y$$

and the speculative demand for money given by

$$L_2 = 250 - 160i$$

Assume that the supply of money is fixed at $M_s = 250$.

(a) Determine the IS and LM schedules and hence the equilibrium levels of Y and i.
(b) How is the equilibrium changed if autonomous investment falls from 75 to 50?
(c) What is the effect on the equilibrium of the original model of changing the money supply from 250 to 275?

5. The model in Question 4 can be extended to include a government sector by changing the goods market equations to

$$C = 15 + 0.8Y_d$$

$$I = 75 - 100i$$

$$Y = C + I + G$$

where Y_d is disposable income, G is government expenditure and the other symbols are as before.

(a) If Y_d is defined by $Y_d = Y - T$, where T is taxes, and T is given by $T = 5 + 0.2Y$, and $G = 300$, determine the IS schedule.

(b) If the transaction-precautionary demand for money is given by

$$L_1 = 0.2Y$$

and the speculative demand for money by

$$L_2 = 750 - 260i,$$

and the supply of money is fixed at $M_s = 950$, determine the LM schedule and hence the equilibrium values of Y and i.

(c) What is the effect on the equilibrium if G increases to 350?

1.13   Determinant notation

We now consider the solution of a general set of simultaneous equations. The general system of two equations in two unknowns is

$$a_{11}x_1 + a_{12}x_2 = b_1 \tag{1}$$

$$a_{21}x_1 + a_{22}x_2 = b_2 \tag{2}$$

where the as and bs are constants. Here, x_1 and x_2 are the unknowns and a_{11} is the coefficient in (1) of x_1 and a_{12} is the coefficient in (1) of x_2 etc. In general, therefore, a_{ij} refers to the coefficient in (i) of x_j.

To obtain the solution for x_1 we eliminate x_2 from the equations by multiplying (1) by a_{22} and (2) by a_{12} to give

$$a_{11}a_{22}x_1 + a_{12}a_{22}x_2 = a_{22}b_1$$

$$a_{12}a_{21}x_1 + a_{12}a_{22}x_2 = a_{12}b_2$$

Subtraction now gives

$$(a_{11}a_{22} - a_{12}a_{21})x_1 = a_{22}b_1 - a_{12}b_2$$

Hence, provided $a_{11}a_{22} - a_{12}a_{21} \neq 0$,

$$x_1 = \frac{a_{22}b_1 - a_{12}b_2}{a_{11}a_{22} - a_{12}a_{21}} \tag{3}$$

and similarly

$$x_2 = \frac{a_{11}b_2 - a_{21}b_1}{a_{11}a_{22} - a_{12}a_{21}} \tag{4}$$

Notice that the denominator in (3) and (4) is the same and that it is a combination of all the a_{ij}-values from (1) and (2). In particular if the pattern of a_{ij}-values from (1) and (2) is reproduced:

$$\begin{matrix} a_{11} & \times & a_{12} \\ a_{21} & & a_{22} \end{matrix}$$

we see that the denominator is the cross-product of $a_{11}a_{22}$ minus the cross-product $a_{12}a_{21}$. We now adopt a special notation and define the *determinant* of the numbers a_{ij} as

$$|\mathbf{A}| = \begin{vmatrix} a_{11} & a_{12} \\ a_{21} & a_{22} \end{vmatrix} = a_{11}a_{22} - a_{12}a_{21}$$

It should be noted that the result of multiplying out or evaluating a determinant is a number. For example, if $a_{11} = 4$, $a_{12} = 2$, $a_{21} = 2$, $a_{22} = 3$, then

$$\begin{vmatrix} 4 & 2 \\ 2 & 3 \end{vmatrix} = (4 \times 3) - (2 \times 2) = 12 - 4 = 8$$

or if $a_{11} = 2$, $a_{12} = 0$, $a_{21} = 1$, $a_{22} = 3$ then

$$\begin{vmatrix} 2 & 0 \\ 1 & 3 \end{vmatrix} = (2 \times 3) - (0 \times 1) = 6 - 0 = 6$$

From the definition of a determinant it will be clear that

$$\begin{vmatrix} b_1 & a_{12} \\ b_2 & a_{22} \end{vmatrix} = a_{22}b_1 - a_{12}b_2$$

and

$$\begin{vmatrix} a_{11} & b_1 \\ a_{21} & b_2 \end{vmatrix} = a_{11}b_2 - a_{21}b_1$$

and hence (3) and (4) can be written in determinant notation:

$$x_1 = \dfrac{\begin{vmatrix} b_1 & a_{12} \\ b_2 & a_{22} \end{vmatrix}}{\begin{vmatrix} a_{11} & a_{12} \\ a_{21} & a_{22} \end{vmatrix}} \quad \text{and} \quad x_2 = \dfrac{\begin{vmatrix} a_{11} & b_1 \\ a_{21} & b_2 \end{vmatrix}}{\begin{vmatrix} a_{11} & a_{12} \\ a_{21} & a_{22} \end{vmatrix}}$$

The solutions to (1) and (2) are seen to be the ratios of determinants. The denominators are the determinants of the coefficients of x_1 and x_2 whilst the numerators are essentially the same determinants with one column, the first for x_1 and the second for x_2, replaced by the column made up of b_1 and b_2. This method of solving simultaneous equations is known as *Cramer's rule*. For example,

$$3x_1 - x_2 = 1$$

$$2x_1 + x_2 = 4$$

$$\begin{vmatrix} a_{11} & a_{12} \\ a_{21} & a_{22} \end{vmatrix} = \begin{vmatrix} 3 & -1 \\ 2 & 1 \end{vmatrix} = (3 \times 1) - [(-1) \times 2] = 3 + 2 = 5$$

$$\begin{vmatrix} b_1 & a_{12} \\ b_2 & a_{22} \end{vmatrix} = \begin{vmatrix} 1 & -1 \\ 4 & 1 \end{vmatrix} = (1 \times 1) - [(-1) \times 4] = 1 + 4 = 5$$

$$\begin{vmatrix} a_{11} & b_1 \\ a_{21} & b_2 \end{vmatrix} = \begin{vmatrix} 3 & 1 \\ 2 & 4 \end{vmatrix} = (3 \times 4) - (1 \times 2) = 12 - 2 = 10$$

Thus the solutions are

$$x_1 = \tfrac{5}{5} = 1, \qquad x_2 = \tfrac{10}{5} = 2$$

It will be clear from this that if the determinant of the coefficients of x_1 and x_2 is zero then there is no unique solution to the equations.

For example,
$$2x_1 + 2x_2 = b_1$$
$$3x_1 + 3x_2 = b_2$$

Here

$$\begin{vmatrix} a_{11} & a_{12} \\ a_{21} & a_{22} \end{vmatrix} = \begin{vmatrix} 2 & 2 \\ 3 & 3 \end{vmatrix} = (2 \times 3) - (2 \times 3) = 6 - 6 = 0$$

and there is no solution. This is because either the two equations are identical and their graphs coincide, or the two equations are contradictory and their graphs are parallel lines. In either case there is no unique point of intersection. Determinants of the kind just discussed, which have two rows and two columns, are known as two by two or (2×2) determinants, and are said to be *second-order* determinants. It is possible to define higher-order determinants such as a (3×3) determinant

$$\begin{vmatrix} a_{11} & a_{12} & a_{13} \\ a_{21} & a_{22} & a_{23} \\ a_{31} & a_{32} & a_{33} \end{vmatrix} = a_{11}\begin{vmatrix} a_{22} & a_{23} \\ a_{32} & a_{33} \end{vmatrix} - a_{12}\begin{vmatrix} a_{21} & a_{23} \\ a_{31} & a_{33} \end{vmatrix} + a_{13}\begin{vmatrix} a_{21} & a_{22} \\ a_{31} & a_{32} \end{vmatrix}$$

which is defined in terms of the three second-order determinants. These are obtained by multiplying a_{11} by the determinant of the coefficients which remains when the first row and first column are eliminated,

$$\begin{vmatrix} a_{11}.\,.\,.a_{12}.\,.\,.a_{13} \\ \vdots \\ a_{21} \quad a_{22} \quad a_{23} \\ \vdots \\ a_{31} \quad a_{32} \quad a_{33} \end{vmatrix} \rightarrow \begin{vmatrix} a_{22} & a_{23} \\ a_{32} & a_{33} \end{vmatrix}$$

and similarly for the other terms. Notice that the pattern of signs is positive on a_{11}, negative on a_{12} and positive on a_{13}. In this example the determinant is said to have been *expanded by the first row* and it is frequently convenient to do this. However, the value of the determinant can also be obtained by expanding by any row or column, using the same principles of multiplying the element from the particular row or column chosen by the second-order determinants formed by deleting the appropriate row and column. The rule for deciding the sign on the coefficient is that if the coefficient for row i, column j is a_{ij} then it is multiplied by $(-1)^{1+j}$ so that if $i + j$ is even the coefficient is positive while if $i + j$ is odd the coefficient is negative.

To illustrate this, consider expanding the general (3×3) determinant by the second column. This gives

$$\begin{vmatrix} a_{11} & a_{12} & a_{13} \\ a_{21} & a_{22} & a_{23} \\ a_{31} & a_{32} & a_{33} \end{vmatrix} = (-1)^{1+2}a_{12}\begin{vmatrix} a_{21} & a_{23} \\ a_{31} & a_{33} \end{vmatrix} + (-1)^{2+2}a_{22}\begin{vmatrix} a_{11} & a_{13} \\ a_{31} & a_{33} \end{vmatrix}$$

$$+ (-1)^{3+2}a_{32}\begin{vmatrix} a_{11} & a_{13} \\ a_{21} & a_{23} \end{vmatrix}$$

$$= -a_{12}(a_{21}a_{33} - a_{23}a_{31}) + a_{22}(a_{11}a_{33} - a_{13}a_{31})$$

$$-a_{32}(a_{11}a_{23} - a_{13}a_{21})$$

This property of determinants is particularly useful when a determinant includes some zero elements. For example, the following determinant can be evaluated easily by expanding by the first column:

$$\begin{vmatrix} 1 & -1 & 2 \\ 0 & 1 & 0 \\ 0 & -2 & 4 \end{vmatrix} = 1\begin{vmatrix} 1 & 0 \\ -2 & 4 \end{vmatrix} - 0 + 0 = (1 \times 4) - (0 \times (-2)) = 4 - 0 = 4$$

The possibility of expanding a determinant by any row or column gives the useful result that if all the elements of any row (or column) are zero then the determinant must also be zero. For example,

$$\begin{vmatrix} 0 & 0 & 0 \\ 3 & 1 & 2 \\ 2 & 4 & 1 \end{vmatrix} = 0\begin{vmatrix} 1 & 2 \\ 4 & 2 \end{vmatrix} - 0\begin{vmatrix} 3 & 2 \\ 2 & 1 \end{vmatrix} + 0\begin{vmatrix} 3 & 1 \\ 2 & 4 \end{vmatrix} = 0$$

Another useful property is that the value of a determinant is unchanged if a constant multiple of any row (or column) is added to (or subtracted from) any other row (or column). This is demonstrated by noting that

$$\begin{vmatrix} 1 & -1 \\ -1 & 2 \end{vmatrix} = (2 \times 1) - [(-1)(-1)] = 2 - 1 = 1$$

while if the second row is added to the first,

$$\begin{vmatrix} 0 & 1 \\ -1 & 2 \end{vmatrix} = (0 \times 2) - [1 \times (-1)] = 0 + 1 = 1$$

Another example of this property can be seen in the evaluation of

$$|\mathbf{A}| = \begin{vmatrix} 10 & 15 & -5 \\ 8 & -9 & 10 \\ -4 & 3 & 2 \end{vmatrix}$$

Here if twice the third column is added to the first column, the result is zero in the first row, first column position. Similarly, adding three times the third column to the second gives another zero in the first row, and it is then straightforward evaluating the determinant by expanding by the first row. That is,

$$|\mathbf{A}| = \begin{vmatrix} 10-10 & 15-15 & -5 \\ 8+20 & -9+30 & 10 \\ -4+4 & 3+6 & 2 \end{vmatrix} = \begin{vmatrix} 0 & 0 & -5 \\ 28 & 21 & 10 \\ 0 & 9 & 2 \end{vmatrix} = -5\begin{vmatrix} 28 & 21 \\ 0 & 9 \end{vmatrix}$$

$$= -5(28 \times 9) = -1260$$

In this example the numbers could have been simplified by using the result that if any row or column is multiplied by a constant, the value of the determinant is also multiplied by the constant. Thus, if a row has a common factor, the factor can be moved outside the determinant. In the previous example row 1 has a common factor of 5 which can be removed:

$$|\mathbf{A}| = \begin{vmatrix} 10 & 15 & -5 \\ 8 & -9 & 10 \\ -4 & 3 & 2 \end{vmatrix} = 5\begin{vmatrix} 2 & 3 & -1 \\ 8 & -9 & 10 \\ -4 & 3 & 2 \end{vmatrix}$$

and now column 1 has a common factor of 2 while column 2 has a common factor of 3. Removing these gives

$$|\mathbf{A}| = 5(2)(3)\begin{vmatrix} 1 & 1 & -1 \\ 4 & -3 & 10 \\ -2 & 1 & 2 \end{vmatrix}$$

and this is easier to evaluate than the original determinant.

If any two rows (or columns) of a determinant are the same, the value of the determinant is zero, since subtracting these will give a row (or column) of zeros. For example,

$$\begin{vmatrix} 1 & 2 & 3 \\ 1 & 2 & 3 \\ 1 & 0 & 4 \end{vmatrix} = \begin{vmatrix} 0 & 0 & 0 \\ 1 & 2 & 3 \\ 1 & 0 & 4 \end{vmatrix} = 0$$

Further discussion of the properties of determinants is deferred until Chapter 2, and this section concludes with the use of third-order determinants in the solution of simultaneous equations.

The general system of three equations in three unknowns can be written

$$a_{11}x_1 + a_{12}x_2 + a_{13}x_3 = b_1$$

$$a_{21}x_1 + a_{22}x_2 + a_{23}x_3 = b_2$$

$$a_{31}x_1 + a_{32}x_2 + a_{33}x_3 = b_3$$

and by analogy with the two-equation system the solution is, using Cramer's rule,

$$x_1 = \frac{|\mathbf{A}_1|}{|\mathbf{A}|}, \ x_2 = \frac{|\mathbf{A}_2|}{|\mathbf{A}|}, \ x_3 = \frac{|\mathbf{A}_3|}{|\mathbf{A}|}$$

where

$$|\mathbf{A}| = \begin{vmatrix} a_{11} & a_{12} & a_{13} \\ a_{21} & a_{22} & a_{23} \\ a_{31} & a_{32} & a_{33} \end{vmatrix}, \quad |\mathbf{A}_1| = \begin{vmatrix} b_1 & a_{12} & a_{13} \\ b_2 & a_{22} & a_{23} \\ b_3 & a_{32} & a_{33} \end{vmatrix}$$

$$|\mathbf{A}_2| = \begin{vmatrix} a_{11} & b_1 & a_{13} \\ a_{21} & b_2 & a_{23} \\ a_{31} & b_3 & a_{33} \end{vmatrix}, \quad |\mathbf{A}_3| = \begin{vmatrix} a_{11} & a_{12} & b_1 \\ a_{21} & a_{22} & b_2 \\ a_{31} & a_{32} & b_3 \end{vmatrix}$$

so that $|\mathbf{A}_i|$ is the value of $|\mathbf{A}|$ with column i replaced by the column of b-values. For example,

$$3x_1 + x_2 - 2x_3 = 2$$

$$x_1 + x_2 + x_3 = 1$$

$$2x_1 + 2x_2 + 3x_3 = 3$$

Here

$$|\mathbf{A}| = \begin{vmatrix} 3 & 1 & -2 \\ 1 & 1 & 1 \\ 2 & 2 & 3 \end{vmatrix} = 3\begin{vmatrix} 1 & 1 \\ 2 & 3 \end{vmatrix} - 1\begin{vmatrix} 1 & 1 \\ 2 & 3 \end{vmatrix} + (-2)\begin{vmatrix} 1 & 1 \\ 2 & 2 \end{vmatrix}$$

$$= 3(3 - 2) - 1(3 - 2) - 2(2 - 2)$$

$$= 3 \times 1 - 1 \times 1 - 2 \times 0 = 2$$

$$|\mathbf{A}_1| = \begin{vmatrix} 2 & 1 & -2 \\ 1 & 1 & 1 \\ 3 & 2 & 3 \end{vmatrix} = 2\begin{vmatrix} 1 & 1 \\ 2 & 3 \end{vmatrix} - 1\begin{vmatrix} 1 & 1 \\ 3 & 3 \end{vmatrix} + (-2)\begin{vmatrix} 1 & 1 \\ 3 & 2 \end{vmatrix}$$

$$= 2(3 - 2) - 1(3 - 3) - 2(2 - 3) = 4$$

$$|A_2| = \begin{vmatrix} 3 & 2 & -2 \\ 1 & 1 & 1 \\ 2 & 3 & 3 \end{vmatrix} = 3\begin{vmatrix} 1 & 1 \\ 3 & 3 \end{vmatrix} - 2\begin{vmatrix} 1 & 1 \\ 2 & 3 \end{vmatrix} + (-2)\begin{vmatrix} 1 & 1 \\ 2 & 3 \end{vmatrix}$$

$$= 3(3-3) - 2(3-2) - 2(3-2) = -4$$

$$|A_3| = \begin{vmatrix} 3 & 1 & 2 \\ 1 & 1 & 1 \\ 2 & 2 & 3 \end{vmatrix} = 3\begin{vmatrix} 1 & 1 \\ 2 & 3 \end{vmatrix} - 1\begin{vmatrix} 1 & 1 \\ 2 & 3 \end{vmatrix} + 2\begin{vmatrix} 1 & 1 \\ 2 & 2 \end{vmatrix}$$

$$= 3(3-2) - 1(3-2) - 2(2-2) = 2$$

Therefore the solution is

$$x_1 = \frac{|A_1|}{|A|} = \frac{4}{2} = 2$$

$$x_2 = \frac{|A_1|}{|A|} = -\frac{4}{2} = -2$$

$$x_3 = \frac{|A_1|}{|A|} = \frac{2}{2} = 1$$

which is easily checked by direct substitution in the original equations.

Cramer's rule is probably most useful for systems of three equations in three unknowns but it can be applied to larger systems. The definition of a (4×4) determinant follows in the obvious way from that of a (3×3) determinant. For larger systems the algebra involved can become tedious and, for numerical problems, computer packages are available which solve such systems without generating serious rounding errors. A related approach is to use matrix methods, which are discussed in Chapter 2.

1.14 Exercises

Use Cramer's rule to find the values of the variables which satisfy the following equations.

1. $2x_1 + 2x_2 = 2$

   $3x_1 - x_2 = 1$

2. $x_1 + 3x_2 = 4$

   $2x_1 + x_2 = 3$

3. $x_1 - x_2 + x_3 = 4$

   $x_1 + x_2 + 3x_3 = 8$

$$x_1 + 2x_2 - x_3 = 0$$

4. $2x_1 + 2x_2 + x_3 = 3$

   $2x_1 - 2x_2 + x_3 = -1$

   $x_1 + x_2 - 2x_3 = 4$

5. $x_1 + x_2 - 2x_3 = 0$

   $2x_1 - 2x_2 - 2x_3 = -4$

   $x_1 + 2x_2 + x_3 = 8$

6. $x_1 + x_2 - 2x_3 = 3$

   $2x_1 - 3x_2 + x_3 = 1$

   $x_1 - x_2 - x_3 = 4.$

1.15 Applications of determinants

In this section we examine some applications of the use of determinants to solve simultaneous equations. The main areas of application are in linear macroeconomic models and in the analysis of equilibria in several markets.

We saw in section 1.10 that a simple three-equation national income model, given by the national income identity

$$Y = C + I + G,$$

a consumption function in which taxes are deducted from gross income,

$$C = a + b(Y - T)$$

and investment is a linear function of income,

$$I = c + dY,$$

can be solved for Y by substituting from the C and I equations into the Y equation and then the solutions for C and I follow.

We now solve this system using determinants. First the equations are written in the standard form, with the variables on the left hand side of the equals sign and the constants on the right hand side. For comparison we also write out the standard equations.

$$Y - C - I = G \qquad\qquad a_{11}x_1 + a_{12}x_2 + a_{13}x_3 = b_1$$

$$-bY + C = a - bT \qquad\qquad a_{21}x_1 + a_{22}x_2 + a_{23}x_3 = b_2$$

$$-dY + I = c \qquad\qquad a_{31}x_1 + a_{32}x_2 + a_{33}x_3 = b_3$$

In the general case the solutions are, using Cramer's rule,

$$x_1 = \frac{|\mathbf{A}_1|}{|\mathbf{A}|}, \; x_2 = \frac{|\mathbf{A}_2|}{|\mathbf{A}|}, \; x_3 = \frac{|\mathbf{A}_3|}{|\mathbf{A}|}$$

For the model the solutions are

$$Y = \frac{|\mathbf{A}_1|}{|\mathbf{A}|}, \; C = \frac{|\mathbf{A}_2|}{|\mathbf{A}|}, \; I = \frac{|\mathbf{A}_3|}{|\mathbf{A}|}$$

where

$$|\mathbf{A}| = \begin{vmatrix} 1 & -1 & -1 \\ -b & 1 & 0 \\ -d & 0 & 1 \end{vmatrix}, \; |\mathbf{A}_1| = \begin{vmatrix} G & -1 & -1 \\ a-bT & 1 & 0 \\ c & 0 & 1 \end{vmatrix}$$

$$|\mathbf{A}_2| = \begin{vmatrix} 1 & G & -1 \\ -b & a-bT & 0 \\ -d & c & 1 \end{vmatrix}, \; |\mathbf{A}_3| = \begin{vmatrix} 1 & -1 & G \\ -b & 1 & a-bT \\ -d & 0 & c \end{vmatrix}$$

By adding the second and third rows to the first row, and expanding by the first row,

$$|\mathbf{A}| = \begin{vmatrix} 1-b-d & 0 & 0 \\ -b & 1 & 0 \\ -d & 0 & 1 \end{vmatrix} = (1-b-d) \begin{vmatrix} 1 & 0 \\ 0 & 1 \end{vmatrix} = (1-b-d)$$

Similarly, $|\mathbf{A}_1| = G + a - bT + c$,

$$|\mathbf{A}_2| = \begin{vmatrix} 1-d & G+c & 0 \\ -b & a-bT & 0 \\ -d & c & 1 \end{vmatrix} = [(1-d)(a-bT)] - [(-b)(G+c)]$$

$$|\mathbf{A}_3| = \begin{vmatrix} 0 & -1 & G \\ 1-b & 1 & a-bT \\ -d & 0 & c \end{vmatrix} = \begin{vmatrix} 1-b & 0 & G+a-bT \\ 1-b & 1 & a-bT \\ -d & 0 & c \end{vmatrix} = \begin{vmatrix} 1-b & G+a-bT \\ -d & c \end{vmatrix}$$

$$= c(1-b) + d(G+a-bT)$$

Therefore,

$$Y = \frac{|\mathbf{A}_1|}{|\mathbf{A}|} = \frac{G + a - bT + c}{(1 - b - d)}$$

$$C = \frac{|\mathbf{A}_2|}{|\mathbf{A}|} = \frac{[(1 - d)(a - bT)] + b(G + c)}{(1 - b - d)}$$

$$I = \frac{|\mathbf{A_3}|}{|\mathbf{A}|} = \frac{c(1-b) + d(G + a - bT)}{(1 - b - d)}$$

These answers agree with the ones obtained at the end of section 1.10.

Our next application of determinants is to the analysis of equilibria in several markets. When different products are competing in the same market, the demand for each product will be affected by the prices of all the products. For example, suppose that there are three types of apple, A, B and C, being sold and their demand and supply equations are:

$$q_A^d = 20 - 2p_A + 4p_B + p_C \qquad q_A^s = -5 + 4p_A$$
$$q_B^d = 10 + 3p_A - 5p_B + 2p_C \qquad q_B^s = -7 + 3p_B$$
$$q_C^d = 70 + 4p_A + 2p_B - 5p_C \qquad q_C^s = -16 + 5p_C$$

Here the supply depends only on the price of the particular type of apple but the demands are interrelated. Equating q^s and q^d in each market, the equilibrium prices satisfy

$$20 - 2p_A + 4p_B + p_C = -5 + 4p_A$$
$$10 + 3p_A - 5p_B + 2p_C = -7 + 3p_B$$
$$70 + 4p_A - 2p_B - 5p_C = -16 + 5p_C$$

and rearranging in the standard form:

$$-6p_A + 4p_B + p_C = -25$$
$$3p_A - 8p_B + 2p_C = -17$$
$$4p_A + 2p_B - 10p_C = -86$$

Using determinants,

$$|\mathbf{A}| = \begin{vmatrix} -6 & 4 & 1 \\ 3 & -8 & 2 \\ 4 & 2 & -10 \end{vmatrix} = -266$$

$$|\mathbf{A_1}| = \begin{vmatrix} -25 & 4 & 1 \\ -17 & -8 & 2 \\ -86 & 2 & -10 \end{vmatrix} = -3990, \quad |\mathbf{A_2}| = \begin{vmatrix} -6 & -25 & 1 \\ 3 & -17 & 2 \\ 4 & -86 & -10 \end{vmatrix} = -3192$$

$$|\mathbf{A_3}| = \begin{vmatrix} -6 & 4 & -25 \\ 3 & -8 & -17 \\ 4 & 2 & -86 \end{vmatrix} = -4522$$

and so the equilibrium prices are $p_A = -3990/-266 = 15$, $p_B = -3192/-266 = 12$ and $p_C = -4522/-266 = 17$. The corresponding quantities of apples are $q_A = 55$, $q_B = 29$ and $q_C = 69$.

1.16 Exercises

1. A simple macroeconomic model is given by

$$Y = C + I + G$$
$$C = 20 + 0.7Y_d$$
$$I = 15 + 0.1Y$$
$$Y_d = Y - T$$
$$T = 5 + 0.3Y$$
$$G = 20$$

(a) By eliminating G, T and Y_d from the equations, reduce the system to three equations in three unknowns and hence use Cramer's rule to find the equilibrium values of Y, C and I. Is there a government budget deficit?

(b) How does the equilibrium change if the government has to balance its budget so that $G = T$?

2. In a simple national income model with a foreign sector the equations are

$$Y = C + I + G + (X - M)$$
$$C = a + bY$$
$$M = c + dY$$
$$T = fM$$

so that imports (M) depend on income and are taxed at a rate f per unit. Assume exports (X) are exogenous and fixed at $X = X_0$ and that $G = T$ and $I = I_0$. Determine the equilibrium levels of Y, C, M.

3. The market for a product has three main brands and the demands for each product are interrelated. For brand A demand (measured in thousands) is $q_A^d = 3 - p_A + p_B$ and supply is $q_s^A = p_A - 2$, for brand B, $q_d^B = 8 - 2p_B + p_c$ and $q_s^B = p_B - 1$ and for brand C, $q_d^C = 6 + 2p_A - p_c$ and $q_s^C = 2p_C - 2$. Find the prices and quantities which result in equilibrium for each brand.

4. Three products compete in a market and their demand and supply functions are

$$q_{d1} = 44 - 2p_1 + p_2 + p_3 \qquad q_{s1} = -10 + 3p_1$$

$$q_{d2} = 25 + p_1 - p_2 + 3p_3 \qquad q_{s2} = -15 + 5p_2$$

$$q_{d3} = 40 + 2p_1 + p_2 - 3p_3 \qquad q_{s3} = -18 + 2p_3$$

where p_1, p_2, and p_3 are the prices. Determine the prices and quantities which result in equilibrium in each market.

1.17   Revision exercises for Chapter 1 (without answers)

1. Sketch the graphs of the following equations for $x = 0$ to $x = 10$.

   (a) $y = 4x - 3$ (b) $y = 30 - 2x$ (c) $y = 15$

   (d) $x = 5$ (e) $y = 20x$

2. If total cost is 100 when output is 20, and total cost is 90 when output is 10, determine the equation of the total cost function, assuming it is linear.

3. If the total cost (y) is related to the level of output (x) by the equation

$$y = 30 + 6x$$

   determine the breakeven point if the price is 9 per unit. What would the net revenue be at an output of 8 units?

4. The supply function for a product is

$$q^s = 20p - 30$$

   where q is the quantity and p is the price. The demand function is

$$q^d = 170 - 5p$$

   (a) Determine the equilibrium price and quantity and sketch the demand and supply equations.

   (b) If the demand equation changes to

$$q^d = 180 - 10p$$

   what is the new equilibrium point?

5. Solve the following simultaneous equations:

   (a) $6x + 6y - 2z = 4$ (b) $2x + 2y - 3z = 4$

   $3x + 3y - z = 2$ $6x - 4y - 5z = 9$

   $x + 4y - z = 3$ $x + y + z = 2$

6. If the supply and demand equations for a product are

$$q^s = 50p - 40$$

$$q^d = 480 - 15p$$

determine the supplier's revenue at the equilibrium point. A flat-rate tax of 4 per unit is imposed. What effect does this have on the supplier's revenue?

7. In the following national income model, Y is national income, C is consumption, Y^d is disposable income, I is investment, G is government expenditure and T is tax revenue:

$$Y = C + I + G$$

$$C = 10 + 0.6Y^d$$

$$Y^d = Y - T$$

$$I = 25$$

$$T = 2 + 0.2Y$$

(a) If $G = 25$, determine the equilibrium level of national income and whether there is a budget deficit.

(b) If $G = T$, determine the equilibrium level of national income.

(c) Does changing the tax function to

$$T = 2 + 0.15Y$$

increase national income?

8. The goods market in an economy is given by

$$C = 25 + 0.7Y$$

$$I = 90 - 40i$$

$$Y = C + I$$

where C is consumption, Y is national income, I is investment and i is the rate of interest. The money market has money supply fixed at $M^s = 500$ and the demand for money is given by

$$M^d = L_1 + L_2$$

where the transactions demand for money is

$$L_1 = 0.25Y$$

and the speculative demand for money is

$$L_2 = 550 - 80i$$

(a) Determine the IS and LM schedules and sketch their graphs.

(b) What are the equilibrium levels of Y and i?

(c) If money supply changes to 550, what effect that this have on the equilibrium value of i?

9. A macroeconomic model of a closed economy is given by the following equations:

$$Y = C + I + G$$

$$C = a + bY^d$$

$$Y^d = Y - T$$

$$T = c + dY$$

$$I = I_0$$

$$G = G_0$$

where Y is national income, C is consumption, I is investment, Y^d is disposable income, T is tax revenue, G is government expenditure and both I_0 and G_0 are constants.

(a) What is the equilibrium level of national income?

(b) If the budget must balance (so that $G_0 = T$) what is the new equilibrium level of national income?

10. There are three main brands in the market for petrol in a particular city and the demand functions are all interrelated. They are:

$$q_1^d = 175 - 3p_1 + 2p_2 + 3p_3$$

$$q_2^d = 20 + p_1 - 2p_2 + 2p_3$$

$$q_3^d = 40 + 2p_1 + p_2 - 2p_3$$

The supply functions are

$$q_1^s = 6p_1 - 60$$

$$q_2^s = 3p_2 - 25$$

$$q_3^s = 3p_3 - 10$$

(a) What are the equilibrium prices and quantities?

(b) If a flat-rate tax of 2 per unit sold is imposed, what is the new equilibrium position?

Chapter 2
Elementary Matrix Algebra

2.1 The matrix notation

A *matrix* is a rectangular array of elements in *rows* and *columns*. Examples of matrices are

$$\begin{bmatrix} 1 & 0 \\ 0 & 1 \end{bmatrix} \quad \begin{bmatrix} a & b & c \\ c & c & b \\ 2a & 2c & b \end{bmatrix} \quad \begin{bmatrix} x_{11} & x_{12} & x_{13} \\ x_{21} & x_{22} & x_{23} \end{bmatrix}$$

which are rectangular arrays with 2 rows and 2 columns, 3 rows and 3 columns and 2 rows and 3 columns respectively. The *order* of a matrix is the number of rows and the number of columns, so that in the above examples the orders are (2×2) or (2 by 2), (3×3), and (2×3) respectively. The elements, which may be numbers or constants or variables, are enclosed between square brackets to signify that they must be considered as a whole and not individually. In contrast to the determinants of Chapter 1, a matrix does not have a single numerical value.

A matrix is often denoted by a single letter in bold-face type and a general matrix of order $m \times n$ is written as

$$\mathbf{X} = \begin{bmatrix} x_{11} & x_{12} & \dots & x_{1n} \\ x_{21} & x_{22} & \dots & x_{2n} \\ \vdots & \vdots & & \vdots \\ x_{m1} & x_{m2} & \dots & x_{mn} \end{bmatrix}$$

where the subscripts identify the row and column in which the element is located. For example

$$x_{12} \text{ is the element in row 1, column 2}$$

$$x_{34} \text{ is the element in row 3, column 4}$$

and $\qquad x_{ij}$ is the element in row i, column j.

43

A *square matrix* has an equal number of rows and columns and in the general case has order $(n \times n)$. If a matrix has only one row it is referred to as a *row vector* and similarly a matrix with only one column is referred to as a *column vector*.

Examples of these are

$$\mathbf{A} = \begin{bmatrix} a_{11} & a_{12} \\ a_{21} & a_{22} \end{bmatrix} \quad \text{is a square matrix of order } (2 \times 2)$$

$$\mathbf{B} = \begin{bmatrix} 3 & 4 & 5 & 1 \end{bmatrix} \quad \text{is a row vector of order } (1 \times 4)$$

$$\mathbf{C} = \begin{bmatrix} c_{11} \\ c_{21} \end{bmatrix} \quad \text{is a column vector of order } (2 \times 1)$$

2.2 Elementary matrix operations

MATRIX EQUALITY

Two matrices are equal if, and only if, the corresponding elements of each matrix are all equal. For example,

$$\begin{bmatrix} x_1 \\ x_2 \end{bmatrix} = \begin{bmatrix} 4 \\ 2 \end{bmatrix} \quad \text{if} \quad x_1 = 4 \quad \text{and} \quad x_2 = 2$$

$$\begin{bmatrix} x_{11} & x_{12} \\ x_{21} & x_{22} \end{bmatrix} = \begin{bmatrix} a_{11} & a_{12} \\ a_{21} & a_{22} \end{bmatrix} \quad \text{if} \quad \begin{aligned} x_{11} &= a_{11}, & x_{12} &= a_{12} \\ x_{21} &= a_{21}, & x_{22} &= a_{22} \end{aligned}$$

and, in general

$$\begin{bmatrix} x_{11} & x_{12} & \cdots & x_{1n} \\ x_{21} & x_{22} & \cdots & x_{2n} \\ \vdots & \vdots & & \vdots \\ x_{m1} & x_{m2} & \cdots & x_{mn} \end{bmatrix} = \begin{bmatrix} a_{11} & a_{12} & \cdots & a_{1n} \\ a_{21} & a_{22} & \cdots & a_{2n} \\ \vdots & \vdots & & \vdots \\ a_{m1} & a_{m2} & \cdots & a_{mn} \end{bmatrix}$$

that is, $\mathbf{X} = \mathbf{A}$, if $x_{ij} = a_{ij}$ for all i and all j, where x_{ij}, a_{ij} represent the elements in the ith row and jth column of the two matrices respectively.

It follows that two matrices can be equal only if they have the same number of elements arranged in corresponding positions, i.e. they must be of the same order. In the above case, both $\mathbf{X}$ and $\mathbf{A}$ are $(m \times n)$.

SCALAR MULTIPLICATION

A *scalar* is a real number, in contrast to a vector or matrix. If k is a scalar then the product of k and a matrix is obtained by multiplying each element of the matrix by k. For example,

$$\mathbf{X} = \begin{bmatrix} x_{11} & x_{12} \\ x_{21} & x_{22} \\ x_{31} & x_{32} \end{bmatrix} \qquad k\mathbf{X} = \begin{bmatrix} kx_{11} & kx_{12} \\ kx_{21} & kx_{22} \\ kx_{31} & kx_{32} \end{bmatrix}$$

$$\mathbf{A} = \begin{bmatrix} 2 & 4 \\ 0 & -1 \end{bmatrix} \qquad 3\mathbf{A} = \begin{bmatrix} 6 & 12 \\ 0 & -3 \end{bmatrix}$$

ADDITION AND SUBTRACTION

If $\mathbf{A}$ and $\mathbf{B}$ are of the same order then $\mathbf{C} = \mathbf{A} + \mathbf{B}$ is found by adding the corresponding elements in $\mathbf{A}$ and $\mathbf{B}$, and $\mathbf{D} = \mathbf{A} - \mathbf{B}$ is found by subtracting the corresponding elements in $\mathbf{B}$ from $\mathbf{A}$. For example,

Let
$$\mathbf{A} = \begin{bmatrix} 1 & 3 \\ 2 & 4 \\ 0 & -1 \end{bmatrix} \qquad \mathbf{B} = \begin{bmatrix} 2 & x \\ 0 & -1 \\ 2 & 2 \end{bmatrix}$$

then
$$\mathbf{C} = \mathbf{A} + \mathbf{B} = \begin{bmatrix} 1+2 & 3+x \\ 2+0 & 4+(-1) \\ 0+2 & -1+2 \end{bmatrix} = \begin{bmatrix} 3 & 3+x \\ 2 & 3 \\ 2 & 1 \end{bmatrix}$$

and
$$\mathbf{D} = \mathbf{A} - \mathbf{B} = \begin{bmatrix} 1-2 & 3-x \\ 2-0 & 4-(-1) \\ 0-2 & -1-2 \end{bmatrix} = \begin{bmatrix} -1 & 3-x \\ 2 & 5 \\ -2 & -3 \end{bmatrix}$$

MATRIX MULTIPLICATION

Two matrices $\mathbf{A}$ and $\mathbf{B}$ can be multiplied together to give $\mathbf{AB}$ only if the number of *columns* in $\mathbf{A}$ is equal to the number of *rows* in $\mathbf{B}$. For example,

if
$$\mathbf{A} = \begin{bmatrix} a_{11} & a_{12} \\ a_{21} & a_{22} \\ a_{31} & a_{32} \end{bmatrix} \qquad \text{and} \qquad \mathbf{B} = \begin{bmatrix} b_{11} & b_{12} & b_{13} \\ b_{21} & b_{22} & b_{23} \end{bmatrix}$$

then $\mathbf{A}$ is (3×2), $\mathbf{B}$ is (2×3) and the product of $\mathbf{A}$ and $\mathbf{B}$ can be found since $\mathbf{A}$ has 2 columns and $\mathbf{B}$ has 2 rows. The product matrix $\mathbf{C}$ is

$$\mathbf{C} = \mathbf{AB} = \begin{bmatrix} a_{11} & a_{12} \\ a_{21} & a_{22} \\ a_{31} & a_{32} \end{bmatrix} \begin{bmatrix} b_{11} & b_{12} & b_{13} \\ b_{21} & b_{22} & b_{23} \end{bmatrix}$$

$$= \begin{bmatrix} a_{11}b_{11} + a_{12}b_{21} & a_{11}b_{12} + a_{12}b_{22} & a_{11}b_{13} + a_{12}b_{23} \\ a_{21}b_{11} + a_{22}b_{21} & a_{21}b_{12} + a_{22}b_{22} & a_{21}b_{13} + a_{22}b_{23} \\ a_{31}b_{11} + a_{32}b_{21} & a_{31}b_{12} + a_{32}b_{22} & a_{31}b_{13} + a_{32}b_{23} \end{bmatrix}$$

$$= \begin{bmatrix} c_{11} & c_{12} & c_{13} \\ c_{21} & c_{22} & c_{23} \\ c_{31} & c_{32} & c_{33} \end{bmatrix}$$

where $c_{11} = a_{11}b_{11} + a_{12}b_{21}$

$c_{12} = a_{11}b_{12} + a_{12}b_{22}$

and the general term is $c_{ij} = a_{i1}b_{1j} + a_{i2}b_{2j}$.

The elements of $\mathbf{C}$ are obtained by multiplying row i from $\mathbf{A}$ by column j from $\mathbf{B}$ to give c_{ij}.

Notice that $\mathbf{A}$ is (3×2), $\mathbf{B}$ is (2×3) and $\mathbf{C}$ is $(3 \times 2)(2 \times 3)$ $= (3 \times 3)$.

Example 1

$$\mathbf{A} = \begin{bmatrix} 2 & 1 \\ 4 & -1 \end{bmatrix} \qquad \mathbf{B} = \begin{bmatrix} 3 \\ 2 \end{bmatrix}$$

$\mathbf{A}$ is (2×2), $\mathbf{B}$ is (2×1) so $\mathbf{C}$ is $(2 \times 2)(2 \times 1) = (2 \times 1)$

$$\mathbf{C} = \begin{bmatrix} 2 & 1 \\ 4 & -1 \end{bmatrix} \begin{bmatrix} 3 \\ 2 \end{bmatrix} = \begin{bmatrix} 2 \times 3 + 1 \times 2 \\ 4 \times 3 + (-1) \times 2 \end{bmatrix} = \begin{bmatrix} 6 + 2 \\ 12 - 2 \end{bmatrix} = \begin{bmatrix} 8 \\ 10 \end{bmatrix}$$

Example 2

$$\mathbf{A} = \begin{bmatrix} 2 & 1 & 4 \\ 3 & 2 & 0 \end{bmatrix}, \qquad \mathbf{B} = \begin{bmatrix} 2 & 3 \\ 1 & 4 \end{bmatrix}$$

Here $\mathbf{A}$ is (2×3) and $\mathbf{B}$ is (2×2) and so the product $\mathbf{AB}$ is not defined since the number of columns in $\mathbf{A}$ does not equal the number of rows in $\mathbf{B}$. However, if

$$\mathbf{B} = \begin{bmatrix} 2 & 3 \\ 1 & 4 \\ 0 & 1 \end{bmatrix}$$

then **B** is (3×2) and the product is of order $(2 \times 3)(3 \times 2) = (2 \times 2)$

$$\mathbf{C} = \mathbf{AB} = \begin{bmatrix} 2 & 1 & 4 \\ 3 & 2 & 0 \end{bmatrix} \begin{bmatrix} 2 & 3 \\ 1 & 4 \\ 0 & 1 \end{bmatrix}$$

$$= \begin{bmatrix} 2 \times 2 + 1 \times 1 + 4 \times 0 & 2 \times 3 + 1 \times 4 + 4 \times 1 \\ 3 \times 2 + 2 \times 1 + 0 \times 0 & 3 \times 3 + 2 \times 4 + 0 \times 1 \end{bmatrix}$$

$$= \begin{bmatrix} 5 & 14 \\ 8 & 17 \end{bmatrix}$$

In this case it is also possible to form the product $\mathbf{D} = \mathbf{BA}$, since **B** is (3×2) and **A** is (2×3) and hence **D** is of order $(3 \times 2)(2 \times 3) = (3 \times 3)$

$$\mathbf{D} = \mathbf{BA} = \begin{bmatrix} 2 & 3 \\ 1 & 4 \\ 0 & 1 \end{bmatrix} \begin{bmatrix} 2 & 1 & 4 \\ 3 & 2 & 0 \end{bmatrix}$$

$$= \begin{bmatrix} 2 \times 2 + 3 \times 3 & 2 \times 1 + 3 \times 2 & 2 \times 4 + 3 \times 0 \\ 1 \times 2 + 4 \times 3 & 1 \times 1 + 4 \times 2 & 1 \times 4 + 4 \times 0 \\ 0 \times 2 + 1 \times 3 & 0 \times 1 + 1 \times 2 & 0 \times 4 + 1 \times 0 \end{bmatrix}$$

$$= \begin{bmatrix} 13 & 8 & 8 \\ 14 & 9 & 4 \\ 3 & 2 & 0 \end{bmatrix}$$

Here **D** is (3×3) whilst **C** is (2×2). Therefore **D** and **C** cannot be equal and we have

$$\mathbf{AB} \neq \mathbf{BA}$$

This is true in this particular case and in fact is usually true even when the products **AB** and **BA** are of the same order.

Example 3

$$\mathbf{A} = \begin{bmatrix} a_{11} & a_{12} \\ a_{21} & a_{22} \end{bmatrix} \qquad \mathbf{X} = \begin{bmatrix} x_1 \\ x_2 \end{bmatrix}$$

A is (2×2) and **X** is (2×1) so the product **AX** is $(2 \times 2)(2 \times 1) = (2 \times 1)$ and is

$$\mathbf{AX} = \begin{bmatrix} a_{11} & a_{12} \\ a_{21} & a_{22} \end{bmatrix} \begin{bmatrix} x_1 \\ x_2 \end{bmatrix} = \begin{bmatrix} a_{11}x_1 + a_{12}x_2 \\ a_{21}x_1 + a_{22}x_2 \end{bmatrix}$$

It follows from this that the set of simultaneous equations

$$a_{11}x_1 + a_{12}x_2 = b_1$$
$$a_{21}x_1 + a_{22}x_2 = b_2$$

can be represented by

$$\mathbf{AX} = \mathbf{b}$$

where

$$\mathbf{b} = \begin{bmatrix} b_1 \\ b_2 \end{bmatrix}$$

THE IDENTITY OR UNIT MATRIX

Corresponding to the number 1 in ordinary algebra there is a matrix **I**, known as the identity or unit matrix, which has the property

$$\mathbf{IA} = \mathbf{A} = \mathbf{AI}$$

(compare with $1 \times a = a = a \times 1$) and **I** is the square matrix

$$\mathbf{I} = \begin{bmatrix} 1 & 0 & 0 \\ 0 & 1 & 0 \\ 0 & 0 & 1 \end{bmatrix}$$

for order (3×3) and for order $(n \times n)$,

$$\mathbf{I} = \begin{bmatrix} 1 & 0 \dots 0 \\ 0 & 1 \dots 0 \\ \vdots & \vdots & \vdots \\ 0 & 0 \dots 1 \end{bmatrix}$$

For example, if

$$\mathbf{A} = \begin{bmatrix} 1 & 2 \\ 4 & 3 \end{bmatrix}, \qquad \mathbf{I} = \begin{bmatrix} 1 & 0 \\ 0 & 1 \end{bmatrix}$$

then

$$\mathbf{IA} = \begin{bmatrix} 1 & 0 \\ 0 & 1 \end{bmatrix}\begin{bmatrix} 1 & 2 \\ 4 & 3 \end{bmatrix} = \begin{bmatrix} 1 \times 1 + 0 \times 4 & 1 \times 2 + 0 \times 3 \\ 0 \times 1 + 1 \times 4 & 0 \times 2 + 1 \times 3 \end{bmatrix}$$

$$= \begin{bmatrix} 1 & 2 \\ 4 & 3 \end{bmatrix} = \mathbf{A}$$

THE TRANSPOSE OF A MATRIX

If the rows and columns of a matrix are interchanged the new matrix is known as the *transpose* of the original matrix. For example,

if
$$\mathbf{A} = \begin{bmatrix} 1 & 2 \\ 3 & 4 \end{bmatrix}$$

the transpose,
$$\mathbf{A}' = \begin{bmatrix} 1 & 3 \\ 2 & 4 \end{bmatrix}$$

where the first row of $\mathbf{A}$, 1 2 becomes the first column of $\mathbf{A}'$, and the second row of $\mathbf{A}$, 3 4 becomes the second column of $\mathbf{A}'$.

If
$$\mathbf{A} = \begin{bmatrix} a_{11} & a_{12} & a_{13} \\ a_{21} & a_{22} & a_{23} \end{bmatrix}$$
then
$$\mathbf{A}' = \begin{bmatrix} a_{11} & a_{21} \\ a_{12} & a_{22} \\ a_{13} & a_{23} \end{bmatrix}$$

If
$$\mathbf{A} = \begin{bmatrix} a & b & c \end{bmatrix}$$
then
$$\mathbf{A}' = \begin{bmatrix} a \\ b \\ c \end{bmatrix}$$

2.3 Exercises

1. If
$$\mathbf{A} = \begin{bmatrix} 2 & 4 \\ 1 & 3 \end{bmatrix} \quad \mathbf{B} = \begin{bmatrix} 2 & 0 \\ -1 & 1 \end{bmatrix} \quad \mathbf{C} = \begin{bmatrix} 3 \\ 1 \end{bmatrix}$$

form the following (if they exist)

(a) $\mathbf{A} + \mathbf{B}$, (b) $\mathbf{A} - 2\mathbf{B}$, (c) $\mathbf{AB}$, (d) $\mathbf{AC}$, (e) $\mathbf{CA}$, (f) $\mathbf{AB}'$, (g) $\mathbf{C}'\mathbf{A}$ and (h) show that $\mathbf{AB} \neq \mathbf{BA}$.

2. Given that

$$A = \begin{bmatrix} 1 & 0 & 2 \\ 1 & 1 & 3 \\ 0 & 2 & -1 \end{bmatrix} \qquad B = \begin{bmatrix} 2 & 1 \\ 1 & -1 \\ 2 & 2 \end{bmatrix} \qquad C = [\, 1 \quad 0 \quad 2 \,]$$

form the following (if they exist):

(a) **AB**, (b) **A'B**, (c) **AC**, (d) **AC'**, (e) **CB**.

2.4   The inverse matrix

In section 2.2 we saw that a set of simultaneous equations could be written in matrix form. For example the equations

$$a_{11}x_1 + a_{12}x_2 + a_{13}x_3 = b_1$$

$$a_{21}x_1 + a_{22}x_2 + a_{23}x_3 = b_2$$

$$a_{31}x_1 + a_{32}x_2 + a_{33}x_3 = b_3$$

can be written as **AX = b**

where $\qquad A = \begin{bmatrix} a_{11} & a_{12} & a_{13} \\ a_{21} & a_{22} & a_{23} \\ a_{31} & a_{32} & a_{33} \end{bmatrix} \qquad X = \begin{bmatrix} x_1 \\ x_2 \\ x_3 \end{bmatrix} \qquad b = \begin{bmatrix} b_1 \\ b_2 \\ b_3 \end{bmatrix}$

The solution of these equations is obtained by expressing the unknowns x_1, x_2, x_3 in terms of the known constants $a_{11}, a_{12}, \ldots, a_{33}, b_1, b_2$ and b_3. For the matrix equation **AX = b** this is equivalent to transferring the **A** to the other side of the equation. To do this we define a matrix A^{-1}, known as the *inverse* of **A**, which has the property

$$AA^{-1} = I = A^{-1}A$$

i.e. the product of a matrix and its inverse is the unit matrix. If the inverse of **A** exists then **A** is said to be *non-singular*.

Since $\qquad\qquad\qquad\qquad$ **AX = b**

multiplying each side by A^{-1}, we have

$$A^{-1}AX = A^{-1}b$$

or $\qquad\qquad\qquad\qquad$ **IX = A⁻¹b**

But **IX = X** and so $\qquad\qquad\qquad$ **X = A⁻¹b**

is the solution to the equation.

Before considering how to obtain A^{-1} from a matrix **A** we recall that in section 1.13 the condition for a solution to exist for a set of simultaneous equations was shown to be that $|\mathbf{A}| \neq 0$, where $|\mathbf{A}|$ is the determinant of the coefficients of the unknowns. In the case of three equations in three unknowns,

$$|\mathbf{A}| = \begin{vmatrix} a_{11} & a_{12} & a_{13} \\ a_{21} & a_{22} & a_{23} \\ a_{31} & a_{32} & a_{33} \end{vmatrix}$$

and this was evaluated as

$$|\mathbf{A}| = a_{11} \begin{vmatrix} a_{22} & a_{23} \\ a_{32} & a_{33} \end{vmatrix} - a_{12} \begin{vmatrix} a_{21} & a_{23} \\ a_{31} & a_{33} \end{vmatrix} + a_{13} \begin{vmatrix} a_{21} & a_{22} \\ a_{31} & a_{32} \end{vmatrix}$$

These (2×2) determinants are the *minors* of the elements of $\mathbf{A}$, so that the minors of a_{11} and a_{12} are respectively

$$\begin{vmatrix} a_{22} & a_{23} \\ a_{32} & a_{33} \end{vmatrix} \quad \text{and} \quad \begin{vmatrix} a_{21} & a_{23} \\ a_{31} & a_{33} \end{vmatrix}$$

and in general the minor of a_{ij} is the determinant of $\mathbf{A}$ with row i and column j eliminated. The *cofactor* (c_{ij}) of the element a_{ij} of the matrix $\mathbf{A}$ is the minor of a_{ij} multiplied by $(-1)^{i+j}$, so that if $i + j$ is even the cofactor and minor are equal, and if $i + j$ is odd, the cofactor is the negative of the minor.

For example if

$$\mathbf{A} = \begin{bmatrix} a_{11} & a_{12} & a_{13} \\ a_{21} & a_{22} & a_{23} \\ a_{31} & a_{32} & a_{33} \end{bmatrix}$$

the minor of a_{22} is

$$\begin{vmatrix} a_{11} & a_{12} & a_{13} \\ a_{21} & \cdots & a_{22} & \cdots & a_{23} \\ a_{31} & a_{32} & a_{33} \end{vmatrix} = \begin{vmatrix} a_{11} & a_{13} \\ a_{31} & a_{33} \end{vmatrix}$$

and the cofactor of a_{22} is

$$c_{22} = (-1)^{2+2} \begin{vmatrix} a_{11} & a_{13} \\ a_{31} & a_{33} \end{vmatrix} = \begin{vmatrix} a_{11} & a_{13} \\ a_{31} & a_{33} \end{vmatrix}$$

For a_{23} the minor is

$$\begin{vmatrix} a_{11} & a_{12} \\ a_{31} & a_{32} \end{vmatrix}$$

and the cofactor is

$$c_{23} = (-1)^{2+3} \begin{vmatrix} a_{11} & a_{12} \\ a_{31} & a_{32} \end{vmatrix} = - \begin{vmatrix} a_{11} & a_{12} \\ a_{31} & a_{32} \end{vmatrix}$$

It follows that for **A** of order 3×3 then

$$|\mathbf{A}| = a_{11}c_{11} + a_{12}c_{12} + a_{13}c_{13}$$

The *cofactor matrix* is the matrix with elements c_{ij}, the cofactors of **A**, and this expansion of the determinant of **A** is known as the *Laplace expansion*. It extends in the obvious way to fourth- and higher- order determinants.

Example 1

$$\mathbf{A} = \begin{bmatrix} 2 & 1 \\ 0 & 4 \end{bmatrix}$$

The minor of a_{11} is
$$\begin{vmatrix} 2 & 1 \\ 0 & 4 \end{vmatrix} = 4$$

The minor of a_{12} is
$$\begin{vmatrix} 2 & 1 \\ 0 & 4 \end{vmatrix} = 0$$

The minor of a_{21} is
$$\begin{vmatrix} 2 & 1 \\ 0 & 4 \end{vmatrix} = 1$$

The minor of a_{22} is
$$\begin{vmatrix} 2 & 1 \\ 0 & 4 \end{vmatrix} = 2$$

The cofactors are $c_{11} = 4$, $c_{12} = 0$, $c_{21} = -1$, $c_{22} = 2$. The cofactor matrix

is
$$C = \begin{bmatrix} c_{11} & c_{12} \\ c_{21} & c_{22} \end{bmatrix} = \begin{bmatrix} 4 & 0 \\ -1 & 2 \end{bmatrix}$$

Example 2

$$A = \begin{bmatrix} 2 & 1 & 1 \\ 0 & 4 & -1 \\ 2 & 2 & 1 \end{bmatrix}$$

The minor of a_{11} is $\begin{vmatrix} 4 & -1 \\ 2 & 1 \end{vmatrix} = 4 + 2 = 6$

The minor of a_{12} is $\begin{vmatrix} 0 & -1 \\ 2 & 1 \end{vmatrix} = 2$

The minor of a_{13} is $\begin{vmatrix} 0 & 4 \\ 2 & 2 \end{vmatrix} = -8$

The minor of a_{21} is $\begin{vmatrix} 1 & 1 \\ 2 & 1 \end{vmatrix} = -1$

The minor of a_{22} is $\begin{vmatrix} 2 & 1 \\ 2 & 1 \end{vmatrix} = 0$

The minor of a_{23} is $\begin{vmatrix} 2 & 1 \\ 2 & 2 \end{vmatrix} = 2$

The minor of a_{31} is $\begin{vmatrix} 1 & 1 \\ 4 & -1 \end{vmatrix} = -5$

The minor of a_{32} is
$$\begin{vmatrix} 2 & 1 \\ 0 & -1 \end{vmatrix} = -2$$

The minor of a_{33} is
$$\begin{vmatrix} 2 & 1 \\ 0 & 4 \end{vmatrix} = 8$$

The cofactor matrix is

$$\mathbf{C} = \begin{bmatrix} 6 & -2 & -8 \\ 1 & 0 & -2 \\ -5 & 2 & 8 \end{bmatrix}$$

One more definition is required before the inverse can be obtained: the *adjoint* of $\mathbf{A}$ is the transpose of the cofactor matrix of $\mathbf{A}$. Thus, if

$$\mathbf{C} = \begin{bmatrix} c_{11} & c_{12} & c_{13} \\ c_{21} & c_{22} & c_{23} \\ c_{31} & c_{32} & c_{33} \end{bmatrix} \qquad \text{adjoint } (\mathbf{A}) = \begin{bmatrix} c_{11} & c_{21} & c_{31} \\ c_{12} & c_{22} & c_{32} \\ c_{13} & c_{23} & c_{33} \end{bmatrix}$$

The inverse of a matrix $\mathbf{A}$ is given by

$$\mathbf{A}^{-1} = \frac{1}{|\mathbf{A}|} \text{ adjoint } (\mathbf{A})$$

This can be shown to be true quite easily when $\mathbf{A}$ is (2×2) and with more difficulty for higher orders.

Let
$$\mathbf{A} = \begin{bmatrix} a_{11} & a_{12} \\ a_{21} & a_{22} \end{bmatrix} \qquad \text{then} \qquad \mathbf{C} = \begin{bmatrix} a_{22} & -a_{21} \\ -a_{12} & a_{11} \end{bmatrix}$$

and hence

$$\text{adjoint } (\mathbf{A}) = \begin{bmatrix} a_{22} & -a_{12} \\ -a_{21} & a_{11} \end{bmatrix}$$

Now

$$|\mathbf{A}| = \begin{vmatrix} a_{11} & a_{12} \\ a_{21} & a_{22} \end{vmatrix} = a_{11}a_{22} - a_{12}a_{21}$$

If $\mathbf{A}^{-1}$ is the inverse then $\mathbf{A}\mathbf{A}^{-1} = \mathbf{I}$.

Now
$$\mathbf{A}\mathbf{A}^{-1} = \frac{\mathbf{A} \text{ adjoint } (\mathbf{A})}{|\mathbf{A}|}$$

$$= \frac{1}{|\mathbf{A}|} \begin{bmatrix} a_{11} & a_{12} \\ a_{21} & a_{22} \end{bmatrix} \begin{bmatrix} a_{22} & -a_{12} \\ -a_{21} & a_{11} \end{bmatrix}$$

$$= \frac{1}{|\mathbf{A}|} \begin{bmatrix} a_{11}a_{22} - a_{12}a_{21} & -a_{11}a_{12} + a_{12}a_{11} \\ a_{21}a_{22} - a_{22}a_{21} & -a_{21}a_{12} + a_{22}a_{11} \end{bmatrix}$$

$$= \frac{1}{|\mathbf{A}|} \begin{bmatrix} a_{11}a_{22} - a_{12}a_{21} & 0 \\ 0 & a_{11}a_{22} - a_{12}a_{21} \end{bmatrix}$$

$$= \mathbf{I}$$

Hence $\mathbf{A}^{-1}$ is the inverse of $\mathbf{A}$.

Example 3

$$2x_1 + 4x_2 = 2$$
$$3x_1 + 4x_2 = 1$$

Let

$$\mathbf{A} = \begin{bmatrix} 2 & 4 \\ 3 & 4 \end{bmatrix} \quad \text{then} \quad \mathbf{C} = \begin{bmatrix} 4 & -3 \\ -4 & 2 \end{bmatrix}$$

and $|\mathbf{A}| = -4$

$$\therefore \quad \mathbf{A}^{-1} = \frac{1}{|\mathbf{A}|} \text{ adjoint } (\mathbf{A}) = \frac{1}{-4} \begin{bmatrix} 4 & -4 \\ -3 & 2 \end{bmatrix} = \begin{bmatrix} -1 & 1 \\ -\frac{3}{4} & -\frac{1}{2} \end{bmatrix}$$

Check: $\mathbf{A}\mathbf{A}^{-1} = \begin{bmatrix} 2 & 4 \\ 3 & 4 \end{bmatrix}\begin{bmatrix} -1 & 1 \\ \frac{3}{4} & -\frac{1}{2} \end{bmatrix} = \begin{bmatrix} 1 & 0 \\ 0 & 1 \end{bmatrix}$

The solution to the equations is

$$\mathbf{X} = \begin{bmatrix} x_1 \\ x_2 \end{bmatrix} = \mathbf{A}^{-1}\mathbf{b} = \begin{bmatrix} -1 & 1 \\ \frac{3}{4} & -\frac{1}{2} \end{bmatrix}\begin{bmatrix} 2 \\ 1 \end{bmatrix}$$

$$= \begin{bmatrix} -1 \\ 1 \end{bmatrix} \quad \text{and so} \quad x_1 = -1, \qquad x_2 = 1$$

Example 4

$$x_1 + x_2 + x_3 = 6$$
$$2x_1 - x_2 + 2x_3 = 6$$
$$x_1 - x_3 = -2$$

Here

$$\mathbf{A} = \begin{bmatrix} 1 & 1 & 1 \\ 2 & -1 & 2 \\ 1 & 0 & -1 \end{bmatrix} \quad \text{and} \quad \mathbf{C} = \begin{bmatrix} 1 & 4 & 1 \\ 1 & -2 & 1 \\ 3 & 0 & -3 \end{bmatrix}$$

and $|\mathbf{A}| = 6$

$$\mathbf{A}^{-1} = \frac{1}{|\mathbf{A}|} \text{ adjoint } (\mathbf{A}) = \frac{1}{6}\begin{bmatrix} 1 & 1 & 3 \\ 4 & -2 & 0 \\ 1 & 1 & -3 \end{bmatrix}$$

Check:

$$\mathbf{A}\mathbf{A}^{-1} = \begin{bmatrix} 1 & 1 & 1 \\ 2 & -1 & 2 \\ 1 & 0 & -1 \end{bmatrix}\frac{1}{6}\begin{bmatrix} 1 & 1 & 3 \\ 4 & -2 & 0 \\ 1 & 1 & -3 \end{bmatrix} = \begin{bmatrix} 1 & 0 & 0 \\ 0 & 1 & 0 \\ 0 & 0 & 1 \end{bmatrix}$$

The solution to the equations is

$$\mathbf{X} = \begin{bmatrix} x_1 \\ x_2 \\ x_3 \end{bmatrix} = \mathbf{A}^{-1}\mathbf{b} = \frac{1}{6}\begin{bmatrix} 1 & 1 & 3 \\ 4 & -2 & 0 \\ 1 & 1 & -3 \end{bmatrix}\begin{bmatrix} 6 \\ 6 \\ -2 \end{bmatrix} = \begin{bmatrix} 1 \\ 2 \\ 3 \end{bmatrix}$$

and so $x_1 = 1$, $x_2 = 2$, $x_3 = 3$.

2.5 Exercises

1. Obtain the inverse of

$$\mathbf{A} = \begin{bmatrix} 2 & 1 \\ 1 & 3 \end{bmatrix}$$

and hence solve the equations

$$2x_1 + x_2 = 4$$
$$x_1 + 3x_2 = 7$$

2. Solve the following equations using the inverse matrix method.

(a)
$$2x_1 + 5x_2 = 2$$
$$3x_1 - 4x_2 = 10$$

(b)
$$2x_1 + 2x_2 + x_3 = 1$$
$$3x_1 + x_2 + x_3 = 2$$
$$x_1 + x_2 + x_3 = 2$$

(c)
$$x_1 - x_2 + x_3 = 3$$
$$2x_1 - 2x_2 + x_3 = 5$$
$$x_1 + x_2 + 2x_3 = 4$$

2.6 Inversion by Gaussian elimination

The inverse of a matrix can also be found by the method known as *Gaussian elimination*. This is particularly suitable when the elements of the matrix are numbers. The method involves augmenting the matrix to be inverted by including the identity matrix next to it. Then row operations are applied to both matrices until the original matrix is transformed into the unit matrix, whereupon the identity matrix will have been transformed into the inverse matrix. The permitted row operations are:

(1) Any row of the matrix may be divided (or multiplied) by any non-zero constant.
(2) A multiple of any row may be subtracted from (or added to) any other row.

For example, to find the inverse of

$$A = \begin{bmatrix} 2 & -1 \\ -1 & 1 \end{bmatrix}$$

write it as $A \mid I$ or

$$\left[\begin{array}{cc|cc} 2 & -1 & 1 & 0 \\ -1 & 1 & 0 & 1 \end{array}\right]$$

The matrix A is to be transformed into the unit matrix so that the first column becomes 1, 0 and the second column, 0, 1. This is best done by working systematically on the first column, and then the second. Now dividing the first row by 2 gives the desired 1 in the first-row, first-column position:

$$\left[\begin{array}{cc|cc} 1 & -0.5 & 0.5 & 0 \\ -1 & 1 & 0 & 1 \end{array}\right]$$

By adding the first row to the second, a 0 will occur in the second-row, first-column position:

$$\left[\begin{array}{cc|cc} 1 & -0.5 & 0.5 & 0 \\ 0 & 0.5 & 0.5 & 1 \end{array}\right]$$

Next, add the second row to the first, to give a 0 in the first-row, second-column position.

$$\left[\begin{array}{cc|cc} 1 & 0 & 1 & 1 \\ 0 & 0.5 & 0.5 & 1 \end{array}\right]$$

Finally, divide the second row by 0.5 to give a 1 in the second-row, second-column position.

$$\left[\begin{array}{cc|cc} 1 & 0 & 1 & 1 \\ 0 & 1 & 1 & 2 \end{array}\right]$$

The result is the unit matrix and so the inverse is

$$\mathbf{A}^{-1} = \begin{bmatrix} 1 & 1 \\ 1 & 2 \end{bmatrix}$$

which can easily be checked by showing $\mathbf{AA}^{-1} = \mathbf{I}$.

For a second example, to find the inverse of

$$\mathbf{B} = \begin{bmatrix} 3 & -1 & 1 \\ -1 & 1 & 0 \\ 2 & 2 & -1 \end{bmatrix}$$

write the matrix next to the (3×3) unit matrix:

$$\begin{bmatrix} 3 & -1 & 1 & \vert & 1 & 0 & 0 \\ -1 & 1 & 0 & \vert & 0 & 1 & 0 \\ 2 & 2 & -1 & \vert & 0 & 0 & 1 \end{bmatrix}$$

The procedure is to transform the first column into 1, 0, 0, the second into 0, 1, 0, and the third into 0, 0, 1. Here we add twice the second row to the first, to give a 1 in the $(1,1)$ position, and also add twice row two to row three to give a 0 in the $(3,1)$ position:

$$\begin{bmatrix} 1 & 1 & 1 & \vert & 1 & 2 & 0 \\ -1 & 1 & 0 & \vert & 0 & 1 & 0 \\ 0 & 4 & -1 & \vert & 0 & 2 & 1 \end{bmatrix}$$

Next, add the first row to the second to give a 0 in the $(2,1)$ position, giving the required 1, 0, 0 in the first column:

$$\begin{bmatrix} 1 & 1 & 1 & \vert & 1 & 2 & 0 \\ 0 & 2 & 1 & \vert & 1 & 3 & 0 \\ 0 & 4 & -1 & \vert & 0 & 2 & 1 \end{bmatrix}$$

To change the $(2,2)$ element to 1, divide the second row by 2:

$$\begin{bmatrix} 1 & 1 & 1 & \vert & 1 & 2 & 0 \\ 0 & 1 & 0.5 & \vert & 0.5 & 1.5 & 0 \\ 0 & 4 & -1 & \vert & 0 & 2 & 1 \end{bmatrix}$$

Subtract row 2 from row 1, and subtract four times row 2 from row 3 to give 0, 1, 0 in the second column:

$$\begin{bmatrix} 1 & 0 & 0.5 & | & 0.5 & 0.5 & 0 \\ 0 & 1 & 0.5 & | & 0.5 & 1.5 & 0 \\ 0 & 0 & -3 & | & -2 & -4 & 1 \end{bmatrix}$$

Divide row 3 by 6, and add the new row to rows 1 and 2:

$$\begin{bmatrix} 1 & 0 & 0 & | & 1/6 & -1/6 & 1/6 \\ 0 & 1 & 0 & | & 1/6 & 5/6 & 1/6 \\ 0 & 0 & -0.5 & | & -1/3 & -2/3 & 1/6 \end{bmatrix}$$

Multiply row 3 by -2, to give 0, 0, 1 in the third column:

$$\begin{bmatrix} 1 & 0 & 0 & | & 1/6 & -1/6 & 1/6 \\ 0 & 1 & 0 & | & 1/6 & 5/6 & 1/6 \\ 0 & 0 & 1 & | & 2/3 & 4/3 & -1/3 \end{bmatrix}$$

The result is that

$$\mathbf{B}^{-1} = \begin{bmatrix} 1/6 & -1/6 & 1/6 \\ 1/6 & 5/6 & 1/6 \\ 2/3 & 4/3 & -1/3 \end{bmatrix}$$

and this can easily be checked by showing that $\mathbf{BB}^{-1} = \mathbf{I}$.

The method of Gaussian elimination can also be used to solve simultaneous linear equations, giving both the inverse of the coefficient matrix and the values of the unknowns. For example, the equations

$$2x - 3y - z = 0$$
$$x + y + z = 4$$
$$-x + y + 2z = 1$$

can be solved by writing the coefficient matrix, the unit matrix and the column of constants thus:

$$\begin{bmatrix} 2 & -3 & -1 & | & 1 & 0 & 0 & | & 0 \\ 1 & 1 & 1 & | & 0 & 1 & 0 & | & 4 \\ -1 & 1 & 2 & | & 0 & 0 & 1 & | & 1 \end{bmatrix}$$

and applying the row operations as before. If the third row is added to the first row and then the second row, this gives a 1 in the (1, 1) position, and a 0 in the (2, 1) position:

$$\begin{bmatrix} 1 & -2 & 1 & 1 & 0 & 1 & 1 \\ 0 & 2 & 3 & 0 & 1 & 1 & 5 \\ -1 & 1 & 2 & 0 & 0 & 1 & 1 \end{bmatrix}$$

Adding the first row to the third makes the first column 1, 0, 0, as required, and adding the second row to the first starts changing the second column to 0, 1, 0:

$$\begin{bmatrix} 1 & 0 & 4 & 1 & 1 & 2 & 6 \\ 0 & 2 & 3 & 0 & 1 & 1 & 5 \\ 0 & -1 & 3 & 1 & 0 & 2 & 2 \end{bmatrix}$$

Adding the third row to the second, and the new second row to the third completes the operations on the second column:

$$\begin{bmatrix} 1 & 0 & 4 & 1 & 1 & 2 & 6 \\ 0 & 1 & 6 & 1 & 1 & 3 & 7 \\ 0 & 0 & 9 & 2 & 1 & 5 & 9 \end{bmatrix}$$

Divide row three by 9:

$$\begin{bmatrix} 1 & 0 & 4 & 1 & 1 & 2 & 6 \\ 0 & 1 & 6 & 1 & 1 & 3 & 7 \\ 0 & 0 & 1 & 2/9 & 1/9 & 5/9 & 1 \end{bmatrix}$$

Subtract four times row 3 from row 1, and subtract six times row 3 from row 2 to give the unit matrix, the inverse and the solutions:

$$\begin{bmatrix} 1 & 0 & 0 & 1/9 & 5/9 & -2/9 & 2 \\ 0 & 1 & 0 & -1/3 & 1/3 & -1/3 & 1 \\ 0 & 0 & 1 & 2/9 & 1/9 & 5/9 & 1 \end{bmatrix}$$

The solutions are $x = 2$, $y = 1$ and $z = 1$, which can be checked in the original equations.

If the inverse matrix is not required row operations can still be used to obtain a solution. The coefficients matrix is still transformed into the unit matrix. For example, the equations

$$x + y = 2$$
$$2x - y = 1$$

give the coefficients matrix and column of constants

$$\begin{bmatrix} 1 & 1 & 2 \\ 2 & -1 & 1 \end{bmatrix}$$

Subtracting twice row 1 from row 2 gives

$$\begin{bmatrix} 1 & 1 & | & 2 \\ 0 & -3 & | & -3 \end{bmatrix}$$

Dividing row 2 by -3, and subtracting the new row 2 from row 1 results in the unit matrix and the solutions:

$$\begin{bmatrix} 1 & 0 & | & 1 \\ 0 & 1 & | & 1 \end{bmatrix}$$

Here, $x = 1$ and $y = 1$. Taking a more complicated example,

$$w + 2x - y + 2z = 3$$
$$2w + x \qquad\quad + z = 3$$
$$w - x + y \qquad\quad = 2$$
$$w + x + y + z = 2$$

The coefficients matrix and column of constants give

$$\begin{bmatrix} 1 & 2 & -1 & 2 & | & 3 \\ 2 & 1 & 0 & 1 & | & 3 \\ 1 & -1 & 1 & 0 & | & 2 \\ 1 & 1 & 1 & 1 & | & 2 \end{bmatrix}$$

Subtracting twice row 1 from row 2, and row 1 from rows 3 and 4 gives

$$\begin{bmatrix} 1 & 2 & -1 & 2 & | & 3 \\ 0 & -3 & 2 & -3 & | & -3 \\ 0 & -3 & 2 & -2 & | & -1 \\ 0 & -1 & 2 & -1 & | & -1 \end{bmatrix}$$

Divide row 2 by -3 and combine the new row 2 with the others to give 0, 1, 0, 0 in the second column

$$\begin{bmatrix} 1 & 0 & 1/3 & 0 & | & 1 \\ 0 & 1 & -2/3 & 1 & | & 1 \\ 0 & 0 & 0 & 1 & | & 2 \\ 0 & 0 & 4/3 & 0 & | & 0 \end{bmatrix}$$

At this stage row 4 can be divided by 4/3 and combined with the other rows to give:

$$\begin{bmatrix} 1 & 0 & 0 & 0 & | & 1 \\ 0 & 1 & 0 & 1 & | & 1 \\ 0 & 0 & 1 & 1 & | & 2 \\ 0 & 0 & 1 & 0 & | & 0 \end{bmatrix}$$

Finally, subtracting row 3 from row 4, adding the new row 4 to row 2 and row 3, and dividing the new row 4 by -1 results in

$$\begin{bmatrix} 1 & 0 & 0 & 0 & | & 1 \\ 0 & 1 & 0 & 0 & | & -1 \\ 0 & 0 & 1 & 0 & | & 0 \\ 0 & 0 & 0 & 1 & | & 2 \end{bmatrix}$$

and $w = 1, x = -1, y = 0, z = 2$. The answer should be checked in the original equations.

2.7 Exercises

1. Obtain the inverse of the following matrices by using Gaussian elimination:

$$A = \begin{bmatrix} 2 & -3 \\ 1 & 1 \end{bmatrix} \quad B = \begin{bmatrix} 0 & 2 \\ -1 & -1 \end{bmatrix} \quad C = \begin{bmatrix} 3 & -4 \\ 1 & 4 \end{bmatrix} \quad D = \begin{bmatrix} 1 & -1 & 0 \\ -2 & 2 & 3 \\ 3 & 0 & 2 \end{bmatrix}$$

$$E = \begin{bmatrix} 2 & 0 & 1 \\ 2 & 3 & -2 \\ -1 & 1 & -1 \end{bmatrix} \quad F = \begin{bmatrix} 3 & -2 & 3 \\ -1 & 0 & 3 \\ 2 & 1 & -1 \end{bmatrix} \quad G = \begin{bmatrix} -1 & 0 & 2 \\ 1 & 2 & 1 \\ 2 & 1 & 3 \end{bmatrix}$$

2. Solve the following equations using Gaussian elimination:

(a)
$$4x - 2y = 6$$
$$2x + y = 5$$

(b)
$$3x - y + 2z = 7$$
$$x + 3y + z = 3$$
$$2x + y + z = 4$$

(c)
$$x + y + z = 2$$
$$4x + 3y + 2z = 7$$
$$x - y - z = 0$$

(d)
$$2w + x + y + z = 3$$

$$w - x - y + z = 3$$
$$w + 2x + 3y - z = 2$$
$$3w + 3x - y + 2z = 4$$

2.8 Linear dependence and rank

For an inverse to exist it is necessary for the original matrix to be square. This has important implications in the solution of equations because a square matrix of coefficients necessarily means that the number of equations is equal to the number of unknowns. If there are more equations than unknowns then some of these must be *redundant* if a solution exists. This means that some of them provide no additional information to what is already known from the other equations. For example,

$$x + y = 2$$
$$2x + y = 3$$
$$2x + 2y = 4$$

In this case there are three equations in two unknowns, but the third one is simply a multiple of the first and provides no further information. A value for x and for y can be obtained from either the first and second equations or the second and third equations taken together.

There is also the possibility that the extra equation is *inconsistent*. For example,

$$x + y = 2$$
$$2x + y = 3$$
$$2x + 2y = 5$$

This set of equations has no solution because the values of x and y obtained by solving the first and second simultaneously do not satisfy the third. In particular, the first and third equations cannot be satisfied simultaneously. (Graphically, they are parallel lines).

In general, with a set of simultaneous equations, when the number of equations is equal to the number of unknowns a unique solution can be obtained providing that no equation is either redundant or inconsistent with the others. In such a case the equations are said to be *linearly independent* and $|\mathbf{A}| \neq 0$. If they are not then there is no unique solution since those that are redundant can be discarded leaving more unknowns than equations.

For example,
$$x_1 + x_2 = 2$$
$$2x_1 + 2x_2 = 4$$

These two equations are obviously identical and are linearly dependent. The value of x_1 can only be determined in terms of x_2.

The *rank* of a matrix is defined as the number of linearly independent rows (or columns) in the matrix. If two equations in two unknowns are linearly independent then the rank of the coefficients matrix is 2. Therefore to check whether a set of equations is independent, the rank of the coefficients matrix could be found. If this equals the number of unknowns then the equations are linearly independent.

There are two ways of checking the rank of a matrix. First, for an $(n \times n)$ square matrix, $\mathbf{A}$, the rank is n if $|\mathbf{A}| \neq 0$, while if $|\mathbf{A}| = 0$ the rank is $n-1$ if a non-zero determinant of size $n-1$ can be found within $\mathbf{A}$. Secondly, the row operations used in the Gaussian elimination method can be applied to the matrix and the rank is the size of the unit matrix that remains in the final matrix.

For example, consider the equations
$$x + 2y + z = 6$$
$$2x + y + z = 5$$
$$4x + 5y + 3z = 17$$

Using the Gaussian elimination method first, the coefficients matrix is

$$\mathbf{A} = \begin{bmatrix} 1 & 2 & 1 \\ 2 & 1 & 1 \\ 4 & 5 & 3 \end{bmatrix}$$

and taking twice row 1 from row 2, and four times row 1 from row three gives

$$\begin{bmatrix} 1 & 2 & 1 \\ 0 & -3 & -1 \\ 0 & -3 & -1 \end{bmatrix}$$

Subtracting row 2 from row 3 and dividing row 2 by -3,

$$\begin{bmatrix} 1 & 2 & 1 \\ 0 & 1 & 1/3 \\ 0 & 0 & 0 \end{bmatrix}$$

Subtracting twice row 2 from row 1

$$\begin{bmatrix} 1 & 0 & 1/3 \\ 0 & 1 & 1/3 \\ 0 & 0 & 0 \end{bmatrix}$$

No further reductions are possible and so the rank is 2, since the final matrix includes a (2×2) identity matrix. Alternatively, the determinant of the coefficients matrix is

$$|A| = \begin{vmatrix} 1 & 2 & 1 \\ 2 & 1 & 1 \\ 4 & 5 & 3 \end{vmatrix} = \begin{vmatrix} 1 & 2 & 1 \\ 0 & -3 & -1 \\ 0 & -3 & -1 \end{vmatrix} = \begin{vmatrix} -3 & -1 \\ -3 & -1 \end{vmatrix} = 0$$

and so the rank is less than 3 and the equations are not linearly independent. Therefore no unique solution for x, y and z is possible. To see if the rank is 2, try to find any non-zero (2×2) determinant within A. Since

$$\begin{vmatrix} 1 & 2 \\ 2 & 1 \end{vmatrix} = 1 - 4 = -3$$

is non-zero the rank is 2. Here the first (2×2) determinant selected turned out to be non-zero. If it had been zero, another attempt at finding a non-zero (2×2) determinant would have been made. While this is not too arduous in this example, for larger matrices it can be time-consuming and in these circumstances the Gaussian elimination method is preferred.

For a second example consider the equations

$$x_1 + 2x_2 + 2x_3 = 7$$
$$x_1 - x_2 + x_3 = 4$$
$$x_1 + x_2 + 3x_3 = 6$$

The matrix of the coefficients is

$$\begin{bmatrix} 1 & 2 & 2 \\ 1 & -1 & 1 \\ 1 & 1 & 3 \end{bmatrix}$$

Subtracting row 1 from row 2 and from row 3,

$$\begin{bmatrix} 1 & 2 & 2 \\ 0 & -3 & -1 \\ 0 & -1 & 1 \end{bmatrix}$$

Adding twice row 3 to row 1, and row 3 to row 2,

$$\begin{bmatrix} 1 & 0 & 4 \\ 0 & -4 & 0 \\ 0 & -1 & 1 \end{bmatrix}$$

Dividing row 2 by -4 and adding the new row 2 to row 3,

$$\begin{bmatrix} 1 & 0 & 4 \\ 0 & 1 & 0 \\ 0 & 0 & 1 \end{bmatrix}$$

Subtracting four times row 3 from row 1 gives the (3×3) unit matrix and so the rank is 3 and the equations are linearly independent. Alternatively, the value of $|\mathbf{A}|$ is

$$|\mathbf{A}| = \begin{vmatrix} 1 & 2 & 2 \\ 1 & -1 & 1 \\ 1 & 1 & 3 \end{vmatrix} = \begin{vmatrix} 3 & 2 & 4 \\ 0 & -1 & 0 \\ 2 & 1 & 4 \end{vmatrix} = - \begin{vmatrix} 3 & 4 \\ 2 & 4 \end{vmatrix} = -4$$

and as this is non-zero the rank is 3.

Our discussion of rank has been concerned with simultaneous equations and square matrices. However, the concept of rank is more general and can be applied to matrices of any size. For example, if

$$\mathbf{A} = \begin{bmatrix} 1 & -2 & 0 & 3 & 5 \\ 2 & 1 & 0 & 2 & 4 \\ 0 & -6 & 2 & 0 & 0 \\ -2 & 2 & 1 & -1 & 4 \end{bmatrix}$$

the maximum possible rank is 4, since there are no (5×5) determinants within $\mathbf{A}$. As mentioned above in the discussion of our first example, a search could now be made for a (4×4) determinant which is non-zero. However, there are 5 possible (4×4) determinants and it would be tedious to evaluate a number of these. Instead, the Gaussian elimination method is used. Combining a multiple of row 1 with each of the others gives

$$\begin{bmatrix} 1 & -2 & 0 & 3 & 5 \\ 0 & 5 & 0 & -4 & -6 \\ 0 & -6 & 2 & 0 & 0 \\ 0 & -2 & 1 & 5 & 14 \end{bmatrix}$$

Subtracting row 4 from row 1, adding row 3 to row 2, and dividing row 3 by 2,

$$\begin{bmatrix} 1 & 0 & -1 & -2 & -9 \\ 0 & -1 & 2 & -4 & -6 \\ 0 & -3 & 1 & 0 & 0 \\ 0 & -2 & 1 & 5 & 14 \end{bmatrix}$$

Combining row 2 with row 3, and row 2 with row 4,

$$\begin{bmatrix} 1 & 0 & -1 & -2 & -9 \\ 0 & -1 & 2 & -4 & -6 \\ 0 & 0 & -5 & 12 & 18 \\ 0 & 0 & -3 & 13 & 26 \end{bmatrix}$$

Multiplying row 2 by -1 and row 3 by -0.2,

$$\begin{bmatrix} 1 & 0 & -1 & -2 & -9 \\ 0 & 1 & -2 & 4 & 6 \\ 0 & 0 & 1 & -2.4 & -3.6 \\ 0 & 0 & -3 & 13 & 26 \end{bmatrix}$$

Combining multiples of row 3 with each of the other rows,

$$\begin{bmatrix} 1 & 0 & 0 & -4.4 & -12.6 \\ 0 & 1 & 0 & -0.8 & -1.2 \\ 0 & 0 & 1 & -2.4 & -3.6 \\ 0 & 0 & 0 & 5.8 & 15.2 \end{bmatrix}$$

Dividing row 4 by 5.8 and combining multiples of the new row with the other rows,

$$\begin{bmatrix} 1 & 0 & 0 & 0 & -1.07 \\ 0 & 1 & 0 & 0 & 0.90 \\ 0 & 0 & 1 & 0 & 2.69 \\ 0 & 0 & 0 & 1 & 2.62 \end{bmatrix}$$

and the rank is 4.

2.9 Exercises

1. Determine the rank of the following matrices using either determinants or the Gaussian elimination method:

(a) $\begin{bmatrix} 2 & -1 \\ 1 & 1 \end{bmatrix}$ (b) $\begin{bmatrix} 1 & 1 & 3 \\ 2 & 1 & 1 \\ 1 & -2 & -1 \end{bmatrix}$ (c) $\begin{bmatrix} 1 & 2 & -4 \\ 2 & -1 & -1 \\ 3 & 1 & -5 \end{bmatrix}$

(d) $\begin{bmatrix} 1 & 0 & 0 \\ 1 & 1 & 1 \\ 0 & 1 & 1 \end{bmatrix}$ (e) $\begin{bmatrix} 1 & 1 & 1 \\ 1 & 0 & 2 \\ 2 & 2 & -1 \end{bmatrix}$

(f) $\begin{bmatrix} 1 & 0 & 0 & -1 \\ 0 & 2 & 2 & 1 \\ 2 & 1 & 3 & 1 \end{bmatrix}$ (g) $\begin{bmatrix} 1 & 0 & -1 & -1 \\ 1 & 1 & -1 & -1 \\ 0 & 1 & -1 & -1 \end{bmatrix}$

2. Test whether the following equations have a unique solution.

(a)
$$x_1 + 3x_2 - 2x_3 = 3$$
$$x_1 - 2x_2 + x_3 = 0$$
$$2x_1 - 2x_2 + 3x_3 = 3$$

(b)
$$x_1 + 2x_2 + x_3 = 2$$
$$2x_1 - x_2 + 2x_3 = -6$$
$$x_1 + x_2 + x_3 = 0$$

(c)
$$x_1 + 3x_2 - x_3 + 2x_4 = 2$$
$$x_1 - x_2 + 2x_3 + x_4 = 3$$
$$x_1 + x_2 + x_3 - 2x_4 = 3$$
$$x_1 + x_2 \qquad\qquad + 5x_4 = 1$$

2.10 Further properties of matrices

For general matrices **A**, **B** and **C**, the following properties of the transform operation can be shown to be valid

$$\begin{aligned} (\mathbf{A}')' &= \mathbf{A} \\ (\mathbf{A} + \mathbf{B})' &= \mathbf{A}' + \mathbf{B}' \\ (\mathbf{AB})' &= \mathbf{B}'\mathbf{A}' \\ (\mathbf{ABC})' &= \mathbf{C}'\mathbf{B}'\mathbf{A}' \end{aligned}$$

assuming that the various matrices are conformable for addition or multiplication. A special case occurs when

$$\mathbf{A}' = \mathbf{A}$$

since this requires that **A** must be a square matrix, because if **A** is $(m \times n)$, **A**' is $(n \times m)$ so $m = n$. Now if the transpose of **A** equals **A**, then **A** is a *symmetric matrix*. For example, if

$$\mathbf{A} = \begin{bmatrix} 1 & 2 & 3 \\ 2 & 1 & 4 \\ 3 & 4 & 1 \end{bmatrix}$$

then $\mathbf{A} = \mathbf{A}'$ and $\mathbf{A}$ is symmetric. Another example of a symmetric matrix is the unit matrix I. More generally, if

$$\mathbf{A} = \begin{bmatrix} a_{11} & a_{12} & a_{13} \\ a_{21} & a_{22} & a_{23} \\ a_{31} & a_{32} & a_{33} \end{bmatrix}$$

then for $\mathbf{A}$ to be symmetric, $a_{ij} = a_{ji}$ for all i, j. If a symmetric matrix $\mathbf{A}$ has the additional property that

$$\mathbf{A}^2 = \mathbf{A}$$

so that multiplication by itself leaves $\mathbf{A}$ unchanged then $\mathbf{A}$ is a *symmetric idempotent matrix*. The unit matrix is a symmetric idempotent matrix, as is the matrix

$$\mathbf{B} = \begin{bmatrix} 1 & 0 & 0 \\ 0 & 0 & 0 \\ 0 & 0 & 0 \end{bmatrix}$$

A useful application of symmetric matrices occurs in statistics and econometrics. If $\mathbf{x}$ is $(n \times 1)$ and $\mathbf{x}' = [x_1\, x_2\, \ldots\, x_n]$

then
$$\mathbf{x}'\mathbf{x} = [x_1\, x_2 \cdots x_n] \begin{bmatrix} x_1 \\ x_2 \\ \cdot\,\cdot \\ x_n \end{bmatrix} = x_1^2 + x_2^2 + \cdots + x_n^2$$

so that $\mathbf{x}'\mathbf{x}$ is the sum of the squares of the elements of $\mathbf{x}$. A weighted sum of squares of $\mathbf{x}$ is obtained by forming $\mathbf{x}'\mathbf{A}\mathbf{x}$ where $\mathbf{A}$ is a $(n \times n)$ *diagonal* matrix (i.e. all $a_{ij} = 0$ for $i \neq j$). For example, if $\mathbf{x}'$ has three elements,

$$\mathbf{x}'\mathbf{A}\mathbf{x} = [x_1\, x_2\, x_3] \begin{bmatrix} a_{11} & 0 & 0 \\ 0 & a_{22} & 0 \\ 0 & 0 & a_{33} \end{bmatrix} \begin{bmatrix} x_1 \\ x_2 \\ x_3 \end{bmatrix}$$

$$= [x_1\, x_2\, x_3] \begin{bmatrix} a_{11}x_1 \\ a_{22}x_2 \\ a_{33}x_3 \end{bmatrix} = a_{11}x_1^2 + a_{22}x_2^2 + a_{33}x_3^2$$

If $\mathbf{A}$ is generalised to be any symmetric matrix, expressions of the type $\mathbf{x}'\,\mathbf{A}\mathbf{x}$ are known as *quadratic forms*. For the (3×3) case,

$$\mathbf{x'Ax} = [x_1 \ x_2 \ x_3] \begin{bmatrix} a_{11} & a_{12} & a_{13} \\ a_{12} & a_{22} & a_{23} \\ a_{13} & a_{23} & a_{33} \end{bmatrix} \begin{bmatrix} x_1 \\ x_2 \\ x_3 \end{bmatrix}$$

$$= [x_1 \ x_2 \ x_3] \begin{bmatrix} a_{11}x_1 + a_{12}x_2 + a_{13}x_3 \\ a_{12}x_1 + a_{22}x_2 + a_{23}x_3 \\ a_{13}x_1 + a_{23}x_2 + a_{33}x_3 \end{bmatrix}$$

$$= a_{11}x_1^2 + a_{22}x_2^2 + a_{33}x_3^2 + 2a_{12}x_1x_2 + 2a_{13}x_1x_3$$
$$+ 2a_{23}x_2x_3$$

Further properties of the inverse operation are that

$$(\mathbf{AB})^{-1} = \mathbf{B}^{-1}\mathbf{A}^{-1}$$
$$(\mathbf{ABC})^{-1} = \mathbf{C}^{-1}\mathbf{B}^{-1}\mathbf{A}^{-1}$$
$$(\mathbf{A'})^{-1} = (\mathbf{A}^{-1})'$$

where here **A**, **B** and **C** are all square matrices of the same size and are non-singular so that their inverses exist. In some applications, it is useful to partition a matrix into sub-matrices. This can be done in several ways and is achieved by effectively adding vertical and horizontal lines to the matrix. For example,

$$\mathbf{A} = \begin{bmatrix} a_{11} & a_{12} & a_{13} & : & a_{14} \\ a_{21} & a_{22} & a_{23} & : & a_{24} \\ a_{31} & a_{32} & a_{33} & : & a_{34} \\ \cdots & \cdots & \cdots & \cdots & \cdots \\ a_{41} & a_{42} & a_{43} & : & a_{44} \end{bmatrix} = \begin{bmatrix} \mathbf{A}_{11} & \mathbf{A}_{12} \\ \mathbf{A}_{21} & \mathbf{A}_{22} \end{bmatrix}$$

where $\mathbf{A}_{11}$ is (3×3), $\mathbf{A}_{12}$ is (3×1), $\mathbf{A}_{21}$ is (1×3) and $\mathbf{A}_{22}$ is (1×1). If two matrices are partitioned they can be added or multiplied only if they are partitioned in a conformable way. Thus for addition, each of the corresponding sub-matrices must have the same dimensions, so that if

$$\mathbf{B} = \begin{bmatrix} \mathbf{B}_{11} & \mathbf{B}_{12} \\ \mathbf{B}_{21} & \mathbf{B}_{22} \end{bmatrix}$$

then $\mathbf{A} + \mathbf{B}$ can be formed only if $\mathbf{A}_{ij}$ and $\mathbf{B}_{ij}$ are the same sizes:

$$\mathbf{A} + \mathbf{B} = \begin{bmatrix} \mathbf{A}_{11} + \mathbf{B}_{11} & \mathbf{A}_{12} + \mathbf{B}_{12} \\ \mathbf{A}_{21} + \mathbf{B}_{21} & \mathbf{A}_{22} + \mathbf{B}_{22} \end{bmatrix}$$

For multiplication, the number of columns in $\mathbf{A}_{ij}$ must equal the number of rows in $\mathbf{B}_{ij}$:

$$\mathbf{A}\mathbf{B} = \begin{bmatrix} \mathbf{A}_{11}\mathbf{B}_{11} + \mathbf{A}_{12}\mathbf{B}_{21} & \mathbf{A}_{11}\mathbf{B}_{12} + \mathbf{A}_{12}\mathbf{B}_{22} \\ \mathbf{A}_{21}\mathbf{B}_{11} + \mathbf{A}_{22}\mathbf{B}_{21} & \mathbf{A}_{21}\mathbf{B}_{12} + \mathbf{A}_{22}\mathbf{B}_{22} \end{bmatrix}$$

The inverse of a partitioned matrix can be found from the above result. Since $\mathbf{A}\mathbf{A}^{-1} = \mathbf{I}$, $\mathbf{B}$ is the inverse of $\mathbf{A}$ if $\mathbf{A}\mathbf{B} = \mathbf{I}$. Partitioning $\mathbf{I}$ so that it conforms to the partitioning of the product $\mathbf{A}\mathbf{B}$,

$$\mathbf{A}\mathbf{B} = \mathbf{I} = \begin{bmatrix} \mathbf{I} & 0 \\ 0 & \mathbf{I} \end{bmatrix}$$

and equating the elements of $\mathbf{A}\mathbf{B}$ gives four equations in the unknowns $\mathbf{B}_{11}$, $\mathbf{B}_{12}$, $\mathbf{B}_{21}$ and $\mathbf{B}_{22}$:

$$\mathbf{I} = \mathbf{A}_{11}\mathbf{B}_{11} + \mathbf{A}_{12}\mathbf{B}_{21}$$
$$0 = \mathbf{A}_{11}\mathbf{B}_{12} + \mathbf{A}_{12}\mathbf{B}_{22}$$
$$0 = \mathbf{A}_{21}\mathbf{B}_{11} + \mathbf{A}_{22}\mathbf{B}_{21}$$
$$\mathbf{I} = \mathbf{A}_{21}\mathbf{B}_{12} + \mathbf{A}_{22}\mathbf{B}_{22}$$

Solving these equations gives

$$\mathbf{B}_{11} = (\mathbf{A}_{11} - \mathbf{A}_{12}\mathbf{A}_{22}^{-1}\mathbf{A}_{21})^{-1}$$
$$\mathbf{B}_{12} = -\mathbf{B}_{11} (\mathbf{A}_{12}\mathbf{A}_{22}^{-1})$$
$$\mathbf{B}_{21} = -\mathbf{A}_{22}^{-1}\mathbf{A}_{21}\mathbf{B}_{11}$$
$$\mathbf{B}_{22} = \mathbf{A}_{22}^{-1} + (\mathbf{A}_{22}^{-1}\mathbf{A}_{21}\mathbf{B}_{11}\mathbf{A}_{12}\mathbf{A}_{22}^{-1})$$

where it is assumed that $\mathbf{A}_{11}$ and $\mathbf{A}_{22}$ are square and have an inverse, and that the multiplied matrices are conformable.

The final topic we consider in this section is the characteristic equation of a square matrix. For a non-singular matrix, $\mathbf{A}$, the matrix $[\mathbf{A} - r\mathbf{I}]$ is called the *characteristic matrix* of $\mathbf{A}$, where r is a scalar. For example, if

$$\mathbf{A} = \begin{bmatrix} 5 & 3 \\ 3 & 5 \end{bmatrix} \text{ then } \mathbf{A} - r\mathbf{I} = \begin{bmatrix} 5 & 3 \\ 3 & 5 \end{bmatrix} - \begin{bmatrix} r & 0 \\ 0 & r \end{bmatrix} = \begin{bmatrix} 5-r & 3 \\ 3 & 5-r \end{bmatrix}$$

Setting the determinant of the characteristic matrix to zero gives the *characteristic equation*. Continuing the example,

$$\begin{vmatrix} 5-r & 3 \\ 3 & 5-r \end{vmatrix} = (5-r)(5-r) - 9 = 16 - 10r + r^2 = 0$$

Here the characteristic equation is a quadratic and using the standard formula for a quadratic equation (see section 3.3 for a full discussion), the two values of r which satisfy the equation are $r = 8$ and $r = 2$. These are referred to as the *characteristic roots*, *latent roots* or *eigenvalues* of $\mathbf{A}$. In general, for an $(n \times n)$ matrix, the characteristic equation will have n roots.

Associated with each characteristic root, r, of $\mathbf{A}$ is a vector, $\mathbf{x}$, say, which satisfies

$$\mathbf{Ax} = r\mathbf{x} \quad \text{or} \quad (\mathbf{A} - r\mathbf{I})\mathbf{x} = \mathbf{O}$$

This vector is called the *characteristic vector*, *latent vector* or *eigenvector* corresponding to this particular root. In the example, for $r = 8$, the equation is

$$\begin{bmatrix} 5-8 & 3 \\ 3 & 5-8 \end{bmatrix} \begin{bmatrix} x_1 \\ x_2 \end{bmatrix} = \begin{bmatrix} 0 \\ 0 \end{bmatrix}$$

or
$$-3x_1 + 3x_2 = 0 \quad \text{or } x_1 = x_2$$
$$3x_1 - 3x_2 = 0 \quad \text{or } x_1 = x_2$$

and so even though there are two equations in two unknowns there is no unique solution. At this point it is convenient to *normalise* the sum of the squares of the elements of $\mathbf{x}$ by making this equal 1. That is,

$$x_1^2 + x_2^2 = 1$$

and with $x_1 = x_2$, the solution is $x_1 = 1/\sqrt{2} = x_2$, and so the characteristic vector is

$$\mathbf{x}' = [1/\sqrt{2} \quad 1/\sqrt{2}]$$

For the other characteristic root, $r = 2$, the equation is

$$\begin{bmatrix} 5-2 & 3 \\ 3 & 5-2 \end{bmatrix} \begin{bmatrix} x_1 \\ x_2 \end{bmatrix} = \begin{bmatrix} 0 \\ 0 \end{bmatrix}$$

or
$$3x_1 + 3x_2 = 0 \text{ or } x_1 = -x_2$$
$$3x_1 + 3x_2 = 0 \text{ or } x_1 = -x_2$$

and again normalising by putting

$$x_1^2 + x_2^2 = 1$$

allows the solutions $x_1 = 1/\sqrt{2}$ and $x_2 = -1/\sqrt{2}$ to be found and the characteristic vector is

$$\mathbf{x}' = [1/\sqrt{2} \ -1/\sqrt{2}]$$

Earlier, a matrix, $\mathbf{A}$, was defined to be symmetric idempotent if

$$\mathbf{A}^2 = \mathbf{A}$$

For any matrix the characteristic vector satisfies

$$\mathbf{A}\mathbf{x} = r\mathbf{x}$$

and if $\mathbf{A}$ is symmetric idempotent, multiplying both sides by $\mathbf{A}$ gives

$$\mathbf{A}^2\mathbf{x} = r\mathbf{A}\mathbf{x}$$
$$= r^2\mathbf{x}$$
$$= \mathbf{A}\mathbf{x}$$

Therefore,
$$r^2\mathbf{x} = r\mathbf{x}$$

or
$$r(r-1)\mathbf{x} = \mathbf{0}$$

Now since $\mathbf{x} \neq \mathbf{0}$, $r(r-1) = 0$ and $r = 0$ or $r = 1$. Thus, for any symmetric idempotent matrix the characteristic roots are either zero or unity.

As an example consider

$$\mathbf{B} = \begin{bmatrix} 0.5 & 0.5 \\ 0.5 & 0.5 \end{bmatrix}$$

which is idempotent since $\mathbf{B}^2 = \mathbf{B}$.

The characteristic equation is

$$|\mathbf{B} - r\mathbf{I}| = 0$$

or

$$\begin{vmatrix} 0.5-r & 0.5 \\ 0.5 & 0.5-r \end{vmatrix} = 0$$

and so $(0.5-r)^2 - 0.25 = 0$ or $r^2 - r = 0$

Therefore the characteristic roots are $r = 1$ and $r = 0$, as expected for a symmetric idempotent matrix.

To obtain the characteristic vectors, taking $r = 1$ first,

$$\mathbf{Bx} = r\mathbf{x}$$

or

$$(\mathbf{B}-r\mathbf{I})\mathbf{x} = \begin{bmatrix} -0.5 & 0.5 \\ 0.5 & -0.5 \end{bmatrix} \begin{bmatrix} x_1 \\ x_2 \end{bmatrix} \begin{bmatrix} 0 \\ 0 \end{bmatrix}$$

This gives $-0.5x_1 + 0.5x_2 = 0$ or $x_1 = x_2$, and normalising by imposing the restriction that $x_1^2 + x_2^2 = 1$ results in $x_1 = 1/\sqrt{2} = x_2$. For the second root, $r = 0$ the characteristic equation is

$$\begin{bmatrix} 0.5 & 0.5 \\ 0.5 & 0.5 \end{bmatrix} \begin{bmatrix} x_1 \\ x_2 \end{bmatrix} = \begin{bmatrix} 0 \\ 0 \end{bmatrix}$$

which gives $x_1 = -x_2$ and $x_1 = 1/\sqrt{2}$, $x_2 = -1\sqrt{2}$.

2.11 Exercises

1. For $\mathbf{A} = \begin{bmatrix} 2 & 1 & 2 \\ 1 & 0 & 1 \end{bmatrix}$ $\mathbf{B} = \begin{bmatrix} 3 & -1 & 1 \\ 2 & -2 & 0 \end{bmatrix}$ $\mathbf{C} = \begin{bmatrix} -1 & 0 \\ 2 & 1 \\ -1 & 0 \end{bmatrix}$

   show that $(\mathbf{A}')' = \mathbf{A}$, $(\mathbf{A} + \mathbf{B})' = \mathbf{A}' + \mathbf{B}'$ and $(\mathbf{BC})' = \mathbf{C}'\mathbf{B}'$.

2. Determine which of the following matrices are idempotent.

$$\mathbf{A} = \begin{bmatrix} 1 & 0 & 0 \\ 0 & 1 & 0 \\ 0 & 0 & 0 \end{bmatrix} \quad \mathbf{B} = \begin{bmatrix} 0.25 & 0.25 & 0.25 & 0.25 \\ 0.25 & 0.25 & 0.25 & 0.25 \\ 0.25 & 0.25 & 0.25 & 0.25 \\ 0.25 & 0.25 & 0.25 & 0.25 \end{bmatrix} \quad \mathbf{C} = \begin{bmatrix} a & 0 \\ 0 & a \end{bmatrix}$$

3. If $\mathbf{A} = \begin{bmatrix} 2 & -3 \\ 1 & 1 \end{bmatrix}$ and $\mathbf{B} = \begin{bmatrix} 2 & 2 \\ -1 & 1 \end{bmatrix}$ show that $(\mathbf{A}')^{-1} = (\mathbf{A}^{-1})'$

and that $(\mathbf{AB})^{-1} = \mathbf{B}^{-1}\mathbf{A}^{-1}$.

4. Determine the characteristic roots and vectors of

$$\mathbf{A} = \begin{bmatrix} 3 & 1 \\ 1 & 3 \end{bmatrix} \text{ and } \mathbf{B} = \begin{bmatrix} 2 & -1 \\ 0 & 1 \end{bmatrix}$$

2.12   Input–output analysis

INTRODUCTION

In any economy there are a number of industries supplying consumer demand. Many of these industries also supply intermediate products which are further processed or utilised by other industries before reaching the final consumer. For example, the glass industry supplies finished products such as mirrors, but also supplies window glass to the building industry to be used in the construction of houses, and toughened glass to the motor industry to be used as windscreens for cars.

The industrial sector of any economy, therefore, consists of a number of interlinked units from which final consumer demand is met. Changes in demand for any one of the final outputs affect the outputs required from many other industries. In order to see the effects that such changes might produce it is necessary to build a model of the system. This was first attempted by Leontief and the theory has been developed under the name *input–output analysis*.

INPUT, OUTPUT AND DEMAND

In building the model Leontief made an important assumption: that the output of one industry which is required to satisfy the demand from another industry is directly proportional to the final output of the latter. This can be written as

$$x_{ij} = a_{ij}X_j$$

where X_j is the total output of industry j,

a_{ij} is a constant of proportionality, and is the proportion of output of industry i that will be required by industry j in order to produce one unit of output. This will depend upon the technology of industry j,

x_{ij} is the output of industry i that will be required by industry j.

For example, let us assume then that the average number of units of glass which is required in the construction of a house is equal to 2. Then this will be the constant of proportionality and the relationship can be expressed as

$$x_{ij} = a_{ij}X_j = 2X_j$$

where X_j is the output of the building industry in terms of the number of house units constructed, and x_{ij} is the output of the glass industry which is required by the building industry for house construction. These figures are, in fact, flows and represent the output per unit of time, e.g. output per year.

If the technology of the industry changes and houses are constructed with much larger expanses of glass or the average size of a house unit increases, then a new constant of proportionality must be used: for example,

$$x_{ij} = 3X_j$$

The constant of proportionality used in this example is an average for all types of houses. It may be possible, and it would certainly be useful, to classify the different class of dwellings by size and by type, e.g. bungalows, semi-detached, detached, etc. and to use a separate constant of proportionality for each sub-category.

An industry may also use some of its own output for further processing or as raw material input and as this demand must be met, it must be included as

$$x_{jj} = a_{jj}X_j$$

In this case the constant of proportionality must be less than one, otherwise the industry would require a greater amount of input of its own output than it was capable of producing. (It is possible to consider the system in terms of net outputs only and to ignore these inputs altogether, but this is not done here.)

The total output of any industry depends upon the demands made upon it by all other industries and also upon the demand for its products made directly by the consumer. This latter, the final demand, is a function of consumers' preferences, relative prices and consumers' income, and also varies with time.

TABLE 2.1

| | Input to | | Level of output |
	Industry 1	Industry 2	
Industry 1	$x_{11} = 200$	$x_{12} = 400$	$X_1 = 1,600$
Industry 2	$x_{21} = 600$	$x_{22} = 100$	$X_2 = 2,700$

For any economy, it is possible to collect data for the flow of goods and services between industries and to specify the final demand for the outputs at any given point in time. In the simple case of a two-industry economy the information relating to the flow of goods which is required to produce a particular level of output might be as presented in Table 2.1.

From this data it is possible to establish the technological coefficients a_{ij} using the formula quoted earlier:

$$x_{ij} = a_{ij} X_j$$

therefore

$$a_{ij} = \frac{x_{ij}}{X_j}$$

From the information given in Table 2.1 it is also possible to determine the final demand which can be supplied. This is given by the level of output less the amounts which are required to meet the demands made by the two industries in producing these outputs. Thus, level of final demand which can be met by Industry 1 is equal to

$$1600 - (200 + 400) = 1000 \text{ units}$$

and for Industry 2 it is equal to

$$2700 - (600 + 100) = 2000 \text{ units}$$

From the information given in Table 2.2 it is possible to determine the final demand which can be met at any other level of output within the range over which the constants of proportionality can be assumed to hold. For example, the information in Table 2.2 tells us that Industry 1 required 0.125 units of its own output to be used as input for each further unit of output. Also Industry 2 requires 0.148 units of the output of Industry 1 for each unit of its own output.

It follows that, in general, if X_1 and X_2 represent the total output of

TABLE 2.2

| | Technological coefficient for | |
	Industry 1	Industry 2
Industry 1	$\dfrac{200}{1,600} = 0.125$	$\dfrac{400}{2,700} = 0.148$
Industry 2	$\dfrac{600}{1,600} = 0.375$	$\dfrac{100}{2,700} = 0.037$

Industries 1 and 2 respectively and C_1 and C_2 are the final demands for the outputs of the industries respectively we can write

$$0.125\,X_1 + 0.148\,X_2 + C_1 \leq X_1$$
and
$$0.375\,X_1 + 0.037\,X_2 + C_2 \leq X_2$$

The first equation states that the total demand for the product of Industry 1 must be less than or equal to the total output of Industry 1. The second equation states the same thing for Industry 2.

If we assume that the total output of each industry is just sufficient to meet all the demands made upon it the inequalities can be replaced by equations as follows:

$$0.125\,X_1 + 0.148\,X_2 + C_2 = X_1$$
$$0.375\,X_1 + 0.037\,X_2 + C_2 = X_2$$

These equations can then be rearranged

$$X_1 - 0.125\,X_1 - 0.148\,X_2 = C_1$$
$$X_2 - 0.375\,X_1 - 0.037\,X_2 = C_2$$

that is
$$(1 - 0.125)\,X_1 - 0.148\,X_2 = C_1$$
and
$$-0.375\,X_1 + (1 - 0.037)\,X_2 = C_2$$

A pair of simultaneous equations such as these can be conveniently written in matrix notation, as was shown in the earlier part of this chapter.

$$\begin{bmatrix} (1 - 0.125) & -0.148 \\ -0.375 & (1 - 0.037) \end{bmatrix} \begin{bmatrix} X_1 \\ X_2 \end{bmatrix} = \begin{bmatrix} C_1 \\ C_2 \end{bmatrix}$$

The condition that this set of equations should have a unique solution in terms of X_1 and X_2 is that the determinant of the matrix of coefficients is not equal to zero.

In this particular type of problem it is necessary to apply the further condition that the values for X_1 and X_2, the output of the industries, must not be negative. This condition is satisfied if the value of the determinant is positive. That is,

$$\begin{vmatrix} (1 - 0.125) & -0.148 \\ -0.375 & (1 - 0.037) \end{vmatrix} > 0$$

If this condition holds it is possible to use the equations to determine either

(a) the final demands which can be met when the output of each industry is fully utilised, i.e. C_1 and C_2 given X_1 and X_2, or
(b) the total outputs which are required to meet a given level of final demand, i.e. X_1 and X_2 given C_1 and C_2.

For example, if the total output of the two industries is 1,600 and 2,700 units respectively, the maximum final demand can be obtained from the following relationships:

$$\begin{bmatrix} 0.125 & 0.148 \\ 0.375 & 0.037 \end{bmatrix} \begin{bmatrix} 1600 \\ 2700 \end{bmatrix} + \begin{bmatrix} C_1 \\ C_2 \end{bmatrix} = \begin{bmatrix} 1600 \\ 2700 \end{bmatrix}$$

or

$$\begin{bmatrix} C_1 \\ C_2 \end{bmatrix} = \begin{bmatrix} 1 - 0.125 & -0.148 \\ -0.375 & 1 - 0.037 \end{bmatrix} \begin{bmatrix} 1600 \\ 2700 \end{bmatrix}$$

$$= \begin{bmatrix} 0.875 \times 1600 - 0.148 \times 2700 \\ -0.375 \times 1600 + 0.963 \times 2700 \end{bmatrix}$$

$$= \begin{bmatrix} 1400 - 400 \\ -600 + 2600 \end{bmatrix} = \begin{bmatrix} 1000 \\ 2000 \end{bmatrix}$$

therefore $C_1 = 1,000$ and $C_2 = 2,000$ which agrees with the information given in Table 2.1.

Let us now assume that the level of final demand for the output of the two industries is reversed, i.e. $C_1 = 2,000$ units and $C_2 = 1,000$ units.

Then we can calculate the levels of output which would be required to meet these final demands from the following relationship:

$$\begin{bmatrix} 0.875 & -0.148 \\ -0.375 & 0.963 \end{bmatrix} \begin{bmatrix} X_1 \\ X_2 \end{bmatrix} = \begin{bmatrix} 2000 \\ 1000 \end{bmatrix}$$

The inverse of the coefficient matrix is

$$\frac{1}{0.7871} \begin{bmatrix} 0.963 & 0.148 \\ 0.375 & 0.875 \end{bmatrix}$$

and hence $\begin{bmatrix} X_1 \\ X_2 \end{bmatrix} = \frac{1}{0.7871} \begin{bmatrix} 0.963 & 0.148 \\ 0.375 & 0.875 \end{bmatrix} \begin{bmatrix} 2000 \\ 1000 \end{bmatrix} = \begin{bmatrix} 2635 \\ 2064 \end{bmatrix}$

i.e. the required output of Industry 1 is 2,635 units and of Industry 2 is 2,064 units (approx.).

2.13 The input–output matrix

In general the technology matrix for a two-sector economy can be written as

$$\begin{bmatrix} a_{11} & a_{12} \\ a_{21} & a_{22} \end{bmatrix}$$

or simply as **A**. This matrix is known as the *input–output matrix*. To satisfy final demand exactly from current output the following equality must be true:

$$\begin{bmatrix} a_{11} & a_{12} \\ a_{21} & a_{22} \end{bmatrix} \begin{bmatrix} X_1 \\ X_2 \end{bmatrix} + \begin{bmatrix} C_1 \\ C_2 \end{bmatrix} = \begin{bmatrix} X_1 \\ X_2 \end{bmatrix}$$

or $\mathbf{AX + C = X}$

These equations can be rearranged in the form

$$\begin{bmatrix} (1 - a_{11}) & -a_{12} \\ -a_{21} & (1 - a_{22}) \end{bmatrix} \begin{bmatrix} X_1 \\ X_2 \end{bmatrix} = \begin{bmatrix} C_1 \\ C_2 \end{bmatrix}$$

or
$$(\mathbf{I} - \mathbf{A})\mathbf{X} = \mathbf{C}$$

where **I** is the unit matrix and $(\mathbf{I} - \mathbf{A})$ is known as the *Leontief matrix*.
For an economy which is divided into n sectors

$$\mathbf{A} = \begin{bmatrix} a_{11} & a_{12} & \dots & a_{1n} \\ a_{21} & a_{22} & \dots & a_{2n} \\ \vdots & \vdots & & \vdots \\ a_{n1} & a_{n2} & \dots & a_{nn} \end{bmatrix}$$

$$(\mathbf{I} - \mathbf{A}) = \begin{bmatrix} (1 - a_{11}) & -a_{12} & \dots & -a_{1n} \\ -a_{21} & (1 - a_{22}) & \dots & -a_{2n} \\ \vdots & & & \\ -a_{n1} & -a_{n2} & \dots & (1 - a_{nn}) \end{bmatrix}$$

The output of each industry which is required to satisfy a given final demand from current production can be found by inverting this matrix.

Since
$$(\mathbf{I} - \mathbf{A})\mathbf{X} = \mathbf{C}$$

$$\mathbf{X} = (\mathbf{I} - \mathbf{A})^{-1}\mathbf{C},$$

and given the technology matrix **A**, along with either **C** or **X**, the other can be found.

The interpretation of this final equation is helped by letting $(\mathbf{I} - \mathbf{A})^{-1} = \mathbf{B}$ so that, for example, taking row i,

$$\mathbf{X}_i = \mathbf{B}_{i1}\mathbf{C}_1 + \mathbf{B}_{i2}\mathbf{C}_2 + \mathbf{B}_{i3}\mathbf{C}_3 + \cdots \mathbf{B}_{in}\mathbf{C}_n$$

and so if $\mathbf{C}_1$ increases by 1, and the other $\mathbf{C}_j$ values are unchanged, $\mathbf{X}_i$ will increase by $\mathbf{B}_{i1}$. That is, $\mathbf{B}_{i1}$ measures the effect that a unit increase in final demand for the output of industry 1 has on the gross output of industry i. But an increase in $\mathbf{C}_1$ also affects the gross output of each of the other industries, and so the total effect on the economy of an increase of 1 in $\mathbf{C}_1$ is found by adding together $\mathbf{B}_{11}, \mathbf{B}_{21}, \dots, \mathbf{B}_{n1}$. Let the result be

$$\mathbf{M}_1 = \mathbf{B}_{11} + \mathbf{B}_{21} + \cdots + \mathbf{B}_{n1}$$

TABLE 2.3

Absorbing sector Producing sector	1	2	3	4	Final demand
1	0.269	0.219	0.246	0.062	3,662
2	0.010	0.239	0.200	0.076	1,259
3	0.008	0.006	0.051	0.230	1,248
4	0.015	0.154	0.074	0.720	4,789

which is called the *output multiplier* for industry 1. For example, from the end of section 2.12, the **B** matrix is

$$\mathbf{B} = (\mathbf{I} - \mathbf{A})^{-1} = \frac{1}{0.7871} \begin{bmatrix} 0.963 & 0.148 \\ 0.375 & 0.875 \end{bmatrix} = \begin{bmatrix} 1.223 & 0.188 \\ 0.476 & 1.112 \end{bmatrix}$$

so that
$$M_1 = 1.223 + 0.476 = 1.699$$

$$M_2 = 0.188 + 1.112 = 1.300$$

Here industry 1 has the larger output multiplier and an increase of 1 in the final demand for the output of industry 1 leads to an increase of 1.699 in total output.

Taking another example, Table 2.3 gives the technology matrix for a four-sector economy.

From this matrix,

$$(\mathbf{I} - \mathbf{A}) = \begin{bmatrix} (1-0.269) & -0.219 & -0.246 & -0.062 \\ -0.010 & (1-0.239) & -0.200 & -0.076 \\ -0.008 & -0.006 & (1-0.051) & -0.230 \\ -0.015 & -0.154 & -0.074 & (1-0.720) \end{bmatrix}$$

and
$$(\mathbf{I} - \mathbf{A})^{-1} = \begin{bmatrix} 1.401 & 0.597 & 0.562 & 0.934 \\ 0.039 & 1.474 & 0.377 & 0.718 \\ 0.038 & 0.233 & 1.195 & 1.053 \\ 0.107 & 0.904 & 0.553 & 4.295 \end{bmatrix}$$

The output of each industry which is required to satisfy the final demand is found by multiplying the inverse matrix by the given final demand vector.

$$\mathbf{X} = (\mathbf{I} - \mathbf{A})^{-1}\mathbf{C} = \begin{bmatrix} 1.401 & 0.597 & 0.562 & 0.934 \\ 0.039 & 1.474 & 0.377 & 0.718 \\ 0.038 & 0.233 & 1.195 & 1.053 \\ 0.107 & 0.904 & 0.553 & 4.295 \end{bmatrix} \begin{bmatrix} 3662 \\ 1259 \\ 1248 \\ 4789 \end{bmatrix}$$

$$= \begin{bmatrix} 11056 \\ 5908 \\ 6967 \\ 22789 \end{bmatrix}$$

$$\therefore x_1 = 11056, \; x_2 = 5908, \; x_3 = 6967, \; x_4 = 22789$$

Here the output multipliers, found by adding the columns in $(\mathbf{I} - \mathbf{A})^{-1}$ are 1.585, 3.208, 2.687 and 7.000, showing that an increase in the final demand for the output of industry 4 will have a large impact on gross output.

If the number of sectors, n, is large it can be awkward obtaining the inverse of the Leontief matrix. In these circumstances it may be convenient to use the fact that

$$(\mathbf{I} - \mathbf{A})(\mathbf{I} + \mathbf{A} + \mathbf{A}^2 + \mathbf{A}^3 + \mathbf{A}^4 + \mathbf{A}^5 + \cdots + \mathbf{A}^m)$$

$$= \mathbf{I}(\mathbf{I} + \mathbf{A} + \cdots + \mathbf{A}^m) - \mathbf{A}(\mathbf{I} + \mathbf{A} + \cdots + \mathbf{A}^m)$$

$$= (\mathbf{I} + \mathbf{A} + \cdots + \mathbf{A}^m) - (\mathbf{A} + \mathbf{A}^2 + \cdots + \mathbf{A}^{m+1})$$

$$= \mathbf{I} - \mathbf{A}^{m+1}$$

Now if the product had resulted in the unit matrix then the expression $(\mathbf{A} + \mathbf{A}^2 + \cdots + \mathbf{A}^m)$ would be the inverse of $\mathbf{I} - \mathbf{A}$. But the presence of $\mathbf{A}^{m+1}$ prevents this. However, since every element of $\mathbf{A}$ is positive and less than 1 – and, more importantly, the total of each column of $\mathbf{A}$ adds up to less than 1 – then as m increases, $\mathbf{A}^{m+1}$ tends to the null matrix. Therefore, approximately, $(\mathbf{I} - \mathbf{A})^{-1} = \mathbf{I} + \mathbf{A} + \mathbf{A}^2 + \mathbf{A}^3 + \mathbf{A}^4 + \cdots + \mathbf{A}^m$ and the size of the last term, $\mathbf{A}^m$, is a guide to how close the approximation is.

As an example, from the technology matrix in Table 2.2,

$$\mathbf{A} = \begin{bmatrix} 0.125 & 0.148 \\ 0.375 & 0.037 \end{bmatrix}$$

and $$(\mathbf{I} - \mathbf{A})^{-1} = \frac{1}{0.7871}\begin{bmatrix} 0.963 & 0.148 \\ 0.375 & 0.875 \end{bmatrix} = \begin{bmatrix} 1.223 & 0.188 \\ 0.476 & 1.112 \end{bmatrix}$$

is the accurate inverse. Here, an approximation is given by

$$\mathbf{I} + \mathbf{A} + \mathbf{A}^2 + \mathbf{A}^3 + \mathbf{A}^4 = \begin{bmatrix} 1.2129 & 0.1870 \\ 0.4547 & 1.1108 \end{bmatrix}$$

and to check how accurate this is the product of $\mathbf{I} - \mathbf{A}$ and the approximate inverse is

$$(\mathbf{I} - \mathbf{A})(\mathbf{I} + \mathbf{A} + \cdots + \mathbf{A}^4) = \begin{bmatrix} 0.9940 & 0.0008 \\ 0.0170 & 0.9996 \end{bmatrix}$$

which is reasonably close to $\mathbf{I}$. The last term included in the approximate inverse is

$$\mathbf{A}^4 = \begin{bmatrix} 0.0065 & 0.0031 \\ 0.0078 & 0.0047 \end{bmatrix}$$

and a closer approximation can be found by including $\mathbf{A}^5$ and further terms. Here the accurate output multipliers are 1.699 and 1.300 while the approximate ones are 1.668 and 1.298.

The conditions that must be satisfied in order that at least one set of final demands can be met from any given set of output levels are

(a)
$$\begin{vmatrix} (1-a_{11}) & -a_{12} & \cdots & -a_{1n} \\ -a_{21} & (1-a_{22}) & \cdots & -a_{2n} \\ \vdots & \vdots & & \vdots \\ -a_{n1} & -a_{n2} & \cdots & (1-a_{nn}) \end{vmatrix} > 0$$

and
(b) all the elements on the main diagonal of the matrix, namely

$(1 - a_{11})$, $(1 - a_{22})$, . . ., $(1 - a_{nn})$, are positive; that is,

$(1 - a_{11}) > 0$, . . ., $(1 - a_{ii}) > 0$, . . ., $(1 - a_{nn}) > 0$

These are known as the *Hawkins–Simon* conditions.

One way of extending basic input–output analysis, which is also useful in regional economics, is to disaggregate final demand ($\mathbf{C}$) into components such as household consumption, government spending, exports and investment. At the same time, households, government and imports can be included as additional sources of inputs. In Table 2.4 an

TABLE 2.4

Outputs produced by	Inputs used by		Final demand			Total output
	Ind. 1	Ind. 2	Households	Govt	Exports	
Industry 1	200	400	400	200	400	1,600
Industry 2	600	100	600	300	1,100	2,700
Payments for Households	400	800	0	0	0	1,200
Government	100	400	0	0	0	500
Imports	300	1,000	0	0	0	1,300
Total input	1,600	2,700	1,000	500	1,500	7,300

extended input–output table is given for a simple two-sector economy. The inter-industry matrix is the same as in Table 2.1 but extra rows are included to indicate payments to households (for supplying labour and as profits on investments), to the government (taxes to meet the costs of services) and for imports (for raw materials). Extra columns are included to allow final demand to be split between consumption by households, spending by the government, and exports of finished goods. The addition of the extra rows results in the total output of each of the industries equalling the total input. In this example, the country exports more than it imports and the government account is balanced.

The technology matrix can also be extended to include this extra information. Thus, households provide 400 units out of the total inputs of 1,600 for agriculture and so the technological coefficient is 400/1,600 or 0.25. The extended matrix is given in Table 2.5 and, because it is not square, it needs to be used with care. The first two rows and columns give the original $\mathbf{A}$ matrix, so that the formulae $\mathbf{C} = (\mathbf{I} - \mathbf{A})\mathbf{X}$ and $\mathbf{X} = (\mathbf{I} - \mathbf{A})^{-1}\mathbf{C}$ can be used as previously. Now, at the end of section 2.12, we saw that the effect of changing the final demands to $\mathbf{C}_1 = 2,000$ and $\mathbf{C}_2 = 1,000$ is to require $\mathbf{X}_1 = 2,635$ and $\mathbf{X}_2 = 2,064$. The payments to households will now be, from Table 2.5,

$$0.2500\mathbf{X}_1 + 0.2963\mathbf{X}_2 = 1270.31,$$

and tax payments will be

$$0.0625\mathbf{X}_1 + 0.1481\mathbf{X}_2 = 470.37,$$

TABLE 2.5

Output produced by	Inputs required by	
	Ind. 1	Ind. 2
Industry 1	0.1250	0.1481
Industry 2	0.3750	0.0370
Payments for Households	0.2500	0.2963
Government	0.0625	0.1481
Imports	0.1875	0.3704

and the value of imports will be

$$0.1875X_1 + 0.3704X_2 = 1258.6$$

These calculations assume that the payments for households, taxes and imports are exogenous so that there is no feedback from, say, increased household income, to consumption by households. This is rather artificial and it might be more sensible to assume that some of the increased income will be consumed. This can be achieved by treating the household sector as an industry rather than as a component of final demand so that it is assumed to require inputs in fixed proportions. The (2×2) $\mathbf{A}$ matrix of Table 2.5 now becomes (3×3) and has an extra column, consisting of 400/1,200, 600/1,200, and 0/1,200, as well as the extra row 0.2500, 0.2963 and 0. The result is that

$$\mathbf{A}^* = \begin{bmatrix} 0.1250 & 0.1481 & 0.3333 \\ 0.3750 & 0.0370 & 0.5000 \\ 0.2500 & 0.2963 & 0.0000 \end{bmatrix}$$

and
$$(I - \mathbf{A}^*)^{-1} = \begin{bmatrix} 1.5620 & 0.4732 & 0.7572 \\ 0.9585 & 1.5176 & 1.0783 \\ 0.6745 & 0.5680 & 1.5088 \end{bmatrix}$$

The new output multipliers are 3.1950, 2.5588 and 3.3443 compared with the previous values of 1.699 and 1.300. The effect of including households as an industry increases the output multipliers substantially.

It is possible to approach the problem slightly differently and to consider the total available capacity of each industry. Let x_i represent the maximum output which can be obtained from industry i. Then we can calculate the value of the vector **C**, (the maximum final demands for each industry), which could be met if every industry produced at its maximum level. This would be obtained from the relationship

$$\mathbf{C} = (\mathbf{I} - \mathbf{A})\mathbf{X}$$

However, we often find that the result is a **C**-vector containing one or more negative elements. These would correspond to a negative final demand which is impossible and means that the economy would not operate with all its industries at their maximum output level.

The problem now becomes one of deciding which industries should operate at less than their maximum output, i.e. to decide where the 'slack' capacity should be. The latter can only be decided in terms of some specified objective. For example, let C_1 be the final demand for the output of industry 1 which is to be satisfied at a price p_1 and in general let C_n be the final demand for the output of industry n which is to be satisfied at a price p_n. One objective may be to maximise the value of final demand and this may be stated as:

Max

$$Z = C_1 p_1 + C_2 p_2 + \cdots + C_n p_n = \sum_{i=1}^{n} C_i p_i$$

subject to the following constraints on the system

$$a_{11}X_1 + a_{12}X_2 + \cdots + a_{1n}X_n + C_1 \leqslant X_1$$
$$a_{21}X_1 + a_{22}X_2 + \cdots + a_{2n}X_n + C_2 \leqslant X_2$$
$$\vdots$$
$$a_{n1}X_1 + a_{n2}X_2 + \cdots + a_{nn}X_n + C_n \leqslant X_n$$

or $$\mathbf{AX} + \mathbf{C} \leqslant \mathbf{X}$$

This is, in fact, a linear-programming type of problem and the method of arriving at an optimum solution in such situations is discussed in Chapter 8.

2.14 Exercises

1. Determine the maximum final demand which can be met in the following situation shown in Table 2.6.

2. What final demand can be met when the level of output of Industry 1 is increased to 2,000 units in a situation which is in all other respects identical to that given in Question 1?

3. Determine the level of output which is necessary to meet final demands of 1,000 and 2,000 respectively when the technological coefficients are given by the following matrices:

TABLE 2.6

| | Input to | | Level of output |
	Industry 1	Industry 2	
Industry 1	200	300	1,500
Industry 2	500	100	2,500

TABLE 2.7

| Output produced by | Inputs required by | | Final demand | |
	Agriculture	Industry	Households	Exports
Agriculture	100	200	300	400
Industry	500	100	600	800
Households	100	800	0	0
Imports	300	900	0	0
Total input	1,000	2,000	900	1,200

(a) $\begin{bmatrix} 0.2 & 0.4 \\ 0.3 & 0.2 \end{bmatrix}$ (b) $\begin{bmatrix} 0.1 & 0.6 \\ 0.4 & 0.1 \end{bmatrix}$

4. The economy of a region can be summarised by the information in Table 2.7.
   (a) Treating households as exogenous, use the (2×2) technology matrix to find the output multipliers for agriculture and industry. What levels of gross output would be needed to meet a final demand for exports of 500 for agriculture and 900 for industry?
   (b) Treating households as an endogenous sector, use the (3×3) technology matrix to find the output multipliers for agriculture and industry, and compare them with the results in part (a). What levels of gross output would be needed to meet a final demand for exports of 500 for agriculture and 900 for industry?
5. The summary input–output table for the USA in 1947 (in billions of dollars) is shown in Table 2.8.
   (a) Determine the technology matrix and show that the inverse of the Leontief matrix is

TABLE 2.8

Output from	Input to Agriculture	Input to Industry	Input to Services	Final demand	Total output
Agriculture	11	19	1	10	41
Industry	5	89	40	106	240
Services	5	37	37	106	185

$$\begin{bmatrix} 1.409 & 0.192 & 0.062 \\ 0.372 & 1.753 & 0.476 \\ 0.286 & 0.367 & 1.351 \end{bmatrix}$$

(b) What gross outputs are needed to satisfy final demands of 15 for agriculture, 120 for industry and 130 for services?

(c) Show that if 'industry' and 'services' are combined into a single sector the result in (b) is basically unchanged. What are the implications of this for an economist who is interested only in the agriculture sector?

2.15 Revision exercises for Chapter 2 (without answers)

1. If $\mathbf{A} = \begin{bmatrix} 1 & 0 & -1 \\ -2 & 1 & 1 \\ 1 & 0 & 2 \end{bmatrix}$ $\mathbf{B} = \begin{bmatrix} 2 & 1 & -1 \\ 0 & 2 & 1 \\ 1 & -1 & 0 \end{bmatrix}$ $\mathbf{C} = \begin{bmatrix} 2 & -2 \\ 1 & 1 \\ 3 & 2 \end{bmatrix}$

   form the following (if they exist): $\mathbf{A} - \mathbf{B}, \mathbf{B} - \mathbf{C}, \mathbf{AB}, \mathbf{BC}, \mathbf{AB}'$ and show that $\mathbf{AB} \neq \mathbf{BA}$.

2. Find the inverse of $\mathbf{A} = \begin{bmatrix} 2 & 0 & -1 \\ -1 & 2 & 2 \\ 0 & 3 & 1 \end{bmatrix}$ and $\mathbf{B} = \begin{bmatrix} a^2 & 1 & a \\ 1 & a & 1 \\ a & 1 & a \end{bmatrix}$

3. Write the following equations in matrix form and solve them by matrix inversion:

   (a) $3x - 2y = 4$ (b) $x + y + z = 3$

   $2x + y = 5$ $2x + 3y - z = 2$

   $3x + 2y + z = 3$

4. Determine the rank of the matrices given in Questions 1 and 2 above.

TABLE 2.9

Output from	Input to Agriculture	Input to Industry	Input to Services	Final demand
Agriculture	50	150	60	90
Industry	250	200	100	150
Services	100	300	100	100

5. Use Gaussian elimination to determine the rank of the following matrices:

$$\mathbf{A} = \begin{bmatrix} 1 & 1 & -1 \\ 2 & 1 & 0 \\ 3 & 1 & 1 \\ 3 & 3 & -1 \end{bmatrix} \quad \mathbf{B} = \begin{bmatrix} 2 & -1 & -2 & 3 & 5 \\ 2 & 1 & 0 & 0 & 4 \\ -1 & 0 & 1 & 1 & 3 \\ 3 & 0 & -1 & 4 & 12 \end{bmatrix}$$

6. Check whether the following equations have a unique solution:

(a) $2x - 4y - 3z = 25$ (b) $5x + 3y + z = 12$

      $6x - 2y + 4z = 118$ $x - y + 3z = 14$

      $3x + 3y + z = 75$ $4x + 2y + 2z = 16$

7. If $\mathbf{x}' = [1\ 2\ 2\ 3]$ and $\mathbf{A} = \begin{bmatrix} 2 & -1 & 0 & 1 \\ 1 & 2 & -1 & 0 \\ 0 & 1 & 2 & -1 \\ 1 & 0 & 1 & 2 \end{bmatrix}$

evaluate the quadratic form $\mathbf{x}'\mathbf{A}\mathbf{x}$.

8. By partitioning the matrix $\mathbf{M}$ after column 2 and row 2 so that it consists of four (2×2) sub-matrices, check that the formulae given in the text for the inverse of a partitioned matrix is correct.

$$\mathbf{M} = \begin{bmatrix} a & 0 & 1 & 0 \\ 0 & a & 0 & 1 \\ 0 & 1 & 0 & a \\ 1 & 0 & a & 0 \end{bmatrix}$$

9. Determine the characteristic roots and vectors of $\mathbf{A}$ and $\mathbf{B}$ in Question 2 above.

10. A simple economy has three industries: agriculture, industry and services, and the input–output table is shown in Table 2.9. What should the total output for each sector be if final demand changes to 100 for agriculture, 200 for industry and 150 for services?

Chapter 3
Non-linear Equations

3.1 The quadratic

In Chapter 1 the relationship between two variables y and x was assumed to be linear and could be written

$$y = a + bx$$

Here a is the *intercept term*, or the value of y when $x = 0$, and b is the *slope* or *gradient*, since if x increases by 1, y increases by b. For a linear function the slope is a constant. If the equation represents a cost function, with x as the level of output (where $x > 0$) and y as the total cost, then b is the marginal cost of an extra unit of output. Here it is assumed that output and costs vary continuously so that the equation is valid for all positive values of x. We saw that the graph of such an equation is a straight line.

There are many applications of algebra in economics and business studies where the assumption of a (possibly approximate) straight-line relationship is reasonably adequate. These include demand and supply functions, some cost functions, consumption functions and production functions. Also, in section 5.15 below, we will see that under very general conditions MacLaurin's theorem allows a non-linear function to be represented as a first approximation by a linear function.

However, there are also many instances where it is clear that the relationship is non-linear and that a linear approximation will be misleading. For example, suppose that the demand curve relating quantity demanded, q, and price, p, is linear, so that

$$p = a - bq \qquad a > 0, b > 0$$

then revenue from selling the quantity q is

$$R = pq = (a - bq)q = aq - bq^2$$

which is a *quadratic function*. In this case the values of R for different values of q are

92

$$q = 1, R_1 = a - b$$
$$q = 2, R_2 = 2a - 4b$$
$$q = 3, R_3 = 3a - 9b$$

and while the change in revenue in going from $q = 1$ to $q = 2$ is $(a - 3b)$, that in going from $q = 2$ to $q = 3$ is $(a - 5b)$. Here the marginal revenue changes with each value of q, while with a linear function it would be a constant.

The general quadratic function can be written

$$y = ax^2 + bx + c$$

where a, b and c are constants, and $a \neq 0$. This is also known as a *second-order polynomial* in x because the highest power to which x is raised is 2. The effect of the squared term is to make the graph curved instead of being a straight line. For example, consider the quadratic function

$$y = 6x^2 + 2x + 3$$

This might be a cost function with x being the level of output and y being the total cost associated with the value of x. In order to plot the graph of this function several points which lie on the graph are needed. Since x is the level of output we will assume that x is positive, that total cost varies continuously with x and that the values of interest are for $x = 0$ to $x = 5$. Let $x = 0$, then substituting into the equation gives

$$y = 6(0)^2 + 2(0) + 3 = 3,$$

while $x = 1$ gives

$$y = 6(1)^2 + 2(1) + 3 = 11,$$

and $x = 2$ gives

$$y = 6(2)^2 + 2(2) + 3 = 31,$$

and $x = 3$ gives

$$y = 6(3)^2 + 2(3) + 3 = 63,$$

and $x = 4$ gives

$$y = 6(4)^2 + 2(4) + 3 = 107$$

and $x = 5$ gives

$$y = 6(5)^2 + 2(5) + 3 = 163$$

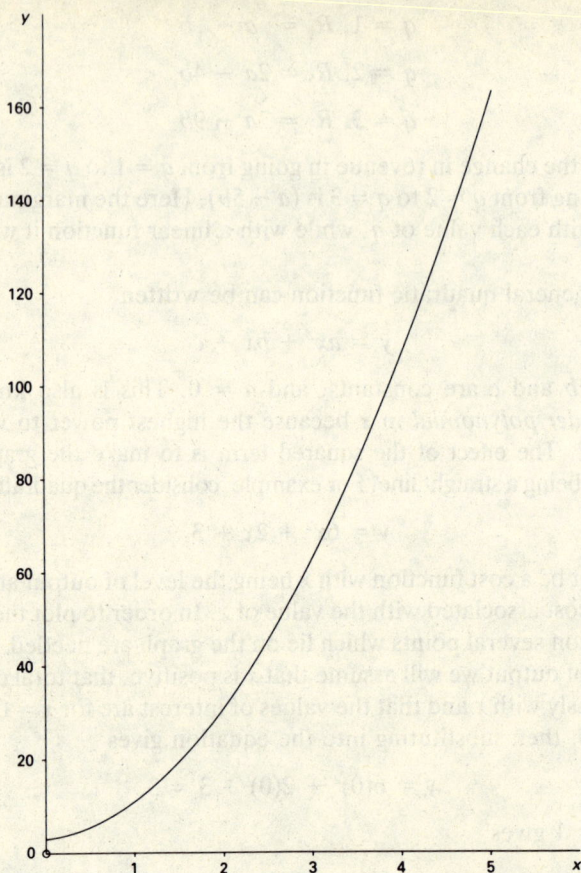

Fig. 3.1

Collecting these values into a table results in

x	0	1	2	3	4	5
y	3	11	31	63	107	163

These values are plotted in Fig. 3.1 and it can be seen that as x increases, so does y, and the rate of increase of y – that is, the gradient – also increases. The result is that the graph curves away from the x-axis. This is because in this example the coefficient on x^2 is positive. If the value of a is negative, the graph will curve towards the x-axis.

For example, let

$$y = 15 + 10x - x^2$$

the values of y for different values of x can be easily obtained by including each term of the quadratic as follows:

x	0	1	2	3	4	5	6	7	8	9	10
constant	15	15	15	15	15	15	15	15	15	15	15
$10x$	0	10	20	30	40	50	60	70	80	90	100
$-x^2$	0	-1	-4	-9	-16	-25	-36	-49	-64	-81	-100
total = y	15	24	31	36	39	40	39	36	31	24	15

The graph is shown in Fig. 3.2. As x increases, y increases, initially by large amounts, but these are decreasing with the result that the value $x = 5$ gives the maximum value of y, and as x increases beyond 5 the graph falls towards the x-axis. The shape of this graph is known as a *parabola* and this is the typical shape of the quadratic function. In the previous example, graphed in Fig. 3.1, since x is the level of output only positive values of x are plotted. However, if the graph is extended in the negative direction, a parabola is also obtained but this has a minimum value (at $x = 0$) rather than a maximum. In general if the coefficient of x^2 is positive (as in Fig. 3.1) there will be a minimum, while if the coefficient is negative (as in Fig. 3.2) there will be a maximum.

The equation of a straight line can be determined if two points which lie on the line are known, since there are two unknown parameters, the intercept term and the slope coefficient. For a quadratic, as there are three parameters, a, b and c, the equation can be determined if three pairs of values of x and y are known. For example, suppose that the following cost information is known:

Output (hundreds)	x	1	2	3
Total cost (£thousands)	y	14	20	28

The fact that x is measured in hundreds and y in thousands can be ignored, as long as this is remembered when the equation is being

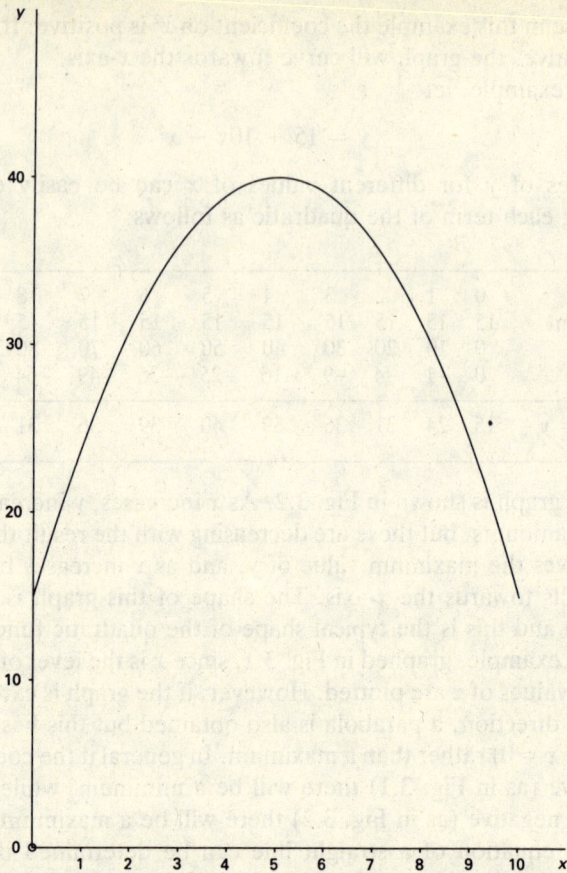

Fig. 3.2

interpreted. Let the function be

$$y = ax^2 + bx + c$$

and again it is assumed that costs vary continously so that this equation
is valid for other values of x. Taking the three values of x for which y
values are known,

$$x = 1 \text{ gives } 14 = \quad a + \quad b + c$$

$$x = 2 \text{ gives } 20 = 4a + 2b + c$$

$$x = 3 \text{ gives } 28 = 9a + 3b + c$$

Solving these by Cramer's rule (see section 1.13), let

$$\det = \begin{vmatrix} 1 & 1 & 1 \\ 4 & 2 & 1 \\ 9 & 3 & 1 \end{vmatrix} = \begin{vmatrix} 1 & 1 & 1 \\ 3 & 1 & 0 \\ 8 & 2 & 0 \end{vmatrix} = 6 - 8 = -2$$

then,

$$a = \frac{\begin{vmatrix} 14 & 1 & 1 \\ 20 & 2 & 1 \\ 28 & 3 & 1 \end{vmatrix}}{\det} = \frac{\begin{vmatrix} 14 & 1 & 1 \\ 6 & 1 & 0 \\ 14 & 2 & 0 \end{vmatrix}}{\det} = \frac{12 - 14}{-2} = \frac{-2}{-2} = 1$$

$$b = \frac{\begin{vmatrix} 1 & 14 & 1 \\ 4 & 20 & 1 \\ 9 & 28 & 1 \end{vmatrix}}{\det} = \frac{\begin{vmatrix} 1 & 14 & 1 \\ 3 & 6 & 0 \\ 8 & 14 & 0 \end{vmatrix}}{\det} = \frac{42 - 48}{-2} = \frac{-6}{-2} = 3$$

$$c = \frac{\begin{vmatrix} 1 & 1 & 14 \\ 4 & 2 & 20 \\ 9 & 3 & 28 \end{vmatrix}}{\det} = \frac{\begin{vmatrix} 1 & 1 & 14 \\ 2 & 0 & -8 \\ 6 & 0 & -14 \end{vmatrix}}{\det} = \frac{28 - 48}{-2} = \frac{-20}{-2} = 10$$

and therefore, $y = x^2 + 3x + 10$

This answer can easily be checked by substituting $x = 1, 2$ and 3 into the equation. Given any value of x, say $x = 4.5$ so that output is 450, the value of y, which here is $y = 43.75$ or £43,750, can be found.

3.2 Exercises

1. Determine the equations of the quadratic cost functions which fit the following data. In each case sketch the graphs for the range $x = 0$ to $x = 10$.

(a) Output	0	2	6
Total cost	4	14	58
(b) Output	4	6	8
Total cost	26	40	58
(c) Output	4	6	8
Total cost	31	37	45
(d) Output	0	5	10
Total cost	20	24.5	30
(e) Output	0	2	4
Total cost	20	22	28

2. Plot the graphs of the following quadratic functions for $x = 0$ to $x = 10$

(a) $y = x^2 + 10x + 100$

(b) $y = x^2 - 10x + 100$

(c) $y = -x^2 + 10x + 100$

(d) $y = -x^2 - 10x + 100$

3.3 The roots of a quadratic equation

The graph of the general quadratic function

$$y = ax^2 + bx + c$$

cuts the x-axis when $y = 0$, and for this value

$$0 = ax^2 + bx + c$$

which is a quadratic equation. The values of x which satisfy this equation are called the *roots* of the quadratic or the *solutions* of the equation.

There are basically three ways of solving this type of equation: by plotting the graph of the quadratic function and finding where $y = 0$; by rearranging the coefficients and finding the factors; and by using the formula which gives the solution. While the first two ways can be useful in some circumstances, they can also be tedious in that a great deal of time might be spent obtaining a large number of pairs of values or trying to find factors before a solution is found. There is also the possibility that there are no solutions, and in this case the search for solutions is futile. However, the third way provides a simple formula which can easily be used and which quickly indicates whether there are any solutions. It is therefore the recommended method for solving a quadratic equation. To obtain the formula, write the general quadratic equation as

$$ax^2 + bx + c = 0$$

Dividing throughout by the coefficient of x^2 and rearranging the equation

$$x^2 + \frac{b}{a}x + \frac{c}{a} = 0$$

The terms containing x can be made into a perfect square by adding the term $(b/2a)^2$ and if we also subtract this term from the left hand side, the equation is unchanged;

$$\left[x^2 + \frac{b}{a}x + \left(\frac{b}{2a}\right)^2 \right] + \frac{c}{a} - \left(\frac{b}{2a}\right)^2 = 0$$

The terms in the square brackets are equal to $(x + b/2a)^2$

$$\therefore \left(x + \frac{b}{2a}\right)^2 = \left(\frac{b}{2a}\right)^2 - \frac{c}{a}$$

$$= \frac{b^2}{4a^2} - \frac{c}{a}$$

$$= \frac{b^2 - 4ac}{4a^2}$$

$$\therefore x + \frac{b}{2a} = \pm \sqrt{\left(\frac{b^2 - 4ac}{4a^2}\right)}$$

$$= \pm \frac{\sqrt{(b^2 - 4ac)}}{2a}$$

Plus and minus signs occur in front of the square root sign. This is because the square of a negative number is always a positive number, e.g. $(-2)^2 = 4$. Therefore the square root of a positive number can be either a positive or negative number, although its absolute value is always the same.

$$\therefore x = \frac{-b}{2a} \pm \frac{\sqrt{(b^2 - 4ac)}}{2a}$$

$$= \frac{-b \pm \sqrt{(b^2 - 4ac)}}{2a}$$

The two roots of the quadratic equation are

$$x_1 = \frac{-b + \sqrt{(b^2 - 4ac)}}{2a} \quad \text{and} \quad x_2 = \frac{-b - \sqrt{(b^2 - 4ac)}}{2a}$$

For example, if $x^2 - 8x + 7 = 0$
then $a = 1$, $b = -8$ and $c = 7$

Therefore,
$$x = \frac{8 \pm \sqrt{(64 - 28)}}{2}$$

$$= 4 \pm 0.5 \sqrt{36} = 4 \pm 3$$

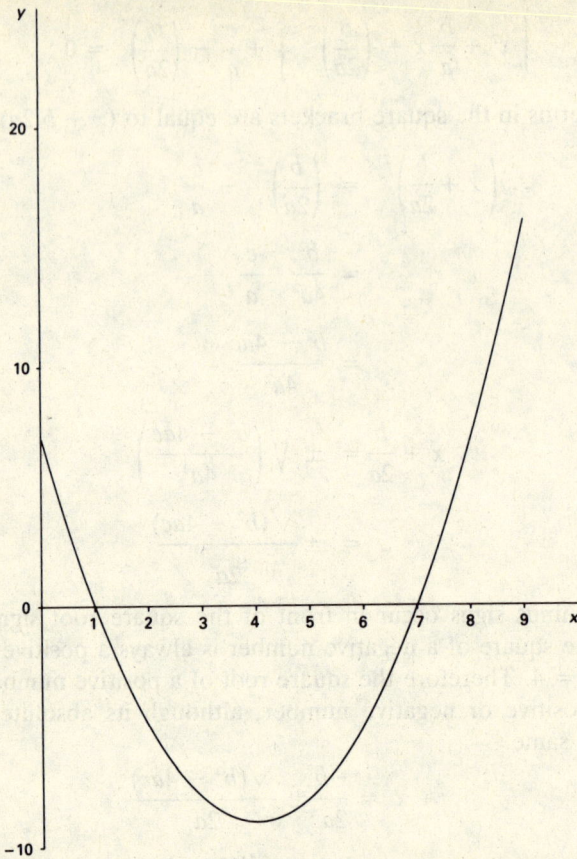

Fig. 3.3

and so $x_1 = 1$ and $x_2 = 7$. Here there are two distinct roots, and this corresponds to the case where the graph of

$$y = x^2 - 8x + 7$$

cuts the x-axis in two places, when $x = 1$ and when $x = 7$. This is shown in Fig. 3.3 where it can be seen that y has a minimum value when $x = 4$. There are also two roots when the graph cuts the x-axis from below, as when

$$y = -x^2 + 8x - 7$$

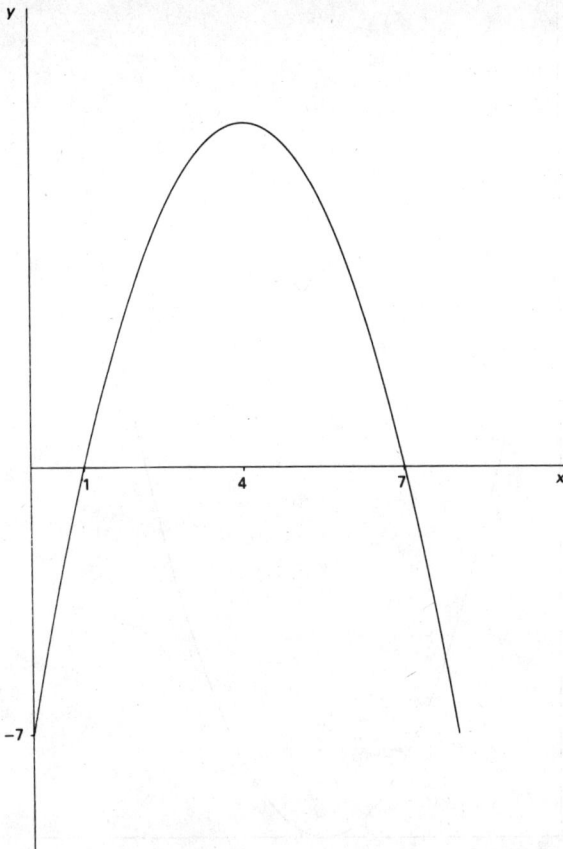

Fig. 3.4

which is shown in Fig. 3.4, where the roots are again $x = 1$ and $x = 7$, and y has a maximum value when $x = 4$. The more general problem of determining whether y has a maximum or minimum value is discussed in detail in Chapter 5.

As a second example consider

$$x^2 - 8x + 16 = 0$$

and using the formula,

$$x = \frac{8 \pm \sqrt{(64 - 64)}}{2} = 4 \pm 0 = 4$$

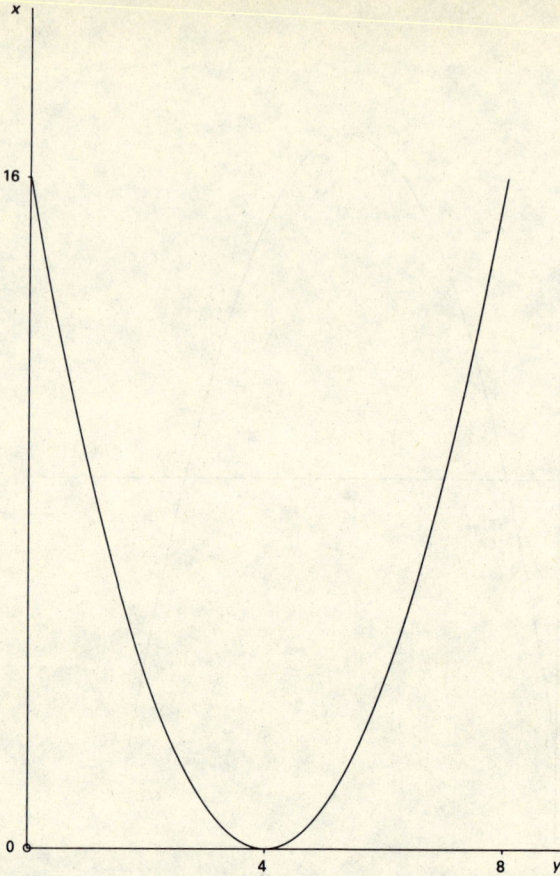

Fig. 3.5

so that there are two equal roots.

We obtain equal roots when the term under the square root sign is equal to zero – that is, when $b^2 = 4ac$. Here the graph of

$$y = x^2 - 8x + 16$$

does not cut the x-axis but touches it when $x = 4$. This is shown in Fig. 3.5 and the minimum value of y is when $x = 4$.

As a third example consider

$$x^2 - 8x + 20 = 0$$

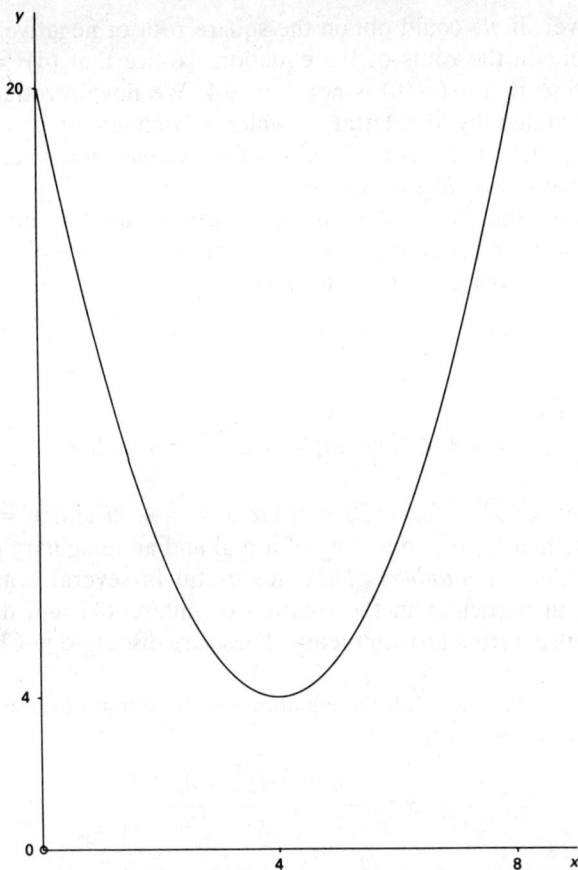

Fig. 3.6

Here, $$x = \frac{8 \pm \sqrt{(64 - 80)}}{2} = 4 \pm 0.5 \sqrt{(-16)}$$

and there is a negative number under the square root sign. This has no real value and there are no real roots in this case. The graph of

$$y = x^2 - 8x + 20$$

is shown in Fig. 3.6 where the value of y reaches a minimum at $x = 4$ so that there is no value of x for which $y = 0$.

However, if we could obtain the square root of negative number, we could obtain the roots of the equation. Notice that $(4)^2 = 16$ and $(-4)^2 = 16$ so that $\sqrt{(-16)}$ is not 4 or -4. We now introduce a new number, denoted by the letter i, which is defined by $i^2 = -1$ or $i = \sqrt{(-1)}$. Since there is no 'real' number which when squared gives -1, is known as an *imaginary number*.

The properties of i are such that it can be used as an ordinary number bearing in mind that $i^2 = -1$, so that $i^3 = -i$ and $i^4 = 1$.

We can now write $-16 = 16(-1) = 16i^2$

Hence $$\sqrt{(-16)} = \sqrt{(16i^2)} = 4i$$

Therefore,

$$x = 4 \pm \tfrac{1}{2}\sqrt{(-16)} = 4 \pm \frac{4i}{2} = 4 \pm 2i$$

and the roots of $x^2 - 8x + 20 = 0$ are $x = 4 + 2i$ and $x = 4 - 2i$. Numbers such as these, consisting of a real and an imaginary part, are known as *complex numbers*. They are useful in several branches of economics, in particular in the solution of differential and difference equations arising from growth theory. These are discussed in Chapters 9 and 10.

The solutions of a quadratic equation can be summarised as follows: If $ax^2 + bx + c = 0$

then $$x = \frac{-b \pm \sqrt{(b^2 - 4ac)}}{2a}$$

and if $b^2 > 4ac$ there are two real and distinct roots

if $b^2 = 4ac$ there are two real and coincident roots

if $b^2 < 4ac$ there are two complex roots

Before leaving quadratic equations we consider some properties of the roots. Let r_1 and r_2 be the roots of

$$ax^2 + bx + c = 0$$

then both $(x - r_1)$ and $(x - r_2)$ are factors of the equation and so

$$(x - r_1)(x - r_2) = 0$$

This can be expanded to give

$$x^2 - (r_1 + r_2) + r_1 r_2 = 0$$

and comparing with

$$x^2 + (b/a)x + (c/a) = 0$$

we see that the sum of the roots,

$$r_1 + r_2 = -b/a$$

and the product of the roots,

$$r_1 r_2 = c/a$$

These can provide useful information when a, b and c are general parameters rather than known numerical values. For example, if c/a is negative then it is clear that, if they are real roots, one of r_1 and r_2 is negative and the other is positive.

3.4 Exercises

1. Use the formula to obtain the roots of the following equations:

   (a) $x^2 - 7x + 12 = 0$ (b) $x^2 + x - 2 = 0$

   (c) $2x^2 + 7x + 3 = 0$ (d) $x^2 - 2x + 1 = 0$

   (e) $x^2 - 1 = 0$ (f) $x^2 + 1 = 0$

   (g) $x^2 - 4x + 5 = 0$ (h) $x^2 + 2x + 2 = 0$

2. By finding the roots of the quadratic, plot the graphs of the following equations over the ranges given:

   (a) $y = x^2 - 10x + 25$ for $x = 0$ to $x = 10$

   (b) $y = -x^2 + 4x - 3$ for $x = 0$ to $x = 4$.

3.5 Other non-linear functions

In discussing the properties of the general quadratic function we saw that the graph could be plotted by taking different values of x and finding the associated values of y, and that the points where the graph cuts the x-axis are the roots of the quadratic equation. Also, the graph is a parabola which has either a maximum value or a minimum value, depending on whether the coefficient of x^2 is negative or positive. Turning now to more general non-linear functions, the same approach can be used to examine their properties.

Both the linear and the quadratic functions are special cases of the general *polynomial function*

$$y = a_n x^n + a_{n-1} x^{n-1} + \cdots + a_1 x + a_0$$

where $a_n, a_{n-1}, \ldots a_0$ are constants and it is assumed that a_n is not zero. Using this notation, for the linear equation, $n = 1$ and

$$y = a_1 x + a_0$$

while for the quadratic $n = 2$ and

$$y = a_2 x^2 + a_1 x + a_0$$

The graph of a polynomial function can be obtained as previously by finding the values of y corresponding to particular values of x. However, with complicated functions it is also useful to:

(a) find where the graph cuts the y-axis and x-axis, and
(b) determine where the maximum and minimum values occur.

Here we will consider (a) and defer a discussion of (b) until Chapter 5.

The graph cuts the y-axis when $x = 0$ and the corresponding value of y is easily evaluated. It is more difficult to find where the graph cuts the x-axis since putting $y = 0$ results in the polynomial equation

$$a_n x^n + a_{n-1} x^{n-1} + \cdots + a_1 x + a_0 = 0$$

The roots of this are required. We saw that the roots of a quadratic equation could be found by using a formula. There also exist formulae for the solution of cubic equations (where $n = 3$) and quartic equations (where $n = 4$) (see Further Reading for details of references) but these are complicated and no formulae apply for $n > 4$. Instead, an alternative method of solution is needed. Here we will consider a simple trial-and-error method and leave a more efficient method to section 5.17.

The general approach is to guess the value of a root, substitute it into the equation and if the value of the equation is not zero, change the guess and repeat the process until the value becomes zero.

For example, consider the problem of plotting the graph of

$$y = x^3 - 8x^2 - 35x + 150$$

for the range $x = 0$ to $x = 12$. If $x = 0$ then $y = 150$, which is where the graph cuts the *y*-axis, but this is not a root of the cubic equation. Trying $x = 1$, $y = 108$ and so while this is also not a root, the value of *y* is smaller so that increasing the value of *x* should be a move in the correct direction – that is, towards the root. For $x = 2$, $y = 56$ and for $x = 3$, $y = 0$ so that $x = 3$ is a root. Therefore $(x - 3)$ is a factor and the equation can be written

$$y = x^3 - 8x^2 - 35x + 150 = (x - 3)Z$$

where *Z* is an unknown quadratic expression. But we can divide both sides of the equation by $(x - 3)$ to give

$$Z = \frac{(x^3 - 8x^2 - 35x + 150)}{x - 3}$$

and this can be simplified by long division. The rules for this are quite straightforward. First, always write any algebraic expression in the order of descending powers of *x*. Second, the first term in the divisor (that is, *x* in $x - 3$) is divided into the first term of the cubic (x^3 here) and the result placed above the cubic, as shown below. Then multiply the $(x - 3)$ by x^2, placing the result under the cubic, and subtract this from the equation. Next, move the rest of the equation down. This gives:

$$
\begin{array}{r}
x^2 \\
\hline
x - 3 \,\big)\, x^3 \;-\; 8x^2 \;-\; 35x \;+\; 150 \\
x^3 \;-\; 3x^2 \\
\hline
0 \;-\; 5x^2 \;-\; 35x \;+\; 150
\end{array}
$$

Next, divide *x* into $-5x^2$, which goes $-5x$ times and so $-5x$ is placed at the top and multiplied by $(x - 3)$ before subtracting this from the equation:

$$
\begin{array}{r}
x^2 \;-\; 5x \\
\hline
x - 3 \,\big)\, x^3 \;-\; 8x^2 \;-\; 35x \;+\; 150 \\
x^3 \;-\; 3x^2 \\
\hline
0 \;-\; 5x^2 \;-\; 35x \;+\; 150 \\
-\; 5x^2 \;+\; 15x \\
\hline
-\; 50x \;+\; 150
\end{array}
$$

Finally dividing $(x - 3)$ into $-50x + 150$ gives -50 and nothing left over. The result of the division is

$$x^3 - 8x^2 - 35x + 150 = (x - 3)(x^2 - 5x - 50)$$

which can be checked by multiplication. Now since $y = 0$,

$$x^2 - 5x - 50 = 0$$

and solving by the formula gives $x = 10$ and $x = -5$. So far we have established that the graph cuts the x-axis when $x = 3$, $x = 10$ and $x = -5$.

Since the range of values which are of interest is $x = 0$ to $x = 12$, we complete the following table:

x	0	1	2	3	4	5	6	7	8	9	10	11	12
y	150	108	56	0	−54	−100	−132	−144	−130	−84	0	128	306

The graph is shown in Fig. 3.7 and has a minimum value at $x = 7$.

As a second example of solving a polynomial equation let $y = 5x^3 + 27x^2 + 17x - 21$ and putting $x = 0$ gives $y = -21$. For $x = 1$, $y = 28$ and so as the sign on the value of y changes there must be at least one root between $x = 0$ and $x = 1$. This is obvious once the graph is considered. At $x = 0$ the value of y is negative while at $x = 1$, y is positive, and so the graph must cross the x-axis. Here the values of y at $x = 0$ and $x = 1$ are approximately equal (ignoring the sign), and so our next guess at a root is $x = 0.5$, which gives $y = -5.125$, indicating that the root is greater than $x = 0.5$. Trying $x = 0.6$ gives $y = 0$ and so $x = 0.6$ is a root. Therefore $(x - 0.6)$ is a factor and dividing through by this:

$$
\begin{array}{r}
5x^2 + 30x + 35 \\
x - 0.6 \overline{\smash{\big)}\ 5x^3 + 27x^2 + 17x - 21} \\
\underline{5x^3 - 3x^2} \\
0 \qquad 30x^2 + 17x - 21 \\
\underline{30x^2 - 18x} \\
0 \qquad 35x - 21 \\
\underline{35x - 21} \\
0
\end{array}
$$

Thus $5x^3 + 27x^2 + 17x - 21 = (x - 0.6)(5x^2 + 30x + 35)$ and setting this to zero results in the quadratic,

$$5x^2 + 30x + 35 = 0$$

which has the roots $x = -3 + \sqrt{2}$ and $x = -3 - \sqrt{2}$

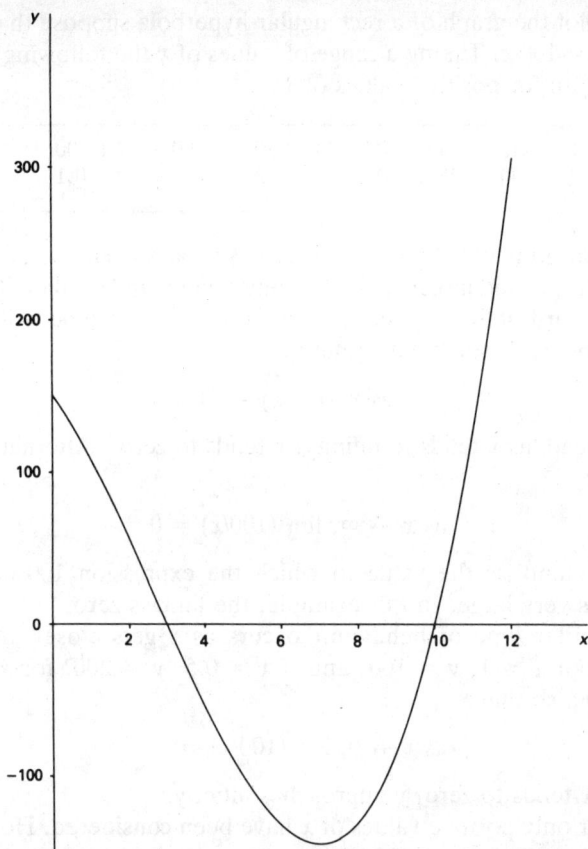

Fig. 3.7

While polynomial functions have many applications in economics and business studies, other types of functions also occur. One of the common ones is the *rectangular hyperbola* which has the general equation

$$xy = c$$

where c is a constant which is neither zero nor infinity. If x is price and y is the quantity sold then xy is the total revenue so that this curve represents a *constant revenue function*. Thus whatever price is charged the revenue is always the same, and so it can be referred to as an iso-revenue or equal-revenue function. In section 5.5 we will see that such a demand function has a constant elasticity of -1.

To plot the graph of a rectangular hyperbola suppose that $c = 100$ so that $y = 100/x$. Taking a range of values of x the following table can be drawn up for positive values of x:

x	1	2	4	10	20	50	100	1,000	10,000
y	100	50	25	10	5	2	1	0.1	0.01

Notice that no matter how large x is, y is always positive and that as x becomes larger and larger, y will become smaller and smaller. That is, as x tends towards infinity, y tends towards zero. This can be written more formally by using some new notation:

$$\text{as } x \to \infty, y \to 0$$

which is read 'as x tends to infinity, y tends to zero'. Alternatively, we can write:

$$\text{as } x \to \infty, \lim (100/x) = 0$$

Thus the 'limit' is the value to which the expression $100/x$ tends as x becomes very large. In this example, the limit is zero.

A similar type of behaviour occurs as x gets closer and closer to zero. For $x = 1$, $y = 100$, and if $x = 0.5$, $y = 200$, for $x = 0.01$, $y = 10,000$, so that

$$\text{as } x \to 0, \lim (100/x) = \infty$$

Thus, as x tends to zero, y approaches infinity.

So far only positive values of x have been considered. However, if x is negative, since $xy = 100$, then y must also be negative. A table of values can be drawn up as before:

x	−1	−2	−4	−10	−20	−50	−100	−1,000	−10,000
y	−100	−50	−25	−10	−5	−2	−1	−0.1	−0.01

Here it can be seen that,

$$\text{as } x \to -\infty, \lim (100/x) = -0$$

so that as x takes a larger negative value, y remains negative but tends to zero (from below). Similarly,

$$\text{as } x \to -0, \lim (100/x) = -\infty$$

so that y becomes infinitely large in the negative direction.

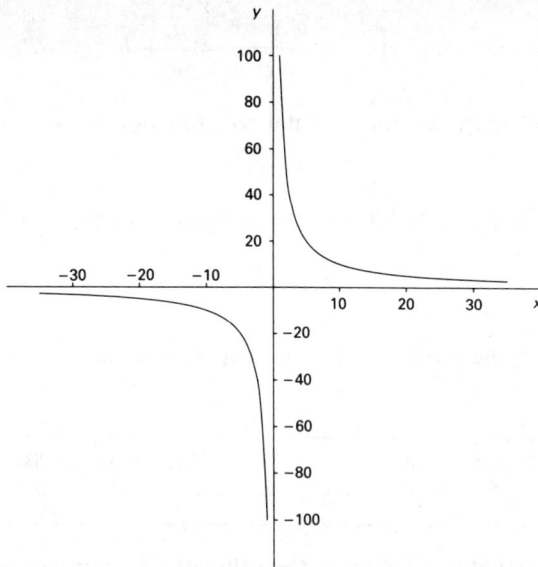

Fig. 3.8

The graph of $xy = 100$ is shown in Fig. 3.8 and consists of two separate curves which do not join. As x tends to zero from the positive side, y tends to infinity and the graph approaches, but does not meet, the y-axis. Thus the graph is said to approach the y-axis *asymptotically*. That is, the y-axis, which is the line $x = 0$, is called an *asymptote* of $xy = 100$. Similarly, on the positive side, the x-axis, which is the line $y = 0$, is an asymptote. It is also clear that when negative values of x are considered the x-axis and the y-axis are asymptotes of $xy = 100$.

The discussion above has been for $xy = 100$, but it obviously applies for any curve $xy = c$ where c is a non-zero constant. If c is negative the two parts of the graph are in the other two quadrants.

Another version of the rectangular hyperbola occurs with average cost functions when the total cost function is linear. Let the total cost function be

$$TC = a + bx$$

where x is the level of output and a and b are constants. We define the *average cost* of production, AC, by

$$AC = \frac{\text{total cost}}{\text{level of output}} = \frac{TC}{x}$$

and here
$$AC = \frac{a + bx}{x} = \frac{a}{x} + b$$

For example, suppose that the total cost function is

$$TC = 100 + 3x$$

so that the fixed cost is 100 and the marginal cost per unit is 3. Then

$$AC = \frac{100}{x} + 3$$

Since here x is the level of output it is expected to be positive. Forming a table of values:

x	1	2	4	10	20	50	100	1,000	10,000
AC	103	53	28	13	8	5	4	3.1	3.01

it can be seen that as x increases the value of AC approaches 3. That is,

$$\text{as } x \to \infty, \lim (AC) = 3$$

The graph is shown in Fig. 3.9 and here the asymptotes are the y-axis ($x = 0$) and the horizontal line $y = 3$. In economic terms the fixed cost is being spread over more and more units of output and so the average cost is approaching the marginal cost.

If the total cost function is non-linear, the behaviour of average cost is more complex. For example, let

$$TC = 100 + 3x + x^2$$

then
$$AC = \frac{100}{x} + 3 + x$$

and as x tends to infinity, AC also increases continuously. However, as x increases, the term $100/x$ becomes smaller and smaller so that AC will be dominated by $(3 + x)$. That is, AC will approach the straight line $y = 3 + x$ but will not cross it and so this will be an asymptote. As previously, the y-axis will also be an asymptote since as $x \to 0, AC \to \infty$. The graph is shown in Fig. 3.10.

This section has been concerned with non-linear functions and, in particular, how to draw their graphs. While it is possible to graph any function by evaluating a large number of points, it is useful to find where

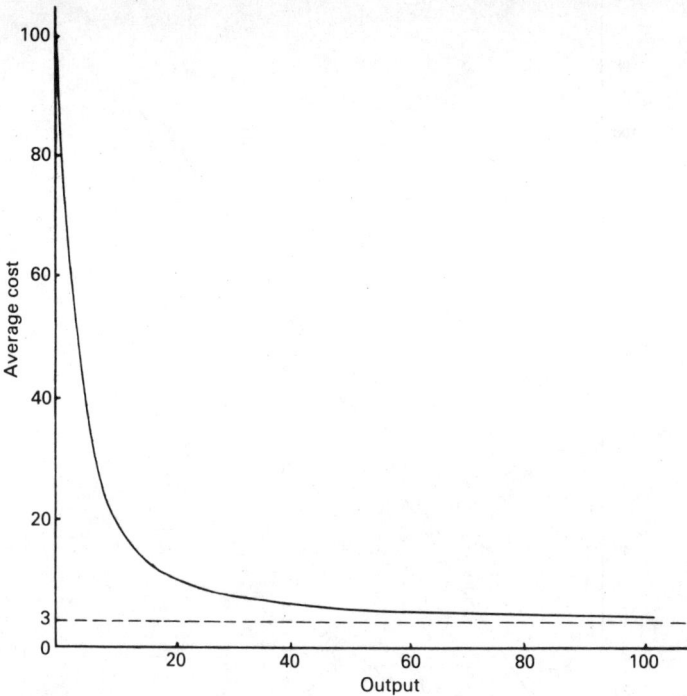

Fig. 3.9

the graph crosses the y-axis (where $x = 0$), and the x-axis (where $y = 0$). For polynomial functions the points where $y = 0$ are the roots of an equation and they can be found by a trial-and-error method. For other non-linear functions, such as the hyperbola, it is important to detect the presence of asymptotes. In Chapter 4 we discuss exponential and logarithmic functions. A detailed treatment of maximum and minimum values is presented in section 5.7 and we return to the problem of plotting graphs in section 5.18.

3.6 Exercises

1. Sketch the graphs of the following functions over the range of values indicated.

   (a) $y = x^2 - 8x + 12$ for $x = 0$ to $x = 6$

   (b) $y = x^3 - 13x^2 - 170x + 600$ for $x = 0$ to $x = 6$

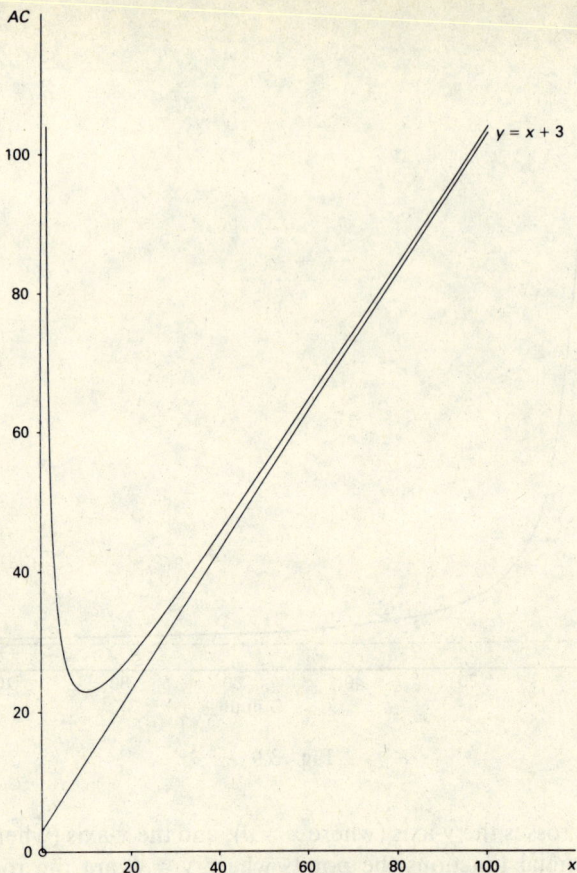

Fig. 3.10

(c) $y = (2x - 3)(x + 1)(x - 5)$ for $x = -2$ to $x = 6$

(d) $y = 32x - 2x^3$ for $x = -4$ to $x = 4$

(e) $y = x^5$ for $x = 0$ to $x = 3$

(f) $xy = 20$ for positive values of x

(g) $xy = 20 + x$ for positive values of x.

2. In a simple production process, if L is the amount of labour input and K is the amount of capital input, the volume of output, Q, is

given by $Q = KL$. If Q is assumed to be fixed then this gives the trade-off between using L and K in producing the output Q. It is known as an isoquant since it shows how different values of K and L can be combined to give a particular value of Q. Sketch the isoquants for $Q_1 = 20$, $Q_2 = 50$ and $Q_3 = 100$. Notice that they do not cross.

3. Sketch the graphs of the following total cost functions and their average cost functions (x is the level of output):

(a) $TC = 100$

(b) $TC = 100 + 5x$

(c) $TC = 100 + 5x + x^2$.

3.7 Breakeven point

In Chapter 1 the breakeven point was defined as that level of output at which *the total cost and the total revenue are equal*. In general, a unique value is obtained when both the cost and revenue functions are linear. However, this is not necessarily the case when one or both of the functions are non-linear.

Let us first consider the case in which the total revenue function is linear, so that the price is independent of the quantity sold, and the total cost function is quadratic. For example,

$$\text{Cost function:} \quad TC = 7 + 2x + x^2$$

$$\text{Revenue function:} \quad TR = 10x$$

where x is the level of output. At the breakeven point,

$$TC = TR \text{ and so}$$

$$7 + 2x + x^2 = 10x$$

or,

$$x^2 - 8x + 7 = 0$$

This quadratic equation has the roots

$$x = \frac{8 \pm \sqrt{(64 - 28)}}{2} = 1 \text{ and } 7$$

and, therefore, either $x = 1$ or $x = 7$ satisfies the equation. This means that there are two points at which total revenue equals total cost and this can be seen to be true by sketching both graphs on the same set of axes as shown in Fig. 3.11.

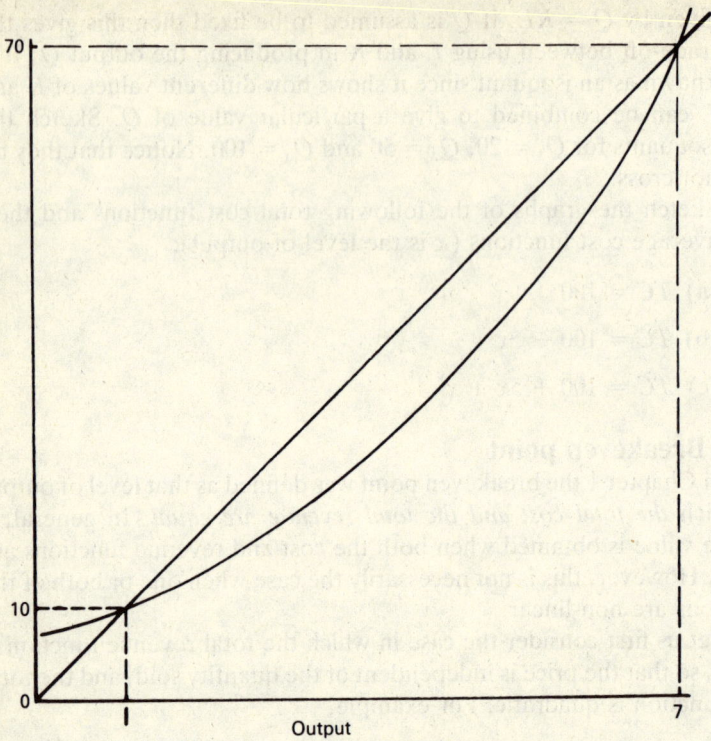

Fig. 3.11

In this case net revenue does not increase continuously with output but has the following pattern:

$x < 1$ Net revenue is negative

$x = 1$ Net revenue is zero

$1 < x < 7$ Net revenue increases at first and then decreases again

$x = 7$ Net revenue is zero

$x > 7$ Net revenue is negative

It therefore has a maximum value for an output somewhere between $x = 1$ and $x = 7$. The output at which this maximum value occurs can be obtained using the differential calculus as we shall see later in Chapter 5. For the present it is sufficient to note that the linear equation and

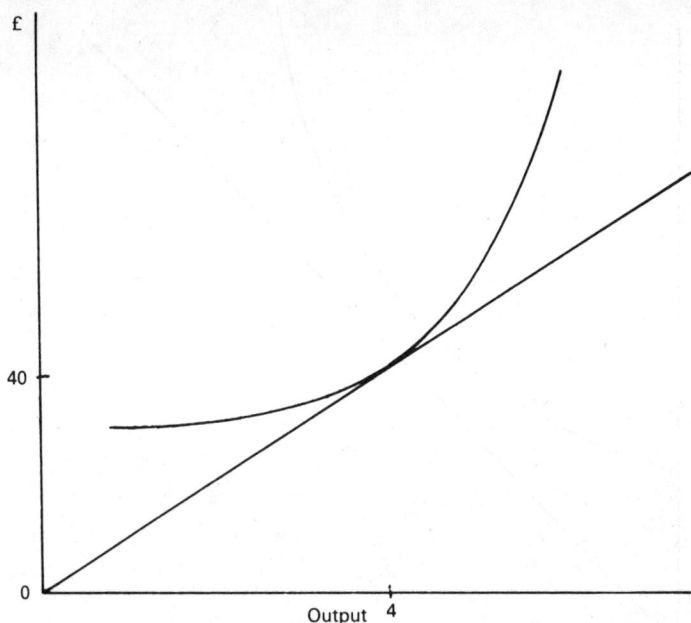

Fig. 3.12

the quadratic equation intersect in two points and the maximum net revenue occurs at an output somewhere between these two points.

But this is not always so. Consider the case where the following equations apply:

Cost function	TC	$=$	$16 + 2x + x^2$
Revenue function	TR	$=$	$10x$
Then	TC	$=$	TR
when	$16 + 2x + x^2$	$=$	$10x$
or	$x^2 - 8x + 16$	$=$	0

and there are repeated roots with just one solution, $x = 4$. The graphs are shown in Fig. 3.12 where it can be seen that there is one point at which the two lines meet. When $x = 4$ the straight line is a *tangent* to the curve. Mathematically the two coincident solutions are $x = 4$. Economically, it is apparent that there is only one point at which the net revenue is non-negative and at no point is it positive. It is therefore likely that

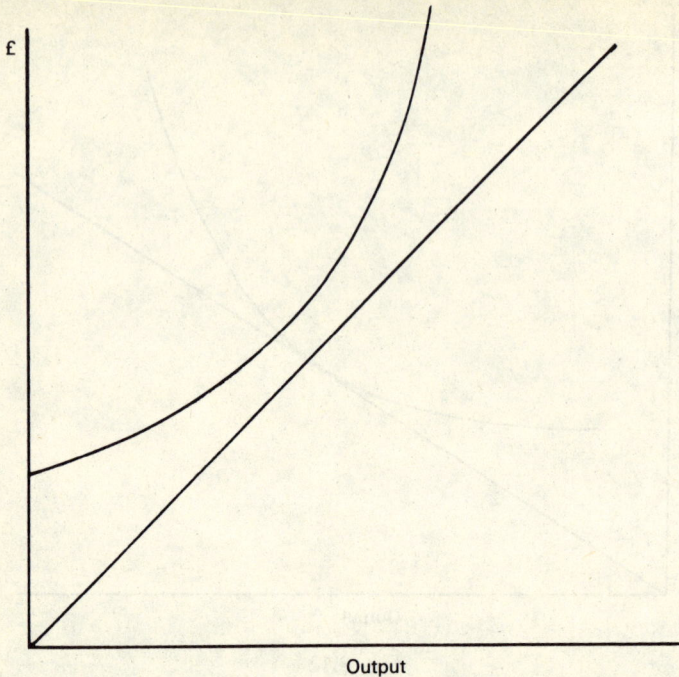

Fig. 3.13

the company will consider either of the two following possibilities.

(a) A price increase if this is possible. The resulting revenue function would then cut the cost function in two places and if the necessary output could be sold at the new price a positive net revenue could be obtained.

(b) A cost reduction in the process. This would again result in the cost and revenue functions intersecting in two points between which there would be an area of positive net revenue.

One further case can occur when the two functions do not intersect at all. This is illustrated graphically in Fig. 3.13

In this case there is no output for which there is a positive net revenue. This situation would arise if the fixed costs of the plant were to increase without a corresponding increase in the price of the product. For example,

Cost function $TC = 20 + 2x + x^2$

Revenue function	$TR = 10x$
Then	$TC = TR$
when	$20 + 2x + x^2 = 10x$
or	$x^2 - 8x + 20 = 0$

This equation has two complex roots and therefore there is no real value of output for which a breakeven point occurs.

3.8 Simultaneous quadratic equations

We have used linear equations to represent demand and supply functions but we are now able to extend the analysis a little further by considering quadratic equations. For example,

$$\text{demand} \quad q_d = f(p) = p^2 - 8p + 15$$

$$\text{supply} \quad q_s = g(p) = 2p^2 + 3p - 3$$

Graphical representation of these two equations on the same set of axes shows that there is a position where the curves intersect. This is the point at which supply and demand are equal (see Fig. 3.14)

The equilibrium values of p and q could be obtained graphically but we can also solve the two quadratic equations simultaneously. For example,

$$q_d = p^2 - 8p + 15$$

$$q_s = 2p^2 + 3p - 3$$

and in equilibrium $q_d = q_s$

$$\therefore \ p^2 - 8p + 15 = 2p^2 + 3p - 3$$

$$\therefore \ p^2 + 11p - 18 = 0$$

$$\therefore \ p = \frac{-11 \pm \sqrt{[11^2 + (4 \times 18)]}}{2} = \frac{-11 \pm \sqrt{193}}{2}$$

$\sqrt{193}$ can be found by reference to tables of square roots or by the use of a calculator. The result is

$$p = \frac{-11 \pm 13.9}{2}$$

and the two roots are $p_1 = 2.9/2 = 1.45$ and $p_2 = -24.9/2 = -12.45$.

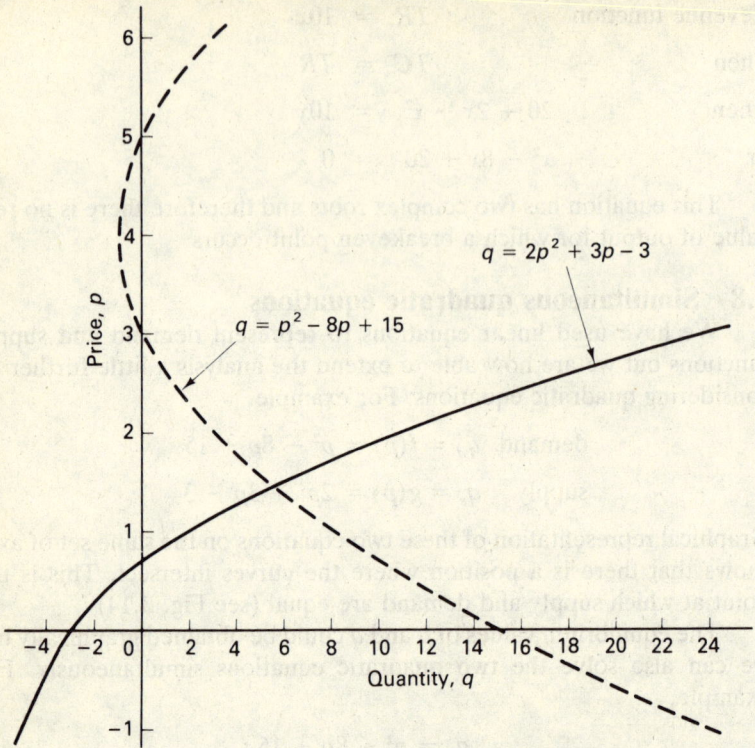

Fig. 3.14

Mathematically the two solutions are both correct because the quadratic curves intersect in two places, but by the very nature of our problem price cannot be negative, and therefore we must take the positive root.

The equilibrium price will be 1.45 units and at this price

$$2.1 - 11.6 + 15 = 5.5 \text{ units are produced and sold}$$

3.9 Net revenue

Quadratic equations often arise in economic theory from situations where the original data can be represented by a linear equation.

Let the demand function for a product be

$$q_D = f(p) = 1000 - 10p$$

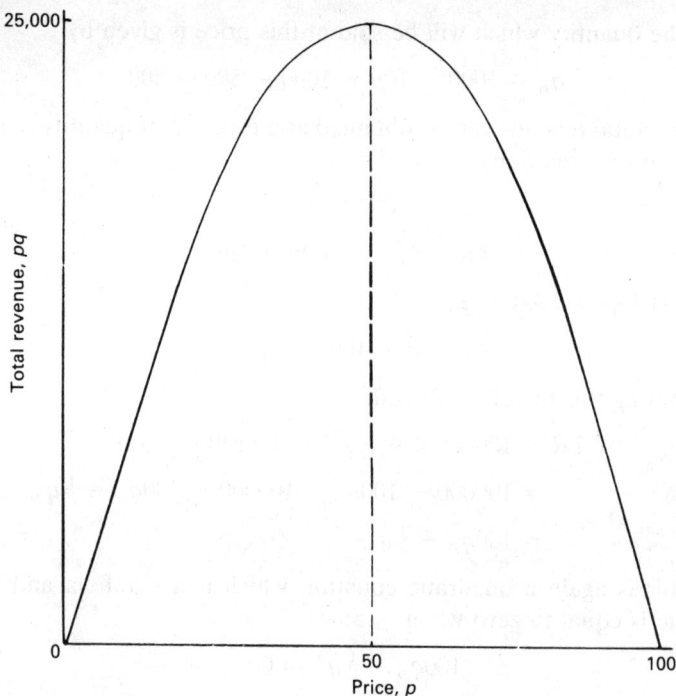

Fig. 3.15

The total revenue, which is obtained by the sale of q items, is obtained by multiplying price by quantity sold.

$$\therefore \text{Total revenue} = pq = p(1000 - 10p)$$
$$= 1000p - 10p^2$$

This is a quadratic equation and a graph of the function shows that the total revenue rises at first and then after reaching some maximum value starts to decline (Fig. 3.15).

This graph is a *parabola*. It is symmetrical about the line parallel to the revenue axis through the point $p = 50$ as shown by the dotted line. The maximum revenue is therefore obtained when $p = 50$ and this is equal to

$$p(1000 - 10p) = 50(1000 - 500) = 25000$$

The quantity which will be sold at this price is given by

$$q_D = 1000 - 10p = 1000 - 500 = 500$$

The total revenue can be obtained as a function of quantity demanded from the equations

$$q_D = 1000 - 10p \tag{1}$$

$$TR = pq = 1000p - 10p^2 \tag{2}$$

from (1) $10p = 1000 - q_D$

$$\therefore p = 100 - \tfrac{1}{10}q_D$$

Substituting this in (2), we obtain

$$TR = 1000(100 - \tfrac{1}{10}q_D) - 10(100 - \tfrac{1}{10}q_D)^2$$

$$= 100000 - 100q_D - 100000 + 200q_D - \tfrac{1}{10}q_D^2$$

$$= 100q_D - \tfrac{1}{10}q_D^2$$

This is again a quadratic equation which is a parabola and total revenue is equal to zero when

$$100q_D - \tfrac{1}{10}q_D^2 = 0$$

$$q_D(100 - \tfrac{1}{10}q_D) = 0$$

$\therefore$ either $$q_D = 0$$

or $$100 - \tfrac{1}{10}q_D = 0$$

that is, $$q_D = 1000$$

The maximum revenue is equal to $500 \times 50 = 25000$. The function can now be represented graphically as shown in Fig. 3.16.

The company will be interested not so much in the total revenue but in the net revenue after the costs of production etc have been met. The line superimposed on the graph corresponds to the linear cost function.

$$TC = 5000 + 15q_D$$

The net revenue for any quantity of output q_D is given by the vertical distance between these two functions at that quantity. Zero net revenue is obtained at the two breakeven points which are given by

$$TR = TC$$

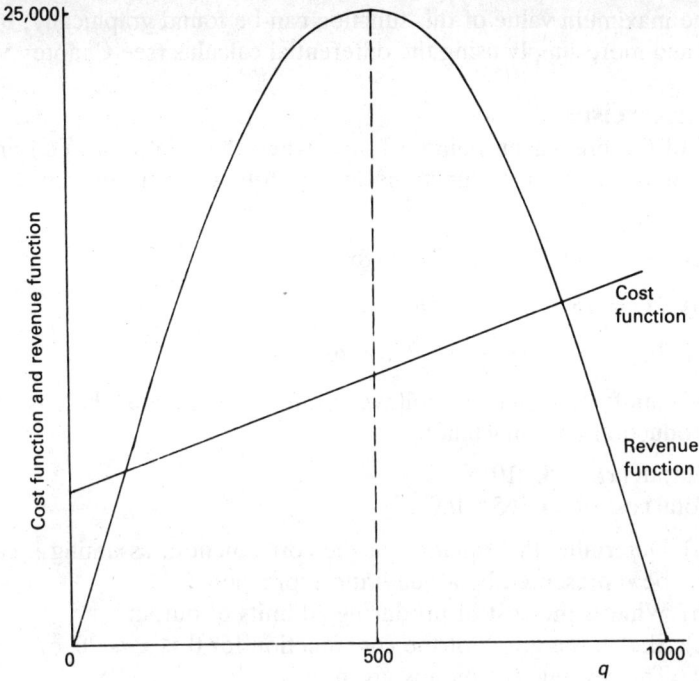

Fig. 3.16

$$100q_D - \tfrac{1}{10}q_D^2 = 5000 + 15q_D$$

i.e. $\tfrac{1}{10}q_D^2 - 85q_D + 5000 = 0$

$$q_D = \frac{85 \pm \sqrt{(85^2 - 4 \times \tfrac{1}{10} \times 5000)}}{2 \times \tfrac{1}{10}} = \frac{85 \pm 72.28}{0.2}$$

That is approximately 64 units and 786 units. Between these two points is the quantity of output which yields the maximum net revenue and it could be obtained from the equation representing the net revenue function:

$$\text{Net revenue} = \text{total revenue} - \text{total cost}$$

$$= 100q_D - \tfrac{1}{10}q_D^2 - (5000 + 15q_D)$$

$$= 85q_D - \tfrac{1}{10}q_D^2 - 5000$$

The maximum value of this function can be found graphically, but is obtained more simply using the differential calculus (see Chapter 5).

3.10 Exercises

1. Find the breakeven points (if any) when the total cost (TC) and total revenue (TR) functions are as follows (x is the level of output):

   (a) $TC = 9 + 2x + x^2$, $TR = 8x$

   (b) $TC = 15 + x + x^2$, $TR = 10x - 3$

   (c) $TC = 10 + 3x + 2x^2$, $TR = 6x$.

2. A manufacturer has the following information about the costs of production on a machine:

Output (x)	5	10	15
Total costs	20	65	160

   (a) Determine the equation of the cost function, assuming it can be represented by a quadratic expression.
   (b) What is the cost of producing 20 units of output?
   (c) Sketch the graph of the cost function for $0 \leqslant x \leqslant 20$.
   (d) The revenue function is given by

   $$y_R = 10x + 15$$

   Show that there are two levels of output at which total revenue equals total costs, and illustrate this graphically.

3. The total cost function for an output of x units of a product is

   $$TC = 250 + 4x$$

   Draw the graph of the average cost function and show that the average cost approaches the value of 4 asymptotically.

4. Show that the equations

   $$xy = 10$$

   and

   $$3y = 32 - 2x$$

   are satisfied for two values of x and sketch the graphs of the equations.

5. Show that the simultaneous equations

   $$y = 20 + 3x + x^2$$

   $$y = 5x + b$$

have the two real solutions when $b = 20$, two real coincident solutions when $b = 19$ and two complex solutions when $b = 18$. Sketch the graphs of these equations to illustrate these three cases.

6. Find the equilibrium price and quantity for the demand curve

$$q = 250 - 4p - p^2$$

and the supply curve

$$q = 2p^2 - 3p - 40$$

and sketch the curves for the range

$$p = 0 \quad \text{to} \quad p = 10$$

7. Determine the equilibrium price and quantity for the following demand and supply functions. Sketch the curves for the range $p = 0$ to $p = 8$

	Demand	*Supply*
(a)	$q - 25 + 3p = 0$	$q = 2p^2 - 40$
(b)	$p^2 + q^2 = 32$	$4 = 3p - 2q$
(c)	$pq = 6$	$q = 3(p - 1)$

8. The demand function for a product is $q_D = 200 - 2p$ and the total cost function is $TC = 20 + 5q_D$. Determine the net revenue function and the breakeven levels of output and price. Sketch the graph of the net revenue function for $q_D = 0$ to $q_D = 200$.

3.11 Discontinuous functions

So far we have looked at those functions for which the graph is a continuous line between the values considered. However, an equation may be valid only between certain limits. For example, the relationship between the total cost of manufacture and the number of units of output of a product may be represented by an equation such as the following:

$$TC = a + bx$$

where a is the fixed cost and b the variable cost of production which in this case is constant for all levels of output. But it is apparent that the volume of output cannot be increased indefinitely for a given amount of fixed costs. For example extra investment in machinery will eventually be needed. We must, therefore, add the condition that the equation is valid only up to a certain quantity of output above which an increase in

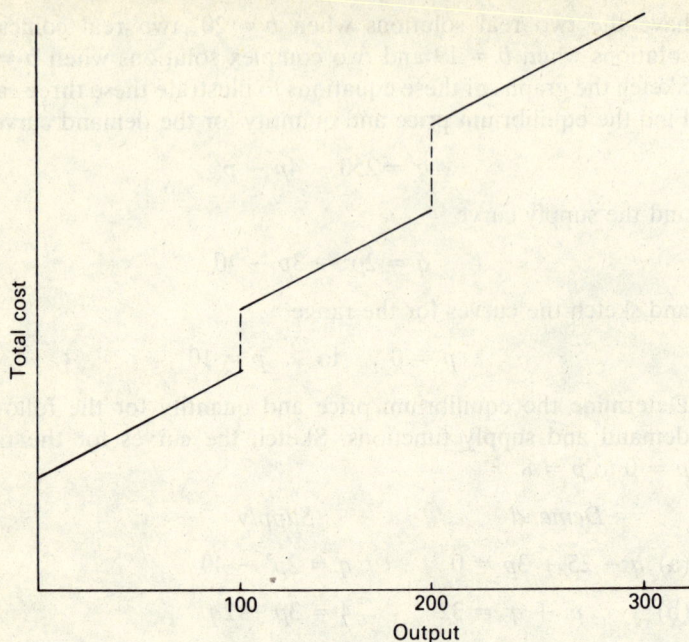

Fig. 3.17

fixed costs must be incurred. By a stepwise construction we might arrive at the following set of equations.

$$TC = a_1 + bx \qquad 0 < x \leqslant 100$$

$$TC = a_2 + bx \qquad 100 < x \leqslant 200$$

$$TC = a_3 + bx \qquad 200 < x \leqslant 300$$

with the condition $a_1 < a_2 < a_3$.

This set of equations can be graphed on the one set of axes as in Fig. 3.17. The graph in this case is not continuous between the limits $x = 0$ and $x = 300$. It consists of three separate sections and at the values $x = 100$ and $x = 200$ there are discontinuities. At these values a very small increase in one variable, quantity, requires a large increase in the other variable, total cost. In this particular example the discontinuity occurs when the value of the constant a in the equation is changed.

Another example of a discontinuity occurs when there are reductions for purchasing large quantities of goods. For example, suppose that the price per unit is 6 for quantities below 100 and that a discount of 20% is given for purchases of 100 to 499 and a discount of 30% on

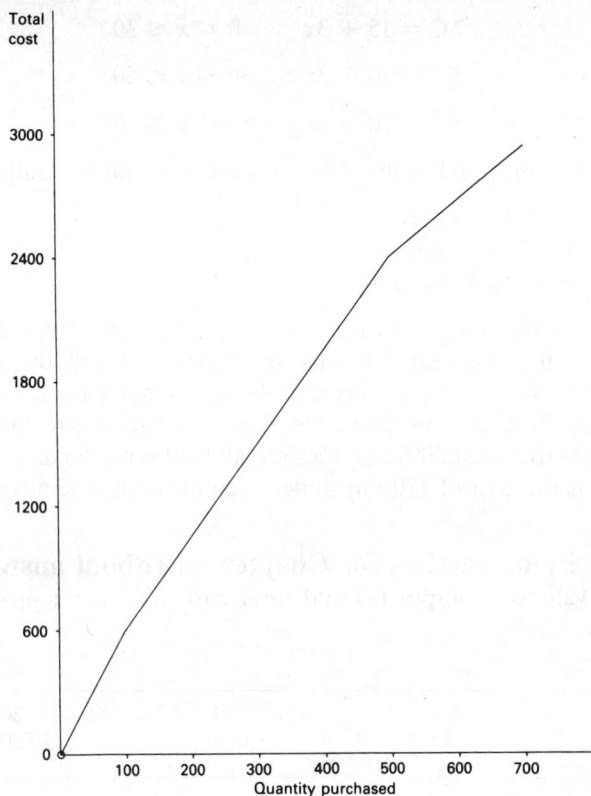

Fig. 3.18

purchases of 500 or more. Let x be the quantity purchased then the total cost to the purchaser is

$$TC = 6x \qquad \text{for } 0 < x < 100$$

$$TC = 6x(0.80) \text{ for } 99 < x < 500$$

$$TC = 6x(0.70) \text{ for } 499 < x$$

The graph is shown in Fig. 3.18 and it can be seen that the slope of the cost curve changes at $x = 100$ and $x = 500$.

3.12 Exercises

1. A firm has a discontinuous cost function and the relationship between output (x) and total cost (TC) is given by

$$TC = 15 + 3x \qquad 0 \leqslant x \leqslant 30$$

$$TC = 40 + 3x \qquad 30 \leqslant x \leqslant 50$$

$$TC = 70 + 3x \qquad 50 \leqslant x \leqslant 100$$

Illustrate this graphically. What is the cost of raising output

(a) from 19 to 21 units?
(b) from 29 to 31 units?
(d) from 49 to 51 units?

2. A canner sells beans at a price of 30p per tin but offers a discount of 15% if more than 100 tins are bought. Sketch the revenue function. What is the total cost of 99 tins? What is the total cost of 101 tins? Noticing this difference, the canner decides to change the discount to one of 20% but does not allow this on the first 100 tins. What is the cost of 120 tins under each of these schemes?

3.13   Revision exercises for Chapter 3 (without answers)

1. The values of output (x) and total cost (TC) for a production process are

x	10	20	30
TC	430	1,060	2,900

Assuming that the total cost function is quadratic, determine its equation. What is the fixed cost? Sketch the average cost function for $x = 0$ to $x = 30$.

2. Find the roots of the following quadratic equations:

(a) $x^2 - 6x + 4 = 0$ (b) $2x^2 + 8x + 8 = 0$

(c) $x^2 + 4x + 5 = 0$ (d) $2x^2 - 3x + 3 = 0$

3. Draw the graphs of the following functions for $x = 0$ to $x = 10$:

(a) $y = x^2 - 6x + 4$ (b) $y = x^2 + 4x + 5$

(c) $xy = 15$ (d) $xy = 15 + 2x$

4. The total cost function for a product is

$$TC = 100 + 4x + x^2$$

Sketch the graph for $x = 0$ to $x = 20$ and add the average cost function. What happens to the average cost as x tends to infinity?

5. What is the breakeven value of output (x) if the total cost function is

$$TC = a + bx + cx^2$$

and the price is p per unit? Sketch the net revenue function.

6. The demand function for a product is

$$q^d = 200 - 2p - 2p^2$$

and the supply function is

$$q^s = 4p - 20$$

What are the equilibrium price and quantity? Sketch the two functions.

7. Determine the equilibrium price and quantity if the demand and supply equations are:

$$q^d = 500 - 4p - p^2$$
$$q^s = p^2 + 6p - 40$$

8. Show that the equations $xy = 30$

$$6x + y = a$$

have no real solution when $a = 20$. For what values of a are there two real solutions?

9. The total cost function for a product is

$$TC = 150 + 3q + q^2$$

If the price is 8 per unit, determine the net revenue function and the breakeven quantity.

10. A machine for producing a component has an installation cost of 500 and the marginal cost per component is 20. The capacity of the machine is 200 components per week and if output needs to be greater than 200 another identical machine can be installed. Write down the total cost function for producing up to 400 components per week and sketch the average cost function.

Chapter 4

Series

4.1 Introduction

This chapter covers some basic applications of series in economics and business studies, mainly in the areas of interest rates and finance, but also extended to exponential and logarithmic functions. We start with the idea of a *progression*. This is a number of terms arranged in a definite order and for which there is a pattern or rule which allows further terms to be identified. For example:

$$100, 101, 102, \quad 103, \quad 104, \ldots$$

$$1, \quad 10, 100, 1000, 10000, \ldots$$

$$0, 100, 100, \quad 200, \quad 400, \ldots$$

are all progressions. In the first example 100 is the *first term* or *initial term* and each subsequent term is obtained by adding 1 to the previous term. In the second example, 1 is the first term and each subsequent term is obtained by multiplying the previous term by 10. In the third example the first term is 0, the second term 100, and each subsequent term is obtained by adding together all the previous terms, so that $0 + 100 = 100$, $0 + 100 + 100 = 200$, $0 + 100 + 100 + 200 = 400$, and so on.

4.2 Arithmetic progressions

An *arithmetic progression* is one in which there is a *constant difference between each term*. The first example above is an arithmetic progression. The difference between any two (consecutive) terms is 1. Another example is the progression

$$100, 90, 80, 70, 60, \ldots$$

where 100 is the first term and the difference between any two terms is -10. In general, any arithmetic progression can be written as

$$a, a + d, a + 2d, a + 3d, \ldots$$

where a is the first term, and d is known as the *common difference*. For example, for the progression

$$100, 101, 102, 103, 104, \ldots$$

$$a = 100, d = 1$$

while for

$$100, 90, 80, 70, 60, \ldots$$

$$a = 100, d = -10.$$

Notice that in the general arithmetic progression the second term is $a + d$, the third term is $a + 2d$, the fourth term is $a + 3d$, and so the nth term is $a + (n - 1)d$. For example for $a = 100$ and $d = 1$, the twentieth term is

$$a + (20 - 1)d = 100 + 19 = 119$$

and for $a = 100$, and $d = -10$, the tenth term is

$$a + (10 - 1)d = 100 + 9(-10) = 10$$

A useful application of arithmetic progressions is in the calculation of simple interest.

For example, suppose £100 is invested at 5 per cent per annum simple interest. Then the interest for the first year is £5 (from £100 × 0.05 = £5). Similarly, the interest for any subsequent year will be £5, and so the value of the investment is as shown in Table 4.1, since $a = 100, d = 5$.

The value at the beginning of the tenth year (i.e. after 9 complete years) is $a + (10 - 1)d = 100 + 9(5) = £145$, and the value of the investment after 20 years (i.e. at the beginning of year 21) is

$$100 + (21 - 1)5 = £200$$

TABLE 4.1

Year	1	2	3	4	....	n
Value at beginning of year	100	100 + 5 = 105	105 + 5 = 110	110 + 5 = 115		100 + (n − 1)5

It is frequently useful to be able to obtain the sum of an arithmetic progression to give an *arithmetic series*. Returning to the general progression $a, a + d, a + 2d, a + 3d, \ldots, a + (n - 1)d$, we see that the sum of the first n terms, S_n, is

$$S_n = a + (a + d) + (a + 2d) + \cdots + [a + (n - 1)d]$$

$$= na + d + 2d + 3d + \cdots + (n - 1)d$$

which can also be written (by reversing the order) as

$$S_n = na + (n - 1)d + (n - 2)d + \cdots + 3d + 2d + d$$

Adding these two expressions together gives

$$2S_n = 2na + nd + nd + \cdots + nd$$

since such terms as $d + (n - 1)d = d + nd - d = nd$ and

$$3d + (n - 3)d = 3d + nd - 3d = nd$$

Hence,

$$2S_n = 2na + (n - 1)nd$$

as there are $(n - 1)$ terms equal to nd.

$$\therefore S_n = \frac{n}{2}[2a + (n - 1)d]$$

Alternatively, this can be written as

$$S_n = n\frac{[a + \{a + (n - 1)d\}]}{2}$$

$$= \text{(number of terms) (average of first and last terms)}.$$

For example, for the progression $100, 101, 102, \ldots$ in which $a = 100$, $d = 1$, the sum of the first n terms is

$$S_n = \frac{n}{2}[200 + (n - 1)]$$

so that the sum of the first 3 terms is

$$S_3 = \tfrac{3}{2}[200 + (3 - 1)] = \frac{3(202)}{2} = 303$$

which can easily be verified. The sum of the first 20 terms is

$$S_{20} = \frac{20}{2}[200 + (20 - 1)] = 2190$$

which is less easily verified.

Example

A man invests £100 per annum in bonds which pay 5% per annum simple interest. What is the value of the investment at the beginning of the 11th year? This value is the sum of the first eleven terms of the arithmetic progression with $a = 100$, $d = 5$ and is given by

$$S_{11} = \frac{11}{2}[200 + (11 - 1)5] = £1375$$

This is verified by looking at the value of the investment for the first few years (Table 4.2). Thus, at the beginning of the 4th year the value is $100 + 105 + 110 + 115$, which is the sum of the first 4 terms of the arithmetic progression with $a = 100$, $d = 5$. Hence the value of the investment at the beginning of the 11th year (i.e. when $n = 11$) is

$$S_{11} = \frac{11}{2}[200 + (11 - 1)5] = £1375.$$

TABLE 4.2

Year	Value at beginning of year
1	100
2	100 + (100 + 5)
3	100 + (100 + 5) + (100 + 10)
4	100 + (100 + 5) + (100 + 10) + (100 + 15)

4.3 Exercises

1. Give the tenth term of each of the following progressions and also determine the sum of the first fifteen terms.

    (a) 1, 3, 5, 7, . . .

    (b) 500, 550, 600, 650, . . .

    (c) 60, 30, 0, − 30, . . .

    (d) 100, 98, 96, 94, . . .

2. What is the value of an investment of £200 after 5 years if simple interest is paid at 10% per annum?

3. A man receives a salary of £8,500 per annum which increases annually by £400. What salary will he receive after 7 years? What will be his total income during the first six years?

4. A government issues savings bonds which sell for £10 and are worth £20 after 15 years. What is the implied rate of simple interest?

4.4 Geometric progressions

A geometric progression is one in which there is a constant ratio between any two consecutive terms. For example in the progression

$$1, \ 10, \ 100, \ 1000, \ 10000, \ \ldots$$

each term is 10 times the previous term so that $100 = 10(10)$ and $10,000 = 10(1,000)$. In the progression $128, 64, 32, 16, 8, \ldots$ each term is 0.5 times the previous term, e.g. $64 = 0.5(128)$, and $32 = 0.5(64)$. In general, any geometric progression can be written as

$$a, \ ar, \ ar^2, \ ar^3, \ \ldots$$

where a is the first term and r is known as the *common ratio*. In the above examples, $a = 1, r = 10$ and $a = 128, r = 0.5$. The nth term in a geometric progression is ar^{n-1}, so that for the progression with $a = 1$, $r = 10$, the sixth term is $1(10^5) = 100,000$, while for the progression with $a = 128, r = 0.5$, the fifth term is $128(0.5^4) = 8$.

One of the applications of geometric progressions is in the calculation of compound interest. Here the sum on which interest is paid includes the interest which has been earned in previous years.

For example, if £100 is invested at 5% per annum compound interest, then after 1 year the interest earned is £5 (from 100×0.05) and the capital invested for the second year is £105. The interest earned by this capital is not £5 but $£105 \times 0.05 = £5.25$. The capital invested for the third year is £105 + £5.25 = £110.25, and the interest earned is $£110.25 \times 0.05 = £5.5125$. This is shown in Table 4.3.

This can be expressed in general terms, as shown in Table 4.4 for an investment of a and an interest rate of i per cent. This shows that the progression for the capital is $a, a(1 + i), a(1 + i)^2, a(1 + i)^3, \ldots$, which is of a geometric form with common ratio $r = 1 + i$.

For example, if $i = 5\% = 0.05$, then $r = 1.05$ and so the nth term in the progression if $a = 100$ is $100(1.05)^{n-1}$. The value of the investment after 8 years is $100(1.05)^{9-1} = £147.75$ (note that $n = 9$ since we want the ninth term in the progression). The evaluation of such quanti-

TABLE 4.3

Beginning of year	Capital	Interest during year
1	100	5
2	105	5.25
3	110.25	5.5125
4	115.7625	5.7881

TABLE 4.4

Beginning of year	Capital	Interest during year
1	a	ai
2	$a + ai = a(1 + i)$	$a(1 + i)i$
3	$a(1 + i) + a(1 + i)i = a(1 + i)^2$	$a(1 + i)^2 i$
4	$a(1 + i)^2 + a(1 + i)^2 i = a(1 + i)^3$	$a(1 + i)^3 i$

ties as 1.05^8 is best done by calculator or by logarithms. A short section on the use of logarithms is included at the end of this chapter (Section 4.14).

The value of a *geometric series*, which is the sum of a geometric progression, is easily obtained by considering the general progression

$$a, ar^2, ar^3, \ldots, ar^{n-1}$$

The sum of the first n terms,

$$S_n = a + ar + ar^2 + ar^3 + \cdots + ar^{n-1}$$

also $$r \cdot S_n = ar + ar^2 + ar^3 + \cdots + ar^{n-1} + ar^n$$

Subtracting gives $S_n - rS_n = a - ar^n$; that is

$$S_n(1 - r) = a(1 - r^n) \quad \text{or} \quad S_n = \frac{a(1 - r^n)}{1 - r}$$

For example, for the progression with $a = 1$, $r = 10$,

$$S_n = \frac{1(1 - 10^n)}{1 - 10}$$

so that the sum of the first 5 terms is

$$S_5 = \frac{1(1 - 10^5)}{1 - 10} = \frac{10^5 - 1}{9} = 11111$$

and the sum of the first 8 terms is

$$S_8 = \frac{1(1 - 10^8)}{1 - 10} = \frac{10^8 - 1}{9} = 11111111$$

A special case arises when $-1 < r < 1$

Then $$S_n = \frac{a}{1 - r} - \frac{ar^n}{1 - r}$$

and as $n \to \infty$, $r^n \to 0$, and the second term will approach zero.

The sum of all the terms in the series or, as it is generally referred to, *the sum to infinity* is given by

$$S_\infty = \frac{a}{1 - r}$$

For example, if $a = 128$, $r = 0.5$

$$S_\infty = \frac{128}{1 - 0.5} = \frac{128}{0.5} = 256$$

This means that adding together the successive terms of the progression 128, 64, 32, 16, 8, 4, 2, 1, 0.5, ... produces a sum which becomes closer and closer to, but never becomes greater than, 256.

It is important to realise that r must satisfy $-1 < r < 1$ for the sum to infinity to be finite. For example, for the series $a = 1$, $r = 10$,

$$S_\infty = \frac{1}{1 - 10} - \frac{1.10^\infty}{1 - 10} = -\frac{1}{9} + \frac{10^\infty}{9}$$

which is obviously infinity.

4.5 Exercises

1. State the sixth term of each of the following progressions and determine the sum of the first 10 terms. Evaluate the sum to infinity if it is finite.

   (a) 10, 30, 90, 270, . . .

   (b) 81, 27, 9, 3, . . .

    (c) 2, −4, 8, −16, . . .

    (d) −1,024, 512, −256, 128, . . .

    (e) 1, 0.1, 0.01, 0.001, . . .

2. What is the value of an investment of £300 at the end of 15 years if compound interest is paid at

    (a) 5% per annum?   (b) 10% per annum?

3. An insurance policy costs £150 per annum for 25 years. What is the final value of the policy if compound interest is earned at a rate of 6% per annum?

4. Government savings bonds are sold for £4 and are worth £5 after 4 years. What is the implied rate of compound interest?

4.6 Discounting

If money is able to earn interest its absolute value in future years will be greater than its current or present value. Conversely money which is to be spent in future years has a present value which is less than its absolute value.

For example, if we have £100 now and we invest this at 6% per annum compound interest, then we will have £106 at the end of one year. If, therefore, we are required to spend £106 in one year's time we would only need to have available at the present time £100 which we could invest at 6%. The present value of a capital sum of £106 which is required in one year's time is therefore equal to £100 when the interest rate is 6% per annum.

A sum of £100 per required in one year's time then has a present value given by

$$PV = \frac{100}{106} \times 100 = \frac{100}{1.06} = \frac{100}{(1 + 6/100)} = £94.3$$

This method of calculating the present value of future sums of money, whether payments or receipts, is known as *discounting*. It can be used to cover any time span as follows:

If future requirements for capital are known with their appropriate time pattern then

$$(PV)_c = a_0 + \frac{a_1}{(1 + i)} + \frac{a_2}{(1 + i)^2} + \cdots + \frac{a_n}{(1 + i)^n}$$

where $(PV)_C$ = present value of capital requirements

a_0 is the immediate requirement

a_1 is the requirement in the first year and discounted as though it appeared at the end of the first year.

a_2 is the requirement in the second year

$\vdots$

a_n is the requirement in the nth year.

If these capital requirements or cash outflows are for a given project then the project will produce a series of receipts, or cash inflows, again over a period of time. These can be discounted using the same compound interest rate.

$$(PV)_R = b_0 + \frac{b_1}{(1 + i)} + \frac{b_2}{(1 + i)^2} + \cdots + \frac{b_n}{(1 + i)^n}$$

where $(PV)_R$ = present value of receipts

b_0 is the immediate return

b_1 is the return in the first year and discounted as though it appeared at the end of the first year

b_2 is the return in the second year

$\vdots$

b_n is the return in the nth year.

For the project to be profitable on a strictly financial basis we must have the condition that

$$(PV)_R > (PV)_C$$

that is, the present value of the receipts from the project must be greater than the present value of the payments.

The above is known as the *present-value method* for assessing investment projects and the *net present value* (*NPV*) is defined as the difference between the discounted values of receipts and payments.

$$NPV = (PV)_R - (PV)_C$$

The value used for i in the discounting process obviously has a considerable influence on the net present value. It is known as the cost of capital and its value is often difficult to establish accurately in many situations. Because of this, preference is often given to a method which allows the cost of capital to be a variable and determines that value

TABLE 4.5

Table of Discount Factors

This table shows the present value of 1 discounted for different numbers of years and at different rates of discount

Rate of discount (%)	8	10	12	14	16
Year					
0	1.000	1.000	1.000	1.000	1.000
1	0.926	0.909	0.893	0.877	0.862
2	0.857	0.826	0.797	0.769	0.743
3	0.794	0.751	0.712	0.675	0.641
4	0.735	0.683	0.636	0.592	0.552
5	0.681	0.621	0.567	0.519	0.476
6	0.630	0.564	0.507	0.456	0.410
7	0.583	0.513	0.452	0.400	0.354
8	0.540	0.467	0.404	0.351	0.305
9	0.500	0.424	0.361	0.308	0.263
10	0.463	0.386	0.322	0.270	0.227
15	0.315	0.239	0.183	0.140	0.108
20	0.215	0.149	0.104	0.073	0.051

which equates the present value of receipts to the present value of payments. That is,

$$(PV)_R = (PV)_C$$

or $a_0 + \dfrac{a_1}{(1 + i)} + \cdots + \dfrac{a_n}{(1 + i)^n} = b_0 + \dfrac{b_1}{(1 + i)} + \cdots + \dfrac{b_n}{(1 + i)^n}$

Since $a_0 - a_n$ and $b_0 - b_n$ are assumed to be known, this equation is a polynomial in i and can be solved by the quadratic formula (when $n = 2$) or by Newton's method (see section 5.17). Because i is a rate of interest it must be positive and normally we expect only one positive value to result. This value is known as the *internal rate of return* (*IRR*) and the higher this is the more profitable the project will be.

Both of the above methods (i.e. *NPV* and *IRR*) require that each year's payments and receipts are discounted by the appropriate factor which is determined by the time period in which these payments and receipts arise. Values of $1/(1 + i)^n$, known as discount factors, are given in Table 4.5. For example, the discount factor at 10% per annum for 5 years is $1/(1 + 0.10)^5 = 0.621$.

4.7 Exercises

1. (a) If £300 is invested for 10 years and is then worth £800, what is the implied rate of compound interest?
   (b) What is the present value of the £800 at 16% per annum compound interest?

2. What is the present value of £1,000 payable after 10 years if the rate of discounting is

   (a) 14%? (b) 8%?

3. (a) How much should be invested at 10% per annum compound interest to give £250 after 5 years?
   (b) What is the value of this investment after 3 years?

4. Two projects are available to a company and the estimated returns are:

End of year	1	2	3
Project *A*	100	200	300
Project *B*	150	300	100

   Which project has the greater present value if the discounting rate is 10%?

5. What is the internal rate of return from a project which has the following costs and receipts?

End of year	1	2	3
Costs	120	120	100
Receipts	100	110	160

4.8 Annuities and sinking funds

We now apply some of these ideas to a number of related practical problems. First we consider annuities, where for a capital payment now an annual income is received for a number of years in the future. The two questions to be answered are: if I pay the capital amount p now what annual income, A, will I receive for n years, and if I want an annual income of A for n years, how much will it cost now? Further applications of the same points arise with perpetual bonds and chief rents. Next, we examine mortgages, where a loan of p now is to be repaid at a rate of A per year, and again the questions concern p (given the value of A) or A (given the value of p). With both annuities and mortgages the capital amount is required immediately and the income or repayments are in the future. The situation is reversed with sinking funds, pension funds and saving schemes. For these, a constant sum A is added to the fund each year and after n years the total value V is returned with interest. Again the questions concern the annual payment, A, and the total value after n years, V.

In the following, two simplifying assumptions will be made. First, it will be assumed that the interest rate in the future will be a known constant. In practice, interest rates fluctuate and future rates are unknown. If in fact we do have some information about future rates, this can be taken into account at the expense of making the various formulae more complicated. If not, we can interpret the assumed rate of interest as being an average of future values. The second assumption is that the payments and receipts occur once a year. In practice, many are made quarterly or monthly. This can be taken account of by adjusting the rate of interest so that an interest rate of 8% per annum, when applied monthly, becomes 8/12 or 0.67% per month. This is discussed further in section 4.10 below.

An *annuity* is a constant annual income which can be bought for cash. For example, an annual income of £10 for 20 years may be purchased for £120. The present value of an annuity of £A payable for n years is given by

$$p = \frac{A}{1 + i} + \frac{A}{(1 + i)^2} + \frac{A}{(1 + i)^3} + \cdots + \frac{A}{(1 + i)^n}$$

$$= \frac{A}{1 + i}\left[1 + \frac{1}{1 + i} + \frac{1}{(1 + i)^2} + \cdots + \frac{1}{(1 + i)^{n-1}} \right]$$

The term in the brackets is the sum of a geometric series with $a = 1$ and $r = 1/(1 + i)$ and so

$$p = \frac{A}{1 + i}\left[\frac{1 - r^n}{1 - r} \right] = \frac{A}{i}\left[1 - \frac{1}{(1 + i)^n} \right]$$

For example, the present value of an annuity of £10 per annum for 20 years if the rate of interest is 7% is

$$p = \frac{10}{0.07}\left[1 - \frac{1}{(1.07)^{20}} \right] = £105.94$$

A special case occurs when the annuity continues indefinitely, since

$$p = \frac{A}{1 + i} + \frac{A}{(1 + i)^2} + \frac{A}{(1 + i)^3} + \cdots = \frac{A/(1 + i)}{1 - 1/(1 + i)} = \frac{A}{i}$$

For example, an annuity of £20 per annum for ever has a present value of

$$p = \frac{20}{0.05} = £400$$

if the rate of interest is 5% per annum, while if the interest rate is 10% per annum the value would be 20/0.10 or £200.

Exactly the same situation arises with perpetual government *bonds*. In the UK there are some special government securities known as 2½% Consols (or Consolidated Stock) which were originally sold for £100 each in return for a promise to pay the owner the fixed amount of £2.50 per annum for ever. The current value of Consols depends on the rate of interest and, if the rate of interest is 9%, is given by

$$p = \frac{2.50}{i} = \frac{2.50}{0.09} = £27.78$$

Another application of the same concept is the idea of a *chief rent*. This is an annual charge paid by some home owners in the UK to the original owner of land which has been used for housing. The landowner accepts an annual payment for the land instead of a lump sum. For example, a plot of land may have an annual chief rent of £25. The present value to the land owner, assuming that the first payment is due one year from now and the interest rate is 8% is

$$p = \frac{A}{i} = \frac{25}{0.08} = 312.5$$

Notice that if there is a payment due immediately then the value is

$$p = A + \frac{A}{1 + i} + \cdots = \frac{A}{1 - \{1/(1 + i)\}} = \frac{A(1 + i)}{i}$$

instead of A/i. For the chief rent example the value would be 337.5 which is £25 more than in the previous case because of the extra payment of this amount.

With *mortgages*, an amount p is borrowed now, usually to pay for a house, and is to be repaid by n equal annual payments of A. Let M_t be the mortgage debt at the end of year t. Then, if p is borrowed and the interest rate is i, the amount owed after 1 year is

$$M_1 = p(1 + i) - A$$

since the first term covers the capital borrowed and the interest charge and the repayment of A is made towards the debt. In the same way,

$$M_2 = M_1(1 + i) - A$$

since M_1 is owed at the end of year 1. But we can express M_1 in terms of p and so, using the expression for M_1,

$$M_2 = \{p(1 + i) - A\}(1 + i) - A$$
$$= p(1 + i)^2 - A \{(1 + i) + 1\}$$

Similarly,

$$M_3 = M_2(1 + i) - A$$
$$= [p(1 + i)^2 - A\{(1 + i) + 1\}](1 + i) - A$$
$$= p(1 + i)^3 - A\{(1 + i)^2 + (1 + i) + 1\}$$

In general,

$$M_t = p(1 + i)^t - A\{(1 + i)^{t-1} + (1 + i)^{t-2} + \cdots + (1 + i) + 1\}$$

The final term can be simplified since it is the sum of t terms of a geometric progression with a common ratio of $(1 + i)$ and so

$$(1 + i)^{t-1} + (1 + i)^{t-2} + \cdots + (1 + i) + 1 = \frac{1(1-(1 + i)^t}{1 - (1 + i)}$$
$$= \frac{(1 + i)^t - 1}{i}$$

Therefore,

$$M_t = p(1 + i)^t - \frac{A\{(1 + i)^t - 1\}}{i}$$

But if the mortgage is to be paid off over n years, M_n must be zero and therefore,

$$0 = p(1 + i)^n - \frac{A\{(1 + i)^n - 1\}}{i}$$

This gives an equation for p in terms of A, i and n,

$$p = \frac{A\{(1 + i)^n - 1\}}{(1 + i)^n i}$$

or for A in terms of p, i and n

$$A = \frac{i(1 + i)^n p}{(1 + i)^n - 1}$$

For example, if $p = £100,000$, the interest rate is 10% and $n = 25$ years,

$$A = \frac{0.1(1 + 0.1)^{25} (100000)}{(1 + 0.1)^{25} - 1}$$

$$= \frac{0.1(10.8347)(100000)}{9.8347} = £11016.8 \text{ per annum}$$

As a second example, suppose a borrower is able to repay £7000 per annum. What amount can be borrowed over 20 years if the interest rate is 12%? Here,

$$p = \frac{A\{(1 + i)^n - 1\}}{(1 + i)^n i} = \frac{7000\{(1.12)^{20} - 1\}}{1.12^{20}(0.12)}$$

$$= \frac{7000(8.6463)}{(9.6463)(0.12)} = £52286$$

and so £52,286 could be paid back over 20 years at a rate of £7,000 per annum.

A *sinking fund* is a fund set up to meet some financial commitment, to which a constant sum is added each year. For example, how much should be invested each year if the rate of interest is 5% per annum to give a capital sum of £600 in 10 years?

Let A be the amount invested. After 1 year, the value of the investment will be $A + A(1 + i) = A + A(1.05)$. After 2 years, the value will be $A + A(1.05) + A(1.05)^2$ and in 10 years the value will be

$$A + A(1.05) + A(1.05)^2 + \cdots + A(1.05)^9$$

This has to be equal to £600.

$$\therefore \quad 600 = A(1 + 1.05 + 1.05^2 + \cdots + 1.05^9)$$

$$= \frac{A[1 - (1.05)^{10}]}{1 - 1.05}$$

since the expression in brackets is a geometric series.

$$\therefore \quad A = \frac{600(1 - 1.05)}{1 - (1.05)^{10}} = £47.7$$

That is, if a sum of £47.7 is invested each year it will accumulate to £600 in 10 years at an interest rate of 5% per year.

The same principle applies to life insurance (savings) policies where, after making a regular payment for a number of years, the policy-holder receives a payout. If it is assumed that the whole of the annual premium (A) is invested (so that any administration charges and payments for life cover are ignored) then it is simply a regular savings scheme as operated by some banks and pension funds. If the interest rate is i then the value in year n is

$$V = A + A(1 + i) + A(1 + i)^2 + \cdots + A(1 + i)^{n-1}$$

$$= \frac{A\{1 - (1 + i)^n\}}{1 - (1 + i)} = \frac{A\{(1 + i)^n - 1\}}{i}$$

For example, if $A = £100$, $i = 8\%$ and $n = 25$ then $V = £7,310.59$. As previously, if A is unknown it can be determined given the values of V, i and n, so that, for example, if a policy is required which will be worth £5,000 after 10 years when the interest rate is 11%, the annual premium would be

$$A = \frac{iV}{(1 + i)^n - 1} = \frac{0.11(5000)}{1.11^{10} - 1} = £299.01$$

4.9 Exercises

1. What is the present value of an annuity of £100 per annum which is to be paid 20 times, commencing one year from now, if the interest rate is 8%?

2. What is the present value of a chief rent of £25 if the rate of interest is 4%?

3. A company buys a machine for £2,000 and estimates that its life will be 15 years. How much should be paid annually into a sinking fund to buy a replacement for the machine in 15 years time if the replacement will cost £2,500 and the rate of interest is 9%?

4. An ex-pupil decides to donate to his school a sum of money to provide an annual prize of £60 for the next ten years.

(a) How much should be donated if the rate of interest is 6%?

(b) How much should be donated if the prize is to be paid indefinitely?

5. A couple borrow £70,000 to buy a house. What would their annual repayments be if the interest rate was 12% and the repayment period was 20 years? They decide this repayment rate is too high, and ask for the period of the loan to be extended to 30 years. What will the revised repayment rate be?

6. An insurance company sells policies for which the premium is £300 per year. If the interest rate is 9% what is the value of a policy after (a) 10 years (b) 20 years?

7. A firm introduces a pension fund for its employees in which it makes annual payments of £250 for each employee and interest is added at 8% per annum. How much would an employee's fund be worth if someone was a member for (a) 25 years or (b) 35 years?

4.10   Interest paid continuously

In our discussion of interest payments we have assumed that interest has been paid annually. However, there are many cases where this is not so. For example, some bank accounts and some government bonds pay interest twice a year, while others pay interest quarterly and generally credit card companies charge interest on the balance outstanding each month. The effect of these more-frequent payments is that the money earns interest for a shorter time before the capital increases by the amount of interest. For example, suppose that the interest rate is 6% per annum compound and £100 is invested. If interest is paid twice a year then after the first six months the interest earned is a half of 6% or £3 since the £100 has been invested for half of the year. Effectively, the interest rate is halved because the investment period is halved. The capital at the beginning of the second six months is £103 and so the interest earned during the second half year is 3 per cent of £103 or £3.09, making the value after one year £106.09 which is £100$(1 + 0.03)^2$. Notice that even though the interest rate is 6% the interest received is at the rate of 6.09% because of the twice-yearly payment.

Now suppose for simplicity that the interest rate is 100% per annum compound and that £1 is invested. If interest is paid just once a year the value at the end of a year is £1 + £1 = £2. If interest is paid twice a year the value after a year is £1$(1 + 0.50)^2$ = £2.25. In the same way, if we assume that the interest is paid at the end of each quarter, the capital value at the end of the first quarter is £(1 + 0.25), at the end of the second quarter £1.25(1 + 0.25), at the end of the third quarter £1.25^2 (1 + 0.25) and at the end of the year £1.25^3(1 + 0.25) = 1.25^4 = 2.44.

In each of these cases the value of £1 at the end of the year is given by

$$(1 + 1/n)^n$$

where n is the number of times the interest is paid during the year.

That is, $n = 1$, $\left(1 + \dfrac{1}{1}\right)^1 = £2$

$n = 2$, $\left(1 + \dfrac{1}{2}\right)^2 = 1.5^2 = £2.25$

$n = 4$, $\left(1 + \dfrac{1}{4}\right)^4 = 1.25^4 = £2.44$

We can draw up a table of values of $(1 + 1/n)^n$ for different values of n.

n	1	2	4	8	10	100	1,000	10,000
$(1 + 1/n)^n$	2.00	2.25	2.44	2.56	2.59	2.70	2.717	2.718

As the value of n increases the value of the investment becomes larger, but never exceeds £2.719. The limit to which the value tends as n tends to infinity (ie becomes indefinitely large) is defined as e,

that is, $$e = \lim_{n \to \infty}\left(1 + \frac{1}{n}\right)^n$$

and e is approximately 2.7183. This is an *irrational* number. Its value cannot be determined exactly no matter how many decimal places of working are used. Also, it cannot be represented by the ratio of two whole numbers.

By allowing n to approach infinity interest is being added to the investment more and more frequently and can be regarded as being added continuously.

Returning to the more general case of £1 invested at $x\%$ per annum compound and the interest being paid n times a year, the value of the investment at the end of the year is given by

$$\left(1 + \frac{x}{n}\right)^n$$

To find the limit of this as n approaches infinity it is useful to let $1/k = x/n$ to that $n = kx$. This means that

$$\left(1 + \frac{x}{n}\right)^n = \left(1 + \frac{1}{k}\right)^{kx} = \left[\left(1 + \frac{1}{k}\right)^k\right]^x$$

and letting n approach infinity means that, since x is unchanged, k must approach infinity. The term in the inner brackets approaches the value e so that the whole expression tends to e^x. Thus e^x is the value of £1 after 1 year when interest is paid at $x\%$ per annum and is compounded continuously. Before discussing how e^x might be evaluated we need to examine the binomial theorem.

4.11   The binomial theorem

An expression of the form $(a + x)^n$ can generally be expanded as a series by applying the normal laws of algebra. For example,

$$(a + x)^2 = (a + x)(a + x) \qquad = a^2 + 2ax + x^2$$
$$(a + x)^3 = (a + x)(a + x)(a + x) = a^3 + 3a^2x + 3ax^2 + x^3$$

The algebra becomes a little tedious as the power of the function is increased but it is possible to obtain the expansion in the general case from the *binomial theorem*. This can be stated as follows:

$$(a + x)^n = {}^nC_0a^n + {}^nC_1a^{n-1}x + {}^nC_2a^{n-2}x^2 + \cdots + {}^nC_ra^{n-r}x^r$$
$$+ \cdots + {}^nC_nx^n$$

where ${}^nC_0 = 1$

$${}^nC_1 = n$$

$${}^nC_2 = \frac{n(n-1)}{1 \times 2} = \frac{n(n-1)}{2!}$$

$${}^nC_3 = \frac{n(n-1)(n-2)}{1 \times 2 \times 3} = \frac{n(n-1)(n-2)}{3!}$$

and in general

$$^nC_r = \frac{n(n-1)(n-2)\ldots(n-r+1)}{r!} = \frac{n!}{r!(n-r)!}$$

$r!$ is known as *factorial r* or *r factorial* and is the product of all the integer values from 1 to r. For example, $4! = 1 \times 2 \times 3 \times 4$. The exception to this is $0!$, which is defined to be equal to 1, so that

$$^nC_n = \frac{n!}{n!(n-n)!} = \frac{n!}{n!0!} = \frac{n!}{n!} = 1$$

The coefficients of the terms in a binomial expansion form a symmetrical pattern, as can be seen in Fig. 4.1, known as Pascal's triangle. Each number is the sum of the two numbers on the row above which are closest to it. Thus, the coefficient 10 is the sum of the two numbers 4 and 6 above it.

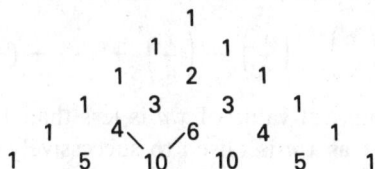

```
              1
           1     1
        1     2     1
     1     3     3     1
  1     4     6     4     1
1     5    10    10     5     1
```

Fig. 4.1

The third line of the triangle gives the coefficients of $(a + x)^2$:

$$(a + x)^2 = \underline{1}a^2 + \underline{2}ax + \underline{1}x^2$$

The fourth line gives the coefficients of $(a + x)^3$:

$$(a + x)^3 = \underline{1}a^3 + \underline{3}a^2x + \underline{3}ax^2 + \underline{1}x^3$$

The fifth line gives the coefficients of $(a + x)^4$:

$$(a + x)^4 = \underline{1}a^4 + \underline{4}a^3x + \underline{6}a^2x^2 + \underline{4}ax^3 + \underline{1}x^4$$

and so on.

The expansion can be divided throughout by the constant a with the result

$$(a + x)^n = a^n \left(1 + \frac{x}{a} \right)^n$$

$$= a^n \left[1 + n\left(\frac{x}{a}\right) + \frac{n(n-1)}{2!}\left(\frac{x}{a}\right)^2 + \cdots \right.$$

$$\left. + \frac{n(n-1)\cdots(n-r+1)}{r!}\left(\frac{x}{a}\right)^r + \cdots + \left(\frac{x}{a}\right)^n \right]$$

The expansion in square brackets is particularly interesting when $n = -1$ since

$$\left(1 + \frac{x}{a}\right)^{-1} = 1 + (-1)\left(\frac{x}{a}\right) + \frac{(-1)(-2)}{1.2}\left(\frac{x}{a}\right)^2$$

$$+ \cdots + \frac{(-1)(-2)\cdots(-r)}{1.2.3\cdots r}\left(\frac{x}{a}\right)^r + \cdots$$

This series contains an infinite number of terms and the coefficient of every term is equal to unity, the signs of these coefficients being alternately positive and negative. That is,

$$\left(1 + \frac{x}{a}\right)^{-1} = 1 - \left(\frac{x}{a}\right) + \left(\frac{x}{a}\right)^2 - \left(\frac{x}{a}\right)^3 + \cdots + (-1)^r\left(\frac{x}{a}\right)^r + \cdots$$

When the numerical value of x/a is less than 1 the terms become smaller and smaller as x/a is raised to successively higher powers and eventually tend to the value zero. That is,

$$(-1)^r\left(\frac{x}{a}\right)^r \to 0 \qquad \text{as} \qquad r \to \infty \qquad \text{for} \qquad \left|\frac{x}{a}\right| < 1$$

In this particular case the sum of the series tends or converges to a finite value as more and more terms are considered. Note the condition $|x/a| < 1$. This means the numerical value, or *modulus*, of x/a must be less than 1. In general it can be shown that a series converges if successive terms tend to zero and also alternate in sign.

It can quite easily be seen that there is an infinite series which converges to a finite value for $(1 - x/a)^{-1}$ when the numerical value of x/a is less than one, since this is a geometric series with a common ratio between -1 and $+1$:

$$\left(1 - \frac{x}{a}\right)^{-1} = 1 + (-1)\left(-\frac{x}{a}\right) + \frac{(-1)(-2)}{1.2}\left(-\frac{x}{a}\right)^2 + \cdots$$

$$= 1 + \left(\frac{x}{a}\right) + \left(\frac{x}{a}\right)^2 + \cdots + \left(\frac{x}{a}\right)^r + \cdots$$

In this case all the terms are positive.

4.12 The exponential series

The binomial theorem can be used to evaluate e. We know that e is defined by

$$e = \lim_{n \to \infty} \left(1 + \frac{1}{n} \right)^n$$

and so using the binomial theorem to expand $(1 + 1/n)^n$, i.e. with $a = 1$, and $x = 1/n$,

$$\left(1 + \frac{1}{n} \right)^n = {}^nC_0 1 + {}^nC_1 \left(\frac{1}{n} \right) + {}^nC_2 \left(\frac{1}{n} \right)^2 + {}^nC_3 \left(\frac{1}{n} \right)^3 + \cdots + {}^nC_n \left(\frac{1}{n} \right)^n$$

$$= 1 + \frac{n}{n} + \frac{n(n-1)}{2!} \frac{1}{n^2} + \frac{n(n-1)(n-2)}{3!} \frac{1}{n^3} + \cdots + \frac{1}{n^n}$$

$$= 1 + 1 + \left(\frac{n}{n} \right) \left(\frac{n-1}{n} \right) \cdot \frac{1}{2!}$$

$$+ \left(\frac{n}{n} \right) \left(\frac{n-1}{n} \right) \left(\frac{n-2}{n} \right) \frac{1}{3!} + \cdots + \frac{1}{n^n}$$

Allowing n to tend to infinity, all the ratios such as $(n-1)/n$, $(n-2)/n, \ldots$ tend to the value 1 and hence

$$e = \lim_{n \to \infty} \left(1 + \frac{1}{n} \right)^n = 1 + 1 + \frac{1}{2!} + \frac{1}{3!} + \frac{1}{4!} + \cdots$$

where the rth term in the expansion is $1/(r-1)!$.

This expansion, known as the *series expansion of e* can be used to evaluate e to any number of decimal places. For example to obtain e correct to 4 decimal places we continue summing the terms of the expansion until a term is less than 0.0001, as in Table 4.6.

The approximate value of e from the first 9 terms of the expansion is 2.71828.

Similarly, e^x can be expanded as a power series to give

$$e^x = 1 + x + \frac{x^2}{2!} + \frac{x^3}{3!} + \frac{x^4}{4!} + \cdots + \frac{x^r}{r!} + \cdots$$

This expansion is known as the *exponential series*. It can be used to evaluate e^x to any degree of accuracy. Alternatively, values of e^x are included in standard books of logarithm tables and are available at the press of a button on many pocket calculators.

TABLE 4.6

Number of terms summed	Last term	Contribution to e
1	1	1.00000
2	1	1.00000
3	$\dfrac{1}{2!}$	0.50000
4	$\dfrac{1}{3!}$	0.16667
5	$\dfrac{1}{4!}$	0.04167
6	$\dfrac{1}{5!}$	0.00833
7	$\dfrac{1}{6!}$	0.00139
8	$\dfrac{1}{7!}$	0.00020
9	$\dfrac{1}{8!}$	0.00002
		Total 2.71828

To illustrate the use of the series expansion of e^x suppose that $x = 0.05$ so that the continuous interest rate is 5%. Then substituting in the series expansion gives

$$e^{0.05} = 1 + 0.05 + \frac{0.05^2}{2!} + \frac{0.05^3}{3!} + \frac{0.05^4}{4!} + \cdots$$

$$= 1.05 + 0.00125 + 0.0000208 + 0.0000002 + \cdots$$

$$= 1.05127$$

correct to five places after the decimal point. Thus if interest is paid continuously at 5% per annum the value of p after one year is $1.05127p$. The value after t years is given by $pe^{0.05t}$ so that the value of £100 after seven years is

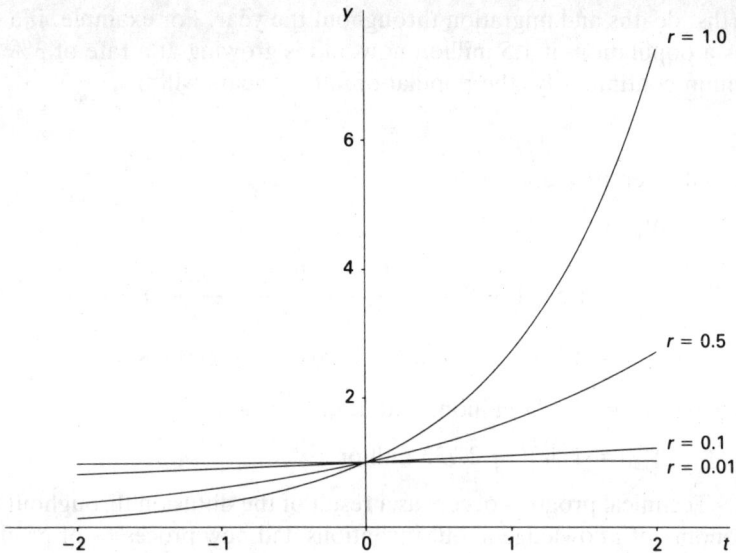

Fig. 4.2

$$pe^{0.05t} = 100e^{0.35}$$

$$= 100 \left\{ 1 + 0.35 + \frac{0.35^2}{2!} + \frac{0.35^3}{3!} + \frac{0.35^4}{4!} + \cdots \right\}$$

$$= 100 \left\{ 1.35 + 0.06125 + 0.0071458 + 0.0006252 + \cdots \right\}$$

$$= 100 \left\{ 1.419021 \right\} = £141.90$$

The general expression for the value of p after t years when the continuous rate of growth is $r\%$ per annum is

$$V_t = pe^{rt}$$

and the graph is shown in Fig. 4.2 for $p = 1$ and r taking the values 0.01, 0.1, 0.5 and 1.0. As r increases the graph becomes steeper.

While the concept of continuous growth has been introduced through continuous compound interest this is not a useful practical application since financial institutions add interest at discrete time intervals. However, there are many variables relevant in economics and business studies which can be assumed to vary continuously. An obvious one is the total population of a country, which changes as a result of births, deaths and migration throughout the year. For example, if a city has a population of 1.5 million now and is growing at a rate of 3% per annum continuously, the population after t years will be

$$V_t = 1.5e^{0.03t}$$

so that after 10 years,

$$V_{10} = 1.5e^{0.3}$$

$$= 1.5 \left\{ 1 + 0.3 + \frac{0.3^2}{2!} + \frac{0.3^3}{3!} + \frac{0.3^4}{4!} + \cdots \right\}$$

$$= 1.5 \left\{ 1.3 + 0.045 + 0.0045 + 0.0003375 + \cdots \right\}$$

$$= 2.025 \text{ million, and after 20 years}$$

$$V_{20} = 1.5e^{0.6} = 2.733 \text{ million.}$$

Technical progress occurs as a result of the diffusion throughout the economy of knowledge about inventions and new processes of production, and results in the growth of national income. This growth is continuous and the previous formula applies. Thus if V_t is the level of national income in year t and r is the rate of technical progress then

$$V_t = pe^{rt}$$

For example, suppose the index of national income is 100 and technical progress occurs at a continuous rate of 2% per annum, then after 8 years national income will be

$$V_8 = 100e^{0.16} = 117.35.$$

In section 4.6 the concept of discounting for discrete annual payments was discussed. If the discounting occurs continuously at the rate $r\%$ per annum the discounted value is

$$V_t = pe^{-rt},$$

so that £2,000 due 10 years from now with the discounting rate 8% per annum has a present value of

$$V_{10} = 2000e^{-0.8} = \text{£}898.66$$

Another application is to rates of decay – that is, negative growth rates. For example, suppose a steel manufacturer has 1,000 tons of steel in stock and this is deteriorating at a continuous rate of 4% per annum.

The remaining stock after t years is given by

$$V_t = 1000e^{-0.04t}$$

and after five years, $V_5 = 818.73$ tons.

Finally, we saw that the value of an amount p after t years when interest is at the rate x and is compounded m times a year is

$$V_{1t} = p(1 + x/m)^{mt}$$

while for interest compounded continuously at the rate $r\%$ per annum,

$$V_{2t} = pe^{rt}$$

and so it is always possible to find a continuous interest rate which is equivalent to a discrete one. For example, if the discrete interest rate is 10%, compounded four times a year, the value after 5 years is

$$V_{15} = p(1 + 0.10/4)^{20}$$

and the equivalent continuous interest value is

$$V_{25} = pe^{5r}$$

For these to be equal, $1.025^{20} = e^{5r}$

This can be solved by taking natural logarithms (see section 4.14 for a discussion),

$$20 \log_e 1.025 = 5r$$

or $$r = 4(0.02469) = 0.0988$$

so that the continuous rate of 9.88% is equivalent to 10% compounded four times a year.

4.13 Exercises

1. £150 can be invested in one of three ways. It can receive 6% per annum simple interest, 5% per annum compound interest, or 4% per annum compound interest with the interest compounded twice a year. Which is the most profitable investment over a period of

   (a) five years   (b) ten years?

2. What is the value of £50 after four years if it is invested at 6% per annum compound interest with the interest compounded three times a year?

3. Use the binomial theorem to expand

   (a) $(1 + x)^5$ (b) $(2 - x)^3$

   (c) $\left(3 + \dfrac{1}{x} \right)^4$ (d) $(2 - x^2)^6$

4. Use the exponential series to obtain approximately the values of

   (a) $e^{0.1}$ (b) $e^{0.5}$

   (c) e^2 and   (d) e^{-1}

   and compare your results with the values given in standard tables of e or with the values given by a pocket calculator.

5. A city has a population of 2 million which is growing at the continuous rate of 4% per annum. What is the projected population of the city after (a) 5 years (b) 25 years? The government decides to introduce policies to try to reduce the growth rate to 2% per annum. What difference will this make to the population projections?

6. Technical progress has increased the national income of a country by an average of 3% per annum. The index of national income is currently 125.0. What will it be in 30 years from now?

7. In a particular year 150,000 new firms are created. It is known that the failure rate is 15% per annum. How many of the firms will survive (a) 4 years, (b) 20 years?

8. What is the continuous rate of interest which is equivalent to 6% per annum compounded twice yearly?

4.14   Logarithms

We know that

$$100 \times 1000 = 100000$$

and since $100 = 10^2$, $1,000 = 10^3$ and $100,000 = 10^5$ then

$$10^2 \times 10^3 = 10^5.$$

That is, the product of 100 and 1,000 can be obtained by writing these numbers as powers of 10 and then adding these powers together. These powers are known as *logarithms* and are referred to as *logarithms to the base 10* because they are powers of 10.

Therefore, $\log_{10} 100 = 2$

$\log_{10} 1000 = 3$

$\log_{10} 100000 = 5$

or, more simply $\log 100 = 2$

$\log 1000 = 3$

$\log 100000 = 5$

Now $10 = 10^1$ and so $\log 10 = 1$

$0.1 = 10^{-1}$ and so $\log 0.1 = -1$

$1 = 10^0$ since $10 \times 0.1 = 1 = 10^1 \times 10^{-1} = 10^0$

and so $\log 1 = 0$.

Logarithms of any number in the range 1–10 can be found by using logarithm tables.

For example, $\log 2 = 0.3010$

$\log 5 = 0.6990$

$\log 7.5 = 0.8751$

$\log 3.001 = 0.4772$

$\log 1.055 = 0.0232$

Logarithms of numbers outside the range 1–10 can be found by expressing the number as the product of a number within this range and ten raised to some power. For example, $500 = 5 \times 10^2$.

Now since $\log 5 = 0.6990$, $5 = 10^{0.6990}$

that is, $\log 500 = 2.6990$,

or $\log 500 = \log 5 + \log 100 = 0.6990 + 2 = 2.6990$

Similarly $0.003001 = 3.001 \times 10^{-3}$ and so $\log (0.003001) = \log (3.001) + \log (10^{-3}) = 0.4772 + (-3) = \bar{3}.4772$.

Notice that we write this as $\bar{3}.4772$ and not as $-3 + 0.4772 = -2.5228$. The reason for this is that it is useful for manipulation if the part of the logarithm after the decimal point is always positive.

The general rules governing logarithms are:

1. $\log ab = \log a + \log b$

We have already used this rule above. For example, $a = 5$, $b = 100$, $ab = 500$.

$$\log 500 = \log 5 + \log 100 = 2.6990.$$

2. $\log \left(\dfrac{a}{b} \right) = \log a - \log b$

For example, $a = 5$, $b = 2$,
$\log \left(\frac{5}{2} \right) = \log 5 - \log 2 = 0.6990 - 0.3010 = 0.3980$
that is, $\log 2.5 = 0.3980$

3. $\log a^n = n \log a$

This follows directly from Rule 1.
For example, $\log (a^2) = \log (a \cdot a) = \log a + \log a = 2 \log a$
Similarly, $\log (5^3) = 3 \log 5 = 3(0.6990) = 2.0970$.

These rules can be used to reduce the arithmetic required in our calculations.

Example 1

Evaluate $450/1.05^6$

By Rule 1, $\log 450 = \log (100 \times 4.5) = \log 100 + \log 4.5$

$$= 2 + 0.6532 = 2.6532$$

By Rule 3, $\log (1.05^6) = 6 \log (1.05) = 6(0.0212)$

$$= 0.1272$$

By Rule 2, $\log \left(\dfrac{450}{1.05^6} \right) = \log 450 - \log (1.05^6)$

$$= 2.6532 - 0.1272 = 2.5260$$

We now need to know the number for which 2.5260 is the logarithm. We can find this either by looking in the body of the logarithm tables for 0.5260, or by using tables of antilogarithms which give the required number directly.

$$\text{antilog } 0.5260 = 3.357$$

and therefore, antilog $2.5260 = 10^2 \times 3.357 = 335.7$; that is,

$$\frac{450}{1.05^6} = 335.7$$

Example 2

Evaluate $50(1.1^{10})/3250$.

We first of all evaluate 1.1^{10}, then $50(1.1^{10})$ and finally divide by 3250.

$$\log (1.1^{10}) = 10 \log 1.1 = 10(0.0414) = 0.4140$$

$$\log 50 = \log (10 \times 5) = \log 10 + \log 5 = 1.6990$$

$$\therefore \quad \log [(50(1.1^{10})] = \log (1.1^{10}) + \log 50 = 0.4140 + 1.6990$$

$$= 2.1130$$

$$\log 3250 = \log (1000 \times 3.25) = 3.5119$$

$$\therefore \quad \log \frac{50(1.1^{10})}{3250} = \log [(50(1.1^{10})] - \log (3250)$$

$$= 2.1130 - 3.5119$$

$$= -1.3989 = -2 + 0.6011 = \bar{2}.6011$$

Hence
$$\left(\frac{50(1.1^{10})}{3250} \right) = \text{antilog } \bar{2}.6011$$

$$= 10^{-2} \text{ antilog } (0.6011)$$

$$= 10^{-2} (3.991)$$

$$= 0.03991$$

Throughout this section all the logarithms used have been to the base 10. This is generally convenient in numerical work. However, in Chapter 5 we are going to see that for most theoretical work it is more convenient to use logarithms to the base e. Such logarithms are known as *natural*, *hyperbolic* or *Naperian logarithms* and are written as $\log_e x$ or *ln x*. The rules of logarithms stated above apply to logarithms with any base. To change the base to which logarithms are measured an additional rule is used:

4. For a positive real number x, $\log_p x = \log_p q \log_q x$

Since we are mainly interested in logarithms to the base 10 and the base e we have the special case

$$\log_e x = \log_e 10 \log_{10} x$$

or
$$\log_{10} x = \log_e x / \log_e 10$$

which allows conversion from one base to the other.

In discussing logarithms we have not mentioned where the tables referred to come from. In section 5.15 we will use Taylor's theorem to show that, for $-1 < x < 1$,

$$\log_e (1 + x) = x - \frac{x^2}{2} + \frac{x^3}{3} - \frac{x^4}{4} + \frac{x^5}{5} + \cdots$$

which is an infinite series with successive terms becoming smaller so that the sum converges rather than tending to infinity. As it stands, this expansion is not particularly useful because of the restriction that $-1 < x < 1$ and we need to be able to find the logarithm of any number. Now if we replace x by $-x$ each time it occurs in the above expansion then,

$$\log_e (1 - x) = -x - \frac{x^2}{2} - \frac{x^3}{3} - \frac{x^4}{4} - \frac{x^5}{5} - \cdots$$

where again $-1 < x < 1$. Therefore by subtraction and using rule 2 for logarithms above,

$$\log_e(1 + x) - \log_e(1 - x) = \log_e\{(1 + x)/(1 - x)\}$$
$$= 2\{ x + \frac{x^3}{3} + \frac{x^5}{5} + \cdots\}$$

since all the even powers of x cancel out. Therefore if we wish to find the natural logarithm of a number, a, say, we find x such that

$$a = \frac{1 + x}{1 - x} \quad \text{and solving,} \quad x = \frac{a - 1}{a + 1}$$

For example, to use the series expansion to evaluate $\log_e 10$ put $a = 10$ so that $x = 9/11$ or 0.81818. Then,

$$\log_e 10 = 2\{0.81818 + 0.18257 + 0.07333 + 0.03506 + 0.01826$$
$$+ 0.01000 + 0.00566 + 0.00329 + 0.00194 + 0.00116$$
$$+ 0.00070 + \cdots\}$$
$$= 2\{1.15015\} = 2.3003$$

Here the calculations are to five places after the decimal point and extra terms were added until the last one was below 0.001, so that the result should be accurate to at least two places after the decimal point. A calculator gives the precise value as 2.3026 and so the result is close. In this example the convergence of the series expansion has been slow.

As another example, to find $\log_e 3$ we have

$$x = (3 - 1)/(3 + 1) = 0.5$$

and $\log_e 3 = 2\{ 0.5 + \dfrac{0.5^3}{3} + \dfrac{0.5^5}{5} + \dfrac{0.5^7}{7} + \dfrac{0.5^9}{9} + \cdots\}$

$\qquad = 2\{0.5 + 0.041667 + 0.00625 + 0.00111 + 0.00022 + \cdots\}$

$\qquad = 2\{0.54924\} = 1.09849$

using the first five terms. A check with a calculator shows that the accurate value is $\log_e 3 = 1.0986$.

Using these results we can obtain

$$\log_{10} 3 = \log_e 3 / \log_e 10 = 1.0986/2.3026 = 0.4771$$

which is the value given in logarithm tables.

The use of natural logarithms helps in some problems concerning continuous growth. For example, suppose that the national income, Y, is growing at a continuous rate of 4% per annum. How long will it take for Y to double?

Here we have $Y_t = Y_0 e^{0.04t}$
where Y_0 is the current value of Y and Y_t is the final value of Y. We require the value of t for which $Y_t/Y_0 = 2$. Therefore,

$$2 = e^{0.04t}$$

Taking natural logarithms,

$$0.69315 = 0.04t$$

since $\log_e e = 1$ and so $t = 17.3$ years.

Finally, since in many applications in economics natural logarithms are used, the graph of $y = \log_e x$ is shown in Fig. 4.3. As x increases, y increases but at a decreasing rate.

4.15 Exercises

1. Use the series expansion of $\log_e\{(1 - x)/(1 + x)\}$ to obtain $\log_e 0.1$ and $\log_e 100$. Hence obtain $\log_{10} 0.1$ and $\log_{10} 100$ and compare the values with those in logarithm tables.
2. The population of a region is 2.4 million and it is growing at a continuous rate of 5% per annum. How many years will it be before the population reaches (a) 3 million (b) 5 million?
3. Last year 100,000 new firms were created in a particular country. If new firms fail at a continuous rate of 12% per annum, how long will it be before there are only 50,000 firms?
4. Pareto's law of distribution of income states that the number of

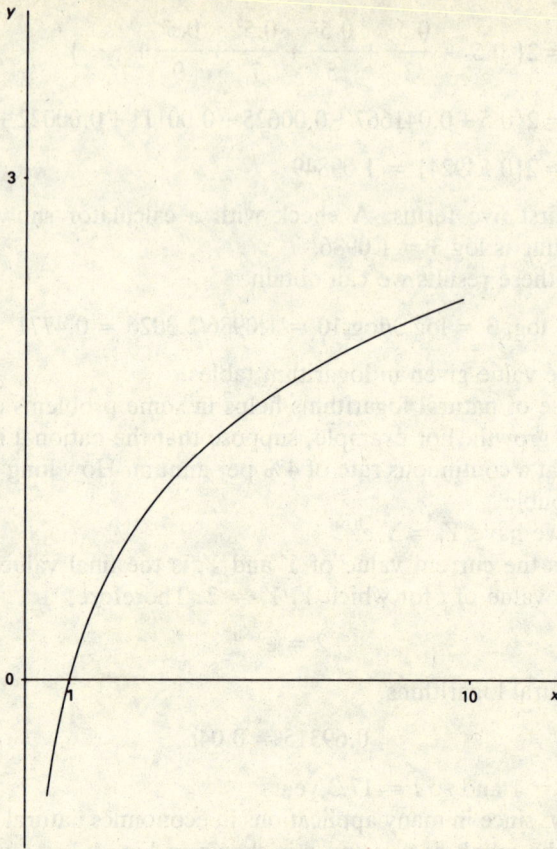

Fig. 4.3

individuals n, from a population of size N, whose income exceeds the value x is given by $n = N/x^{1.5}$. The population of the UK is approximately 55,000,000. According to this law, how many individuals are expected to have incomes above (a) £10,000 (b) £100,000?

4.16 Revision exercises for Chapter 4 (without answers)

1. For the following progressions, identify the type of series, give the expression for the nth term, obtain the sum of the first 20 terms and, if it is finite, the sum to infinity:

    (a) 100, 105, 110, 115, 120, . . .
    (b) 100, 99, 98, 97, 96, . . .
    (c) 1, 2, 4, 8, 16, . . .
    (d) 256, 128, 64, 32, 16, . . .
    (e) 100, −90, 81, −72.9, 65.61, . . .

2. The sum of £1,000 can be invested for 15 years with
    (a) simple interest at the rate of 10% per annum
    (b) compound interest at the rate of 8% per annum, added once
    each year
    (c) compound interest at the rate of 6% per annum, added every
    six months
    (d) continuously added interest at the rate of 5% per annum.
    Which of these gives the highest return?

3. A firm can borrow money at 15% per annum compound interest
    and has the opportunity of investing in a project which has the
    following costs and receipts:

End of year	*1*	*2*	*3*	*4*
Costs	50,000	20,000	5,000	5,000
Receipts	10,000	25,000	55,000	20,000

    Is it worthwhile investing in the project? What is the internal rate
    of return of the project?

4. A lady decides to pay £100 into a fund each year on her grand-
    daughter's birthday (starting at age 5). If compound interest is
    received at 7% per annum, how much would the granddaughter
    receive if given the fund on her twenty-first birthday?

5. What would be the annual payments on a housing loan of £75,000
    payable back over 25 years if interest is charged at 11% per
    annum? If repayments were made at the rate of £6,000 per
    annum, when would the debt be paid off?

6. A firm has a computer which is expected to cost $10,000 to
    replace in 8 years' time. If the interest rate is 9%, how much
    should be saved each year in order to be able to replace the
    computer?

7. Use the binomial theorem to expand $(2 + 3x)^9$ and $(x + 1/x)^{10}$ and
    hence evaluate 2.3^9 and 4.25^{10}.

8. Use the first six terms of the exponential series to evaluate approximately $e^{1.5}$, $e^{-1.5}$ and e^{-4} and compare the results with the values given by a calculator.

9. In 1971 the index of retail prices in the UK was 100. In 1985 the value was 463.3. What is the implied rate of inflation, assuming (a) it is compounded annually and (b) it occurs continuously?

10. Use the series expansion of $\log_e\{(1 - x)/(1 + x)\}$ to obtain $\log_e 5$ and $\log_e 10$.

Chapter 5
Differential Calculus

5.1 Introduction

In Chapters 1 and 3 we saw that an equation relating two variables could be represented by a graph. For example, if y is the total cost of production and x the quantity produced then the equation

$$y = 100 + 3x$$

can be represented by a straight line, whilst

$$y = 100 + 2x + \tfrac{1}{10}x^2$$

can be represented by a curve. The graphs of these equations allow the cost to be determined for permissible levels of production.

But the equations also provide additional information which is very important. They tell us the change in the costs which must be incurred for any given change in production level. This can be determined directly from the graphs by considering a change of x from x_1 to x_2.

In the case of the linear cost function the cost changes from y_1 to y_2 (Fig 5.1). The average cost of each extra unit of production between the values x_1 and x_2 is given by the quotient

$$\frac{y_2 - y_1}{x_2 - x_1}$$

This is the slope of the line and it does not depend upon the value of x_1 or x_2, because the slope is constant for all values of x for a linear function.

This is not so for the quadratic function as can be seen from Fig 5.2. It is clear that with this curve the slope between P_1 and P_2 depends on where P_1 and P_2 are, and that even if P_1 is held constant, the slope will depend on the position of P_2. If the cost changes from y_3 to y_4 when the quantity produced changes from x_1 to x_2, then the average cost of each extra unit is

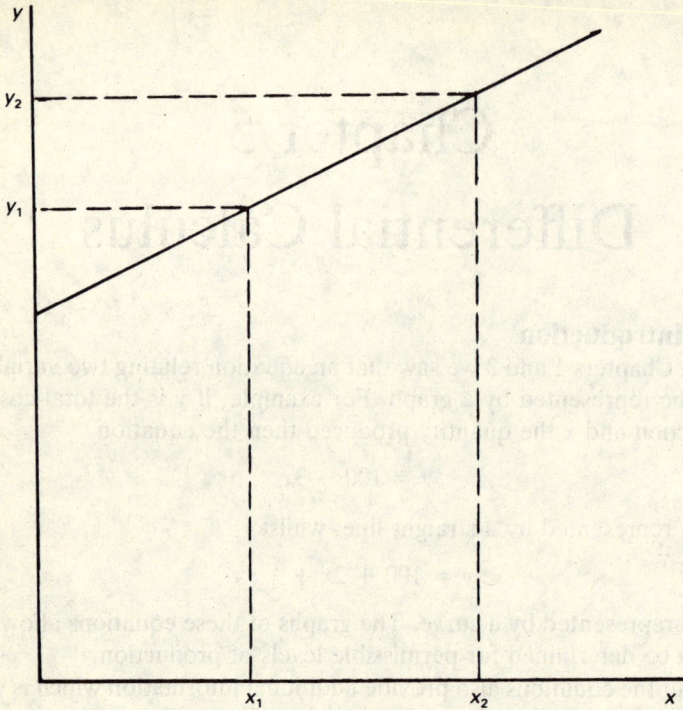

Fig. 5.1

$$\frac{y_4 - y_3}{x_2 - x_1}$$

This is the slope of the straight line joining P_1 and P_2 and is not that of the curve. We therefore define the *slope at a point* on a curve as being a tangent to the curve at that point, where a *tangent* is a line which touches but does not cross the curve. The tangent at P_1 is shown in Fig 5.3, which is an enlargement of Fig 5.2 with P_1 and P_2 close together.

To indicate that only small changes in x and y are being represented we use the symbols δx and δy, which are *not* δ times x and δ times y. The slope of the straight line joining P_1 and P_2 is

$$\frac{\delta y}{\delta x}$$

But this is not the slope of the curve at P_1 (the tangent at P_1). However, if δx becomes smaller then the slope of the straight line $P_1 P_2$ becomes very close to the slope of the tangent at P_1.

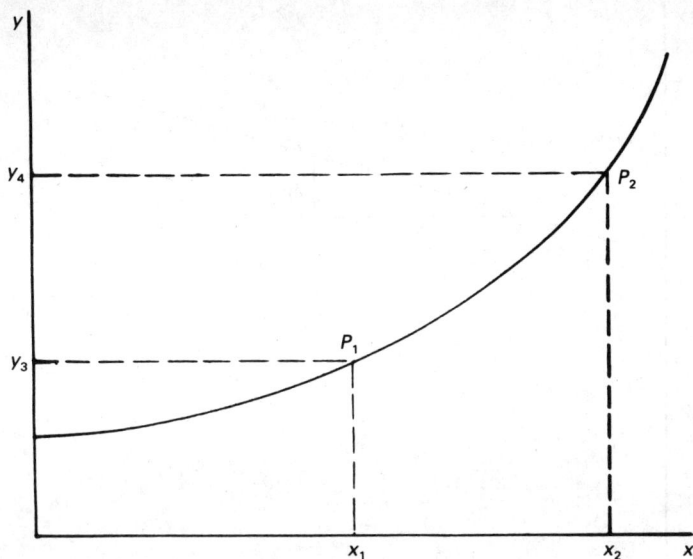

Fig. 5.2

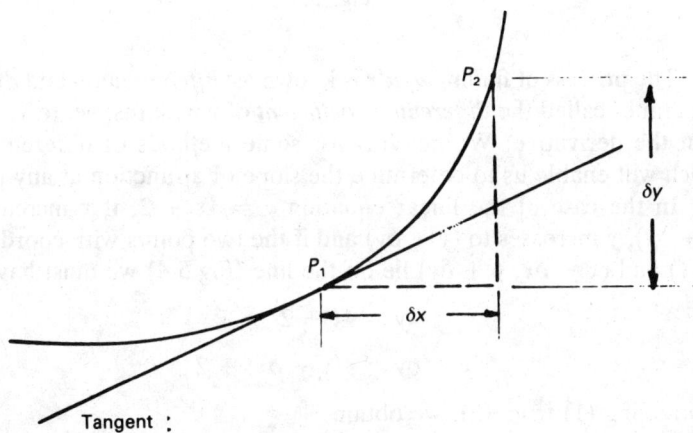

Tangent :

Fig. 5.3

As $\delta x \to 0$, we define limit $(\delta y/\delta x)$ as the *derivative* of y with respect to x, and we denote it by the symbol dy/dx. That is,

$$\frac{dy}{dx} = \lim_{\delta x \to 0}\left(\frac{\delta y}{\delta x}\right)$$

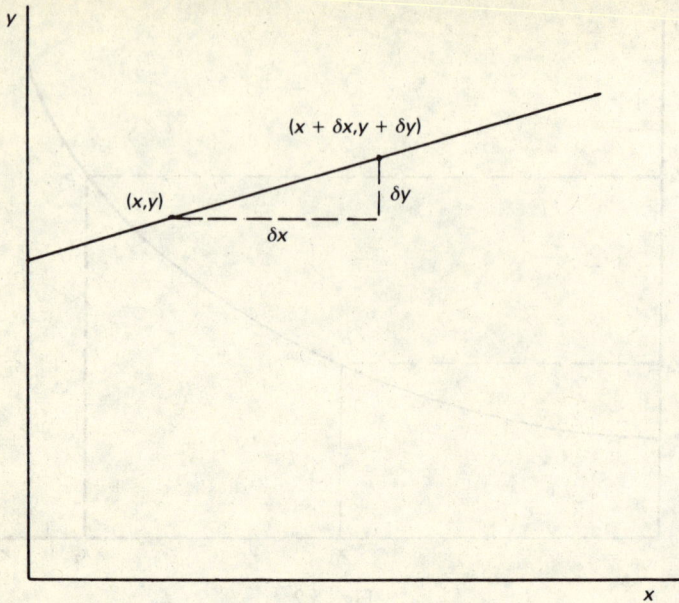

Fig. 5.4

The process of finding dy/dx is known as *differentiation* and dy/dx is sometimes called the *differential coefficient* of y with respect to x, rather than the derivative. We now discuss some methods of differentiation which will enable us to determine the slope of a function at any point.

In the case of the linear equation $y = 4x + 2$, if x increases to $(x + \delta x)$, y increases to $(y + \delta y)$ and if the two points with coordinates (x, y) and $(x + \delta x, y + \delta y)$ lie on the line (Fig 5.4) we must have

$$y = 4x + 2 \tag{1}$$

$$y + \delta y = 4(x + \delta x) + 2 \tag{2}$$

Subtracting (1) from (2), we obtain

$$(y + \delta y) - y = 4(x + \delta x) + 2 - (4x + 2)$$

$$\delta y = 4x + 4\,\delta x + 2 - 4x - 2$$

$$= 4\,\delta x$$

$$\therefore \quad \frac{\delta y}{\delta x} = 4$$

and
$$\frac{dy}{dx} = \lim_{\delta x \to 0} \left(\frac{\delta y}{\delta x} \right) = 4$$

This is the expected result, because we know that the slope is equal to 4 for all positions along the line. That is, it is independent of the value of x.

In the case of a quadratic equation the same procedure can be adopted.

For example if

$$y = x^2 + 2x + 1 \tag{3}$$

then
$$y + \delta y = (x + \delta x)^2 + 2(x + \delta x) + 1 \tag{4}$$

Subtracting (3) from (4)

$$y + \delta y - y = (x + \delta x)^2 + 2(x + \delta x) + 1 - (x^2 + 2x + 1)$$
$$= x^2 + 2x\,\delta x + (\delta x)^2 + 2x + 2\,\delta x + 1 - x^2 - 2x - 1$$
$$= 2x\,\delta x + (\delta x)^2 + 2\,\delta x$$
$$\therefore \quad \delta y = (2x + 2)\,\delta x + (\delta x)^2$$
$$\frac{\delta y}{\delta x} = (2x + 2) + \delta x$$

Now as δx decreases in size the second term on the right-hand side of this equation becomes of less and less significance, and in the limit

$$\frac{dy}{dx} = \lim_{\delta x \to 0} \left(\frac{\delta y}{\delta x} \right) = 2x + 2$$

This means that the slope of the curve at any point is a function of the point itself. For example,

when $x = 1$ $\qquad\qquad \dfrac{dy}{dx} = 2 + 2 = 4$

when $x = 2$ $\qquad\qquad \dfrac{dy}{dx} = 4 + 2 = 6$

5.2 Some general rules

The above procedure can be used to develop general rules for obtaining derivatives. For example, in the case of the general quadratic

$$y = a + bx + cx^2$$

$$(y + \delta y) = a + b(x + \delta x) + c(x + \delta x)^2$$

$$(y + \delta y) - y = a + bx + b\,\delta x + cx^2 + 2cx\,\delta x + c(\delta x)^2 - (a + bx + cx^2)$$

$$\therefore \quad \delta y = b\,\delta x + 2cx\,\delta x + c(\delta x)^2$$

$$\frac{\delta y}{\delta x} = b + 2cx + c\,\delta x$$

$$\therefore \quad \frac{dy}{dx} = b + 2cx$$

that is, if $y = a + bx + cx^2$, then $dy/dx = b + 2cx$.

The derivative is in fact the sum of the derivatives of the three terms taken separately: that is,

if
$$y = a + bx + cx^2$$

then
$$\frac{dy}{dx} = \frac{d}{dx}(a) + \frac{d}{dx}(bx) + \frac{d}{dx}(cx^2)$$

and in general the derivative of any polynomial in x is equal to the sum of the derivatives of each term considered separately.

It can be seen that the derivative of the constant a is zero, that of the term bx is b and that of cx^2 becomes $2cx$. In general the derivative of x^n is equal to nx^{n-1}, which can be proved as follows:

If
$$y = ax^n,$$

$$y + \delta y = a(x + \delta x)^n$$

$$= a(x^n + {}^nC_1 x^{n-1}\,\delta x + \cdots + {}^nC_n(\delta x)^n)$$

from the binomial expansion (section 4.11).

$$y + \delta y - y = a[x^n + {}^nC_1 x^{n-1}\,\delta x + \cdots + {}^nC_n(\delta x^n)] - ax^n$$

or
$$\delta y = a[{}^nC_1 x^{n-1}\,\delta x + {}^nC_2 x^{n-2}\,\delta x^2 + \cdots]$$

Hence
$$\frac{\delta y}{\delta x} = a^n C_1 x^{n-1} + {}^nC_2 x^{n-2}\,\delta x + \cdots$$

$$\therefore \quad \frac{dy}{dx} = \lim_{\delta x \to 0} \frac{\delta y}{\delta x} = anx^{n-1} + 0$$

So if $y = ax^n$, $dy/dx = anx^{n-1}$.

Example 1

If $y = 4x^6$, $n = 6$, $a = 4$, so

$$\frac{dy}{dx} = 4(6)x^5 = 24x^5$$

Example 2

If $y = x^{-2}$, $n = -2$, $a = 1$, so

$$\frac{dy}{dx} = (-2)x^{-3} = \frac{-2}{x^3}$$

THE PRODUCT RULE

If $y = uv$, where u and v are both functions of x, then the derivative

$$\frac{dy}{dx} = v\frac{du}{dx} + u\frac{dv}{dx}$$

This is obtained by the same method as used above. Since u and v are functions of x, a change in the value of x to $x + \delta x$ changes both u and v. Let the new values be $u + \delta u$ and $v + \delta v$. The value of y also changes, and we let this change be δy.

Hence, $\qquad\qquad\qquad y = uv$

and $\qquad\qquad y + \delta y = (u + \delta u)(v + \delta v)$

or $\qquad\qquad y + \delta y = uv + v\,\delta u + u\,\delta v + \delta u\,\delta v$

so that $\qquad\qquad \delta y = v\,\delta u + u\,\delta v + \delta u\,\delta v$

and $\qquad\qquad \dfrac{\delta y}{\delta x} = v\dfrac{\delta u}{\delta x} + u\dfrac{\delta v}{\delta x} + \dfrac{\delta u}{\delta x}\delta v$

Now as $\delta x \to 0$, $\qquad \lim\dfrac{\delta y}{\delta x} = \dfrac{dy}{dx} \quad \lim\dfrac{\delta u}{\delta x} = \dfrac{du}{dx}$

$$\lim\frac{\delta v}{\delta x} = \frac{dv}{dx} \quad \lim \ \delta v \to 0$$

Taking limits of the expression gives

$$\frac{dy}{dx} = v\frac{du}{dx} + u\frac{dv}{dx} + \frac{du}{dx}(0) = v\frac{du}{dx} + u\frac{dv}{dx}$$

An example of the use of this is now given.

$$y = (x + 2)(x^2 + 3)$$

Let $u = x + 2$ and $v = x^2 + 3$

then $$\frac{du}{dx} = 1 \quad \text{and} \quad \frac{dv}{dx} = 2x$$

so that $$\frac{dy}{dx} = (x^2 + 3)(1) + (x + 2)(2x)$$

$$= x^2 + 3 + 2x^2 + 4x = 3x^2 + 4x + 3$$

THE QUOTIENT RULE

If $y = u/v$ where both u and v are functions of x, then

$$\frac{dy}{dx} = \frac{v(du/dx) - u(dv/dx)}{v^2}$$

This can be shown by allowing x to change to $x + \delta x$, and letting the corresponding changes in u, v and y be δu, δv and δy. Then

$$y = \frac{u}{v}$$

$$y + \delta y = \frac{u + \delta u}{v + \delta v}$$

Subtracting, we have

$$\delta y = \frac{u + \delta u}{v + \delta v} - \frac{u}{v} = \frac{vu + v\,\delta u - uv - u\,\delta v}{(v + \delta v)v}$$

$$= \frac{v\,\delta u - u\,\delta v}{v^2 + v\,\delta v}$$

Dividing by δx, we obtain

$$\frac{\delta y}{\delta x} = \frac{v(\delta u/\delta x) - u(\delta v/\delta x)}{v^2 + v\,\delta v}$$

As $\delta x \to 0$, $\quad \dfrac{\delta y}{\delta x} \to \dfrac{dy}{dx}, \quad \dfrac{\delta u}{\delta x} \to \dfrac{du}{dx}, \quad \dfrac{\delta v}{\delta x} \to \dfrac{dv}{dx}, \quad v\,\delta v \to 0.$

Hence $\dfrac{dy}{dx} = \dfrac{v(du/dx) - u(dv/dx)}{v^2}$

Example 3

$$y = \frac{x + 2}{2x}$$

Let $u = x + 2$ and $v = 2x$

$$\frac{du}{dx} = 1 \qquad \frac{dv}{dx} = 2$$

$$\frac{dy}{dx} = \frac{(2x)(1) - (x + 2)(2)}{(2x)^2}$$

$$= \frac{2x - 2x - 4}{4x^2} = \frac{-4}{4x^2} = \frac{-1}{x^2}$$

THE FUNCTION OF A FUNCTION OR CHAIN RULE

This rule states that if y is a function of u, and u is a function of x,

then $\dfrac{dy}{dx} = \left(\dfrac{dy}{du} \right) \left(\dfrac{du}{dx} \right)$

The proof of this rule is as follows

Let $\qquad y = f(u) \qquad$ and $\qquad u = g(x)$

then $\qquad y + \delta y = f(u + \delta u) \qquad$ and $\qquad u + \delta u = g(x + \delta x)$

where δu is assumed to be non-zero.

Hence $\qquad \dfrac{\delta y}{\delta x} = \dfrac{f(u + \delta u) - f(u)}{\delta x}$

$$= \left(\frac{f(u + \delta u) - f(u)}{\delta u} \right) \frac{\delta u}{\delta x}$$

and $\qquad \displaystyle\lim_{\delta x \to 0} \left(\dfrac{\delta y}{\delta x} \right) = \dfrac{dy}{dx} = \left(\dfrac{dy}{du} \right) \left(\dfrac{du}{dx} \right)$

Example 4

$$y = (1 + x)^4$$

In this case, if $u = 1 + x$, then we have $y = u^4$ and $u = 1 + x$, so that y is a function of u and u is a function of x.

Now $$\frac{dy}{du} = 4u^3 \qquad \frac{du}{dx} = 1$$

$$\frac{dy}{dx} = \frac{dy}{du}\frac{du}{dx} = (4u^3)(1) = 4(1 + x)^3$$

LOGARITHMS

Functions involving logarithms can be differentiated using some of the rules of section 4.14. Initially we will not specify the base to which the logarithm is taken. To differentiate $y = \log x$, we allow x to increase by δx and let the corresponding increase in y be δy so that

$$y + \delta y = \log (x + \delta x)$$

Subtracting, we obtain

$$y + \delta y - y = \log (x + \delta x) - \log (x)$$

or $$\delta y = \log \frac{(x + \delta x)}{x} = \log \left(1 + \frac{\delta x}{x} \right)$$

Now we divide by δx, and also multiply the term on the right hand side by (x/x) or by 1. The latter leaves the value of the right hand side unchanged.

$$\frac{\delta y}{\delta x} = \frac{1}{\delta x} \cdot \frac{x}{x} \log \left(1 + \frac{\delta x}{x} \right) = \left(\frac{x}{\delta x} \right) \frac{1}{x} \log \left(1 + \frac{\delta x}{x} \right)$$

$$= \frac{1}{x} \left[\log \left(1 + \frac{\delta x}{x} \right)^{x/\delta x} \right]$$

Let us consider the behaviour of the expression in the outer brackets. It is convenient to let $n = x/\delta x$. Since x is fixed, as $\delta x \to 0$, $n \to \infty$

Now $$\frac{dy}{dx} = \frac{1}{x} \left[\log \left(1 + \frac{1}{n} \right)^n \right]$$

Taking the limit as $\delta x \to 0$ gives

$$\frac{dy}{dx} = \frac{1}{x} \log \left[\lim_{n \to \infty} \left(1 + \frac{1}{n} \right)^n \right]$$

But the limiting value of this expression is defined as e, and so

$$\frac{dy}{dx} = \frac{1}{x} \log (e)$$

If we now specify that we are using logarithms to the base e, since log $(e) = 1$ we have

$$\frac{dy}{dx} = \frac{1}{x} \quad \text{when} \quad y = \log_e (x)$$

In general, from using the function of a function rule,

if
$$y = \log_e (u)$$

$$\frac{dy}{dx} = \frac{1}{u} \frac{du}{dx}$$

since
$$\frac{dy}{du} = \frac{1}{u} .$$

Example 5

$$y = \log_e (2x + 3)$$

$$u = 2x + 3, \qquad du/dx = 2$$

Hence
$$\frac{dy}{dx} = \frac{1}{(2x + 3)} (2) = \frac{2}{2x + 3}$$

We will assume for the rest of this book that whenever we are concerned with differentiation, logarithms use the base e. If this were not the case, a factor $\log_n (e)$ would need to be included, where n is the base of the logarithms.

EXPONENTIAL FUNCTIONS

These can be differentiated most easily by converting to logarithms.

For example, $y = e^x$

Taking logarithms, we obtain

$$\log y = \log (e^x) = x \log e = x$$

The derivative of log y is

$$\frac{d}{dx} (\log y) = \frac{1}{y} \left(\frac{dy}{dx} \right)$$

This arises from the function of a function rule, since y is a function of x, and log y is a function of y.

Hence,
$$\frac{1}{y} \frac{dy}{dx} = 1$$

or
$$\frac{dy}{dx} = y$$

This shows that if $y = e^x$, $dy/dx = e^x$ which is a rather surprising result. The same result is obtained by differentiating the series expansion of e^x (section 4.12) term by term.

Other exponential functions can be differentiated by the same method. In particular,

if
$$y = e^{ax}$$

$$\log y = ax \log e = ax$$

Differentiating, we have

$$\frac{1}{y}\frac{dy}{dx} = a$$

Hence
$$\frac{dy}{dx} = ay = ae^{ax}$$

Example 6

$$y = e^{4x}$$

We have $a = 4$, and so

$$\frac{dy}{dx} = 4e^{4x}$$

TRIGONOMETRIC FUNCTIONS

A number of these are differentiated from first principles in Appendix A, Section A.5 and we summarise the results in Table 5.1.

5.3 Exercises

Find dy/dx for the following functions:

1. $y = 3x - 4$

2. $y = x^2 - 3x + 3$

3. $y = 3x^4 - x^3 + x^2 + 25x$

4. $y = 6x^3 - \dfrac{x^4}{4} + \dfrac{x^3}{3} - \dfrac{1}{x^2}$

5. $y = (2x^2 + 3x - 1)\left(\dfrac{1}{x} - 3x + 2\right)$

TABLE 5.1
Summary table of derivatives

y	$\dfrac{dy}{dx}$	y	$\dfrac{dy}{dx}$
a	0	$\log u$	$\dfrac{1}{u}\dfrac{du}{dx}$
ax^n	anx^{n-1}	e^{ax}	ae^{ax}
uv	$v\dfrac{du}{dx} + u\dfrac{dv}{dx}$	$\sin ax$	$a\cos ax$
$\dfrac{u}{v}$	$\dfrac{v\,(du/dx) - u\,(dv/dx)}{v^2}$	$\cos ax$	$-a\sin ax$
$f(u)$	$\dfrac{dy}{du}\dfrac{du}{dx}$	$\tan ax$	$\dfrac{a}{\cos^2 ax}$

a is a constant, u and v are function of x.

6. $y = (2x + 4)\left(3x + \dfrac{2}{x^2}\right)$

7. $y = \dfrac{4x^2 + 4}{(x^2 + 3x + 2)}$

8. $y = (2x + 3)^5$

9. $y = \log(2x + 3)$

10. $y = x\log(2x^2 + 3x - 5)$

11. $y = 3e^{2x} + 4e^{-3x} + x^2e^x$

12. $y = \sin 2x + 3\cos 5x - \tan 3x$.

5.4 Elasticity of demand

The quantity demanded of some goods is much more sensitive to changes in price than is the quantity demanded of others and this difference is clearly very important. A measure of sensitivity to price change can be obtained by expressing the percentage change in the quantity demanded of a good in terms of the percentage change in price. For very small changes in price this ratio is the *price elasticity of demand*:

$$E_D = \frac{(dq/q) \times 100}{(dp/p) \times 100} = \frac{dq}{dp} \frac{p}{q}$$

If the demand curve is known in the form $q = f(p)$, then it is a simple matter to differentiate and to substitute the values of dq/dp in the above equation. For example, if

$$q = f(p) = 100 - p - p^2$$

$$\frac{dq}{dp} = -1 - 2p = -(1 + 2p)$$

$$\therefore \quad E_D = -(1 + 2p) \frac{p}{q}$$

If the present price is £5 then the quantity demanded is

$$q = 100 - 5 - 25 = 70$$

and
$$E_D = -(1 + 10) \frac{5}{70}$$

$$= -0.8 \text{ approx.}$$

This means that the demand decreases by approximately 0.8% for a 1% increase in price or, conversely, increases by 0.8% for a 1% decrease in price. It follows that, in this case, a decrease in price results in a decrease in total revenue because

$$\text{total revenue} = \text{price} \times \text{quantity} = pq$$

In this case the demand is said to be *inelastic* at a price of £5.

If, however, the present price is £6,

then
$$q = 100 - 6 - 36 = 58$$

and
$$E_D = -(1 + 12) \frac{6}{58}$$

$$= -1.3 \text{ approx.}$$

Thus a 1% decrease in price results in a 1.3% increase in the quantity demanded and if this is supplied it results in an increase in the total revenue. The demand is then said to be *elastic* at a price of £6.

When the elasticity of demand is unity, a rise or fall in price results in the same percentage decrease or increase in quantity demanded and, therefore, leaves the total revenue the same.

Other elasticities are of importance in economics and these are determined in a similar way to the above, e.g. we define

income elasticity of demand $= \dfrac{\text{percentage change in quantity demanded}}{\text{percentage change in income}}$

which can be expressed mathematically as

$$\frac{(dq/q) \times 100}{(dI/I) \times 100} \quad \text{or} \quad \frac{dq}{dI}\frac{I}{q}$$

To determine this elasticity it is necessary to know the relationship between the quantity demanded of a good and the income, that is,

$$q = f(I)$$

It is, therefore, possible to determine other forms of elasticity if the relationship between the variables is known.

One special case which is of interest in economics occurs when the elasticity is a *constant*. For a demand curve, we know that if total revenue is unchanged when the price changes then the price elasticity is -1. But for total revenue to be a constant we have $pq = c$, where c is a constant. Therefore,

$$q = c/p = cp^{-1}$$

and $\dfrac{dq}{dp} = (-1)cp^{-2}$ so that $E_D = \dfrac{dq}{dp}\dfrac{p}{q} = \dfrac{(-1)cp^{-2}p}{cp^{-1}} = -1$

More generally, if the demand curve is $qp^n = c$ then the price elasticity is $-n$. This can be shown using the same method as above, and alternatively by taking logarithms (to the base e) to give

$$\log q + n \log p = \log c$$

or,

$$\log q = \log c - n \log p$$

Differentiating with respect to p,

$$\frac{1}{q}\frac{dq}{dp} = 0 - \frac{n}{p}$$

and hence $$E_D = \frac{dq}{dq}\frac{p}{q} = -n$$

This method also indicates that if q is raised to a power, for example, $q^m p^n = c$ then taking logarithms gives

$$m \log q + n \log p = \log c$$

and the elasticity is $-n/m$.

5.5 Marginal analysis

We have, in the differential calculus, a method for determining the slope of a function. In terms of the total cost curve example, the slope at any point tells us the marginal or extra cost incurred if the volume of production is increased by a very small amount or, conversely, the cost saved if the volume of production is decreased by a very small amount.

The same principles also apply in other uses of marginal concepts, such as marginal revenue (obtained from a total revenue function), the marginal productivity of labour (obtained from a production function), and the marginal propensity to consume (obtained from a consumption function).

For the linear cost function

$$y = 100 + 3x \qquad \frac{dy}{dx} = 3$$

The marginal cost is equal to 3 and is independent of the volume of production.

This can be checked by calculating the cost at any two levels of production which are one unit apart. For example,

when	$x = 3,$	$y = 100 + 9 = 109$
	$x = 4,$	$y = 100 + 12 = 112$
or when	$x = 15,$	$y = 100 + 45 = 145$
	$x = 16,$	$y = 100 + 48 = 148$

Both these sets of costs differ by 3 units.

For the quadratic function

$$y = 100 + 2x + \tfrac{1}{10}x^2$$

$$\frac{dy}{dx} = 2 + \tfrac{1}{5}x$$

In this case the marginal cost is a function of the output. For example,

when $\qquad\qquad x = 50, \qquad \dfrac{dy}{dx} = 2 + 10 = 12$

and when $\qquad x = 100, \qquad \dfrac{dy}{dx} = 2 + 20 = 22$

Therefore, the marginal cost is positive and increasing as output increases.

However, if, as is often the case, more efficient use can be made of the variable factors of production as the level of output is increased, then it is reasonable to expect the marginal cost to be positive but to decrease with output. To satisfy this requirement the derivative could be of the form

$$\frac{dy}{dx} = a - bx$$

where a and b are positive constants. Then as x is increased the marginal cost decreases.

Applying the same ideas to total revenue (TR), and using the fact that $TR = pq$ where p and q are related through the demand function, marginal revenue is found by differentiating TR with respect to q. Notice that the demand function allows p to be replaced by a function of q so that TR is expressed solely in terms of q. It is also possible to use the demand function to replace q by a function of p, so that TR depends on price, but when marginal revenue is referred to it is normally in the context of a change in output rather than price. To illustrate the determination of marginal revenue, suppose the demand function is

$$p = 250 - 2q$$

then $\qquad\qquad TR = pq = (250 - q)q = 250q - q^2$

and so $\qquad\qquad MR = \dfrac{dTR}{dq} = 250 - 2q$

Therefore, as q increases, MR falls and if $q > 125$ MR is negative so that as output increases above this level, total revenue declines.

It should be clear that marginal revenue and elasticity of demand are related. Since $TR = pq$ then differentiating gives

$$\frac{dTR}{dq} = p + q\frac{dp}{dq} = p\left(1 + \frac{q}{p}\frac{dp}{dq}\right)$$

But marginal revenue is given by

$$MR = \frac{dTR}{dq}$$

and elasticity of demand

$$E_D = \frac{p}{q}\frac{dq}{dq}$$

and hence

$$MR = p\left(1 + \frac{1}{E_D}\right)$$

When $E_D = -1$, $MR = 0$, that is, revenue is constant for changes in q. When $E_D = \infty$, $MR = p$, that is, when the firm is a competitor in the product market and the demand is perfectly elastic, the marginal revenue equals price.

Bringing together revenue and costs allows net revenue (*NR*) or profits (π), to be defined by

$$NR = \pi = TR - TC$$

Continuing with the previous example, and assuming the cost function

$$TC = 100 + 3q + 0.5q^2$$

then

$$NR = (250q - q^2) - (100 + 3q + 0.5q^2)$$

$$= 247q - 1.5q^2 - 100$$

The marginal *NR* or marginal profit is

$$\frac{dNR}{dq} = 247 - 3q$$

and this is positive when $q < 82.33$.

A production function relates the level of output from a production process to the level of input. Let q be the level of output and L be the number of units of labour used; then, if other inputs such as capital services and energy are ignored the production function is $q = f(L)$. As previously, the marginal product is found by differentiating q with respect to L. For example, suppose that the production function is

$$q = 4L^2 - 20L$$

then the marginal product of labour is

$$\frac{dq}{dL} = 8L - 20$$

and in this example the marginal product is positive when $L > 2.5$.

Our final application of marginal analysis is with the consumption function which relates consumption expenditure to the level of income. For example, a simple consumption function which is commonly used in basic economics is

$$C = a + bY$$

where a and b are constants. The marginal propensity to consume (*mpc*), which is the amount out of a unit of extra income which is spent on consumption, is given by

$$mpc = \frac{dC}{dY} = b$$

and here this is a constant. The average propensity to consume (*apc*) is C/Y and here,

$$apc = \frac{a + bY}{Y} = \frac{a}{Y} + b$$

Therefore, for this linear consumption function, if a is positive the *apc* is greater than the *mpc*.

In this section we have illustrated some of the many uses of marginal analysis in economics. Perhaps the most important application, however, is the determination of maximum and minimum values and this is the next topic we consider.

5.6 Exercises

1. Find the marginal-cost and average-cost function from the following total-cost functions:

   (a) $TC = 4q^3 + 2q^2 - 25q$

   (b) $TC = (q^3 - 3q)(16 + 5q)$

   (c) $TC = 25 + 6qe^{2q}$

(d) $TC = (3 \log q + 5)(q^2 - q)$

2. Find the elasticity of demand when $p = 10$ for the following equations:

   (a) $p + 2q = 50$

   (b) $qp^2 = 400$

   (c) $q - 35 = -3p$

3. Determine the marginal-revenue function associated with each of the following demand curves (p = price, q = quantity demanded):

   (a) $p = 200 - 3q$ (b) $pq = 100$ (c) $p = 250 - e^{4q}$

4. A firm has a cost function $TC = 2q^2 + 20q + 300$ and its demand function is $4p + q = 260$. Determine (a) the marginal-cost function (b) the marginal-revenue function (c) the marginal-profit function (d) the value of q for which marginal revenue equals marginal cost.

5. The demand function for a product is $p = 177 - q^2$ and the cost function is $TC = 30 + 2q + 2q^2$. Determine the profit function and hence the marginal-profit function. For what level of output is marginal profit zero?

6. If a consumption function is given by $C = a + bY + cY^2$ determine the average propensity to consume and the marginal propensity to consume.

5.7 Maxima and minima

A total-cost function of the form

$$y = 100 + 2x + \tfrac{1}{10}x^2$$

tells us the total cost of manufacture in terms of the total output. The average cost per unit can be obtained by dividing throughout by the quantity produced:

$$\text{average cost per unit} = \frac{\text{total cost}}{\text{quantity}} = \frac{100 + 2x + \tfrac{1}{10}x^2}{x}$$

$$= \frac{100}{x} + 2 + \tfrac{1}{10}x$$

This is an equation showing that the average cost is very high when only a small number of units are made. Initially this value decreases with an

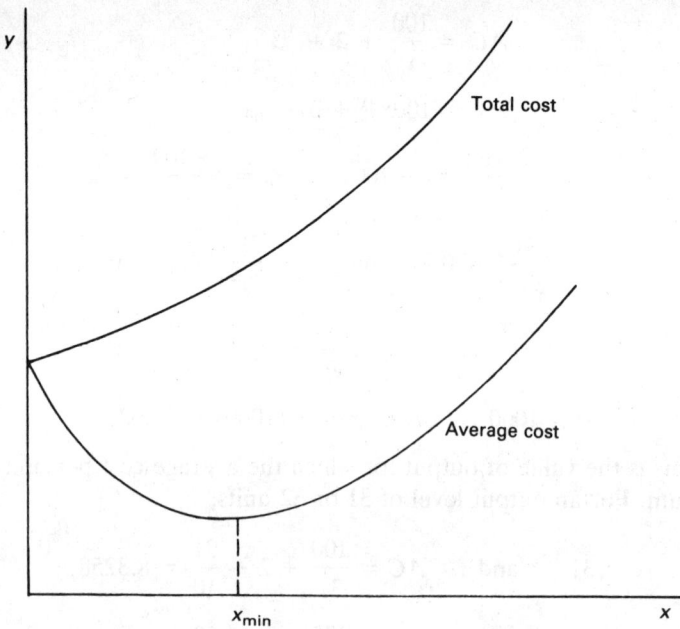

Fig. 5.5

increase in output, but eventually the increasing marginal costs cause the average cost per unit to rise. There is, therefore, some value of output for which the average cost per unit is a minimum.

A rough sketch of the graphs of the functions shows up the situation more clearly (Fig 5.5).

Then for

$$x < x_{min} \qquad \frac{dAC}{dx} < 0 \qquad \text{(i.e. the slope is negative)}$$

$$x > x_{min} \qquad \frac{dAC}{dx} > 0 \qquad \text{(i.e. the slope is positive)}$$

$$x = x_{min} \qquad \frac{dAC}{dx} = 0$$

Therefore, the value of output for which the average cost per unit is a minimum can be found by the differential calculus.

$$AC = \frac{100}{x} + 2 + \tfrac{1}{10}x$$

$$= 100x^{-1} + 2 + \tfrac{1}{10}x$$

$$\therefore \quad \frac{d(AC)}{dx} = -100x^{-2} + \tfrac{1}{10} = \frac{-100}{x^2} + \tfrac{1}{10}$$

$$\frac{d(AC)}{dx} = 0 \quad \text{when} \quad -\frac{100}{x^2} + \tfrac{1}{10} = 0$$

that is,
$$\frac{100}{x^2} = \tfrac{1}{10}$$

and $\qquad x^2 = 1000 \qquad$ or $\qquad x = \sqrt{(1000)} = 31.623$

This is the value of output for which the average cost per unit is a minimum. For an output level of 31 or 32 units,

$$x = 31 \quad \text{and} \quad AC = \frac{100}{31} + 2 + \frac{31}{10} = 8.3258$$

$$x = 32 \quad \text{and} \quad AC = \frac{100}{32} + 2 + \frac{32}{10} = 8.3250$$

If it were possible to obtain an output of 31.623 units then the average cost per unit would be

$$AC = \frac{100}{31.623} + 2 + \frac{31.623}{10}$$

$$= 3.1623 + 2 + 3.1623 = 8.3246$$

The shape of the curve is, therefore, fairly flat about the minimum point and for most practical purposes the difference between the average cost per unit at these levels of output would be considered to be insignificant.

It is interesting to note that at the minimum the term $100/x$ is equal in value to the term $x/10$. The former is equal to the share of the fixed costs which must be carried by each unit of output and, therefore, decreases with output. The latter corresponds to that part of the marginal cost function which depends on the quantity produced. One term, therefore, decreases with output and the other increases, and the minimum average cost occurs when the two terms are equal. At this point the average cost per unit is equal to the marginal cost.

$$\text{Total cost, } y = 100 + 2x + \tfrac{1}{10}x^2$$

$$\text{marginal cost, } \frac{dy}{dx} = 2 + \tfrac{1}{5}x$$

For an output of 31.623 units

$$\text{marginal cost} = 2 + \tfrac{1}{5} \times 31.623 = 8.3246$$

which is the value of the minimum average of cost per unit.

This follows from the fact that for values of output less than x_{min} the average cost is falling and this can only happen if the cost of producing each extra unit is less than average cost of producing all previous units; that is, when

$$0 < x < x_{min}, \qquad MC < AC$$

When the average cost is increasing, this is due to the fact that the cost of producing each extra unit is greater than the average cost of producing all previous units; that is, when

$$x > x_{min}, \qquad MC > AC$$

Then at the minimum the two must be equal; that is, when

$$x = x_{min}, \qquad MC = AC$$

In this example the minimum value was determined by setting the derivative equal to zero. For a curve which reaches a maximum value the same sort of arguments apply with the result that:

for $\qquad\qquad x < x_{max}$ the slope is positive

$$x > x_{max} \text{ the slope is negative}$$

$$x = x_{max} \text{ the slope is zero}$$

and at the maximum value the derivative is zero. It is therefore necessary to decide whether a point for which the derivative is zero corresponds to a maximum or a minimum value. For the average cost function the graph shows that there is a minimum value. However, it is not always easy to sketch the graph and it is not necessary since the differential calculus can be used to distinguish between values corresponding to maximum and minimum positions.

To do this it is simplest to differentiate the function a second time

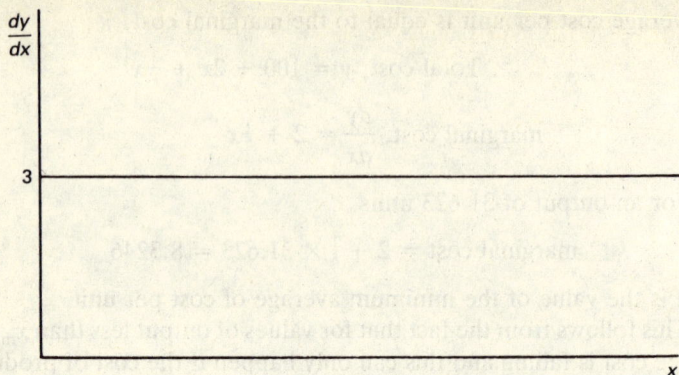

Fig. 5.6

following the same procedure as was used to obtain the first derivative dy/dx. This can best be illustrated by reference to two examples.

In the case of the linear function

$$y = 100 + 3x, \qquad \frac{dy}{dx} = 3$$

This is constant for all values of x and a graph of dy/dx against x is as shown in Fig. 5.6.

The slope of this line is zero at all points. The derivative of dy/dx is therefore zero. If we write y' for dy/dx, then

$$y' = \frac{dy}{dx} = 3 \qquad \text{and} \qquad \frac{dy'}{dx} = 0$$

The derivative of y' is then the second derivative of y and can be written in the alternative forms

$$\frac{dy'}{dx} = \frac{d}{dx}\left(\frac{dy}{dx}\right) = \frac{d^2y}{dx^2}$$

If we let $y'' = d^2y/dx^2$, then the third derivative can be formed in a similar way:

$$\frac{dy''}{dx} = y''' = \frac{d}{dx}\left(\frac{d^2y}{dx^2}\right) = \frac{d^3y}{dx^3}$$

and in general the nth derivative is

$$\frac{d^n y}{dx^n}$$

In the case of the quadratic equation

$$y = a + bx + cx^2$$

then

$$\frac{dy}{dx} = b + 2cx$$

$$\frac{d^2 y}{dx^2} = 2c$$

and $d^3 y/dx^3$ and all higher-order derivatives are equal to zero.

In general a polynomial of degree n, i.e. an equation in which the highest power of the variable is n, has n derivatives not equal to zero and the $(n + 1)$th is always equal to zero.

For the purposes of the analysis we can restrict ourselves to the first two derivatives because they are usually sufficient to determine whether a function has a maximum or minimum value and to distinguish between the two.

For the quadratic

$$y = a + bx + cx^2$$

the slope at any point of the curve is given by

$$\frac{dy}{dx} = b + 2cx$$

This will be positive or negative depending upon the values of b, c and x. Let us consider the two cases of the general quadratic equation over the whole range of possible values of x:

(a) when c is positive and b is negative
(b) when c is negative and b is positive

The general shape of these function is shown in Fig 5.7.

Fig 5.7(a) has a minimum point, i.e. for $x = x_{min}$ the function has a value which is lower than the value for all other values of x.

Fig 5.7(b) has a maximum point, i.e. for $x = x_{max}$ the function has a value which is higher than the value for all other values of x.

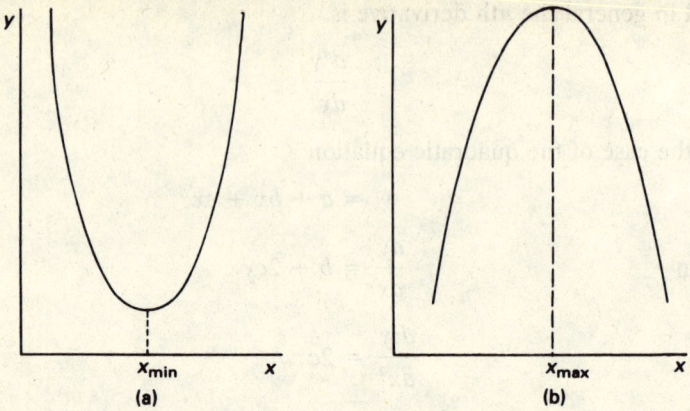

Fig. 5.7

The slope of these curves depends upon the value of x and the sign of the slope depends on the sign of the constant c. This can be seen more clearly if the graph of dy/dx against x is drawn (Fig 5.8).

The slope of the curve at a minimum and a maximum point is zero and therefore the graph of dy/dx crosses the axis of x at this point. But in Fig 5.7 (a) for

$$x < x_{min}, \qquad \frac{dy}{dx} < 0$$

$$x > x_{min}, \qquad \frac{dy}{dx} > 0$$

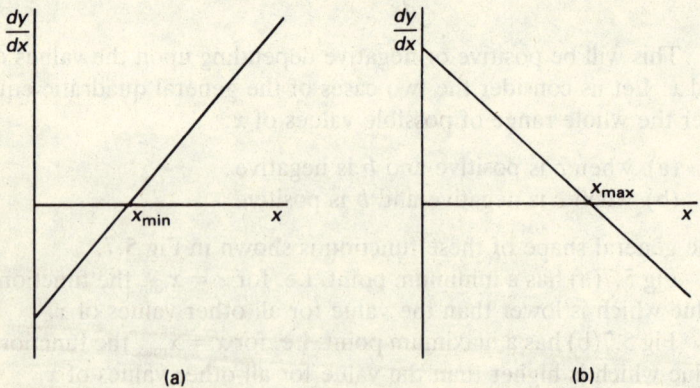

Fig. 5.8

The slope of the graph of dy/dx against x is therefore positive. In Fig 5.7 (b), for

$$x < x_{max}, \qquad \frac{dy}{dx} > 0$$

$$x > x_{max}, \qquad \frac{dy}{dx} < 0$$

Therefore, the slope of dy/dx against x is negative.

The slope of these two graphs can, of course, be determined by differentiating the linear function

$$\frac{dy}{dx} = b + 2cx$$

with the result

$$\frac{d^2y}{dx^2} = 2c$$

The graph of this is illustrated in Fig 5.9 for both signs of the constant c.

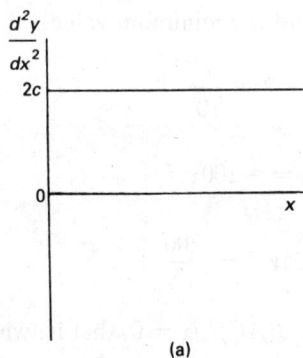

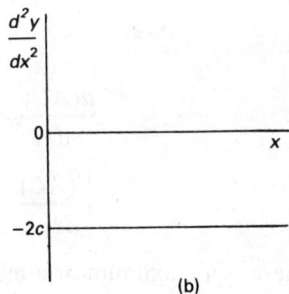

Fig. 5.9

It follows that a quadratic function has a maximum or minimum at the point where

$$\frac{dy}{dx} = 0$$

and that this point is a maximum if d^2y/dx^2 is negative and a minimum if d^2y/dx^2 is positive.

The use of these rules is demonstrated in the following examples:

Example 1

$$y = 100 + 2x - \tfrac{1}{10}x^2$$

$$\frac{dy}{dx} = 2 - \tfrac{1}{5}x$$

and

$$\frac{d^2y}{dx^2} = -\tfrac{1}{5}$$

This is negative and the function has a maximum at the point where

$$\frac{dy}{dx} = 2 - \tfrac{1}{5}x = 0$$

that is, where $x = 10$ and the maximum value of y is given by $y = 100 + 20 - \frac{100}{10} = 110$.

Example 2

Given that total cost $y = 100 + 2x + \tfrac{1}{10}x^2$, find the value of x which minimises the average cost, and hence find the minimum value.

$$\text{Average cost} = AC = \frac{100}{x} + 2 + \frac{x}{10}$$

$$\frac{d(AC)}{dx} = -\frac{100}{x^2} + \tfrac{1}{10} = -100x^{-2} + \tfrac{1}{10}$$

$$\frac{d^2(AC)}{dx^2} = (-100)(-2)x^{-3} = \frac{200}{x^3}$$

There is a maximum or a minimum when $d(AC)/dx = 0$, that is, when

$$\frac{-100}{x^2} + \tfrac{1}{10} = 0$$

or

$$x = 31.623 \text{ (taking the positive root)}$$

At this value of output

$$\frac{d^2(AC)}{dx^2} = \frac{200}{x^3} = \frac{200}{(31.623)^3}$$

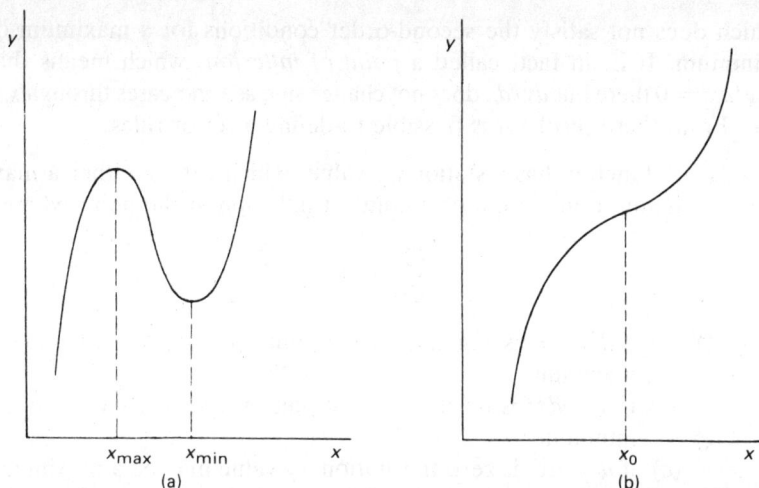

Fig. 5.10

This is positive and therefore the value of $x = 31.623$ gives the minimum value of the average cost function

$$AC = \frac{100}{31.623} + 2 + \frac{31.623}{10} = 8.3246$$

Example 3

The cubic equation

$$y = a + bx + cx^2 + dx^3$$

can have a number of forms depending upon the sign and size of the coefficients of the variable x. The graphs in Fig 5.10 are examples of two of the forms which it might take.

Fig 5.10 (a) has no true maximum or minimum value in terms of our previous definition because y increases indefinitely as x increases and y decreases indefinitely as x decreases. The values of the function corresponding to the value x_{max} and x_{min} are, however, called *local maximum and local minimum*. If we apply the graphical reasoning to this function that was applied to the quadratic function it is easily seen that these two points fulfil the conditions required for maxima and minima.

Fig 5.10 (b) has no local maximum or minimum values but has a value x_0 for which the function has a slope of zero. This, therefore, satisfies the condition $dy/dx = 0$. But it gives the value $d^2y/dx^2 = 0$, which does not satisfy the second-order conditions for a maximum or minimum. It is, in fact, called a *point of inflexion*, which means that $d^2y/dx^2 = 0$ there but dy/dx does not change sign as x increases through x_0.

From these results it is possible to define a set of rules.

(1) A function has a stationary value, which can be either a maximum, a minimum or a point of inflexion at the point where

$$\frac{dy}{dx} = 0$$

(2) (a) if d^2y/dx^2 is negative at that point, the stationary value is a maximum
     (b) if d^2y/dx^2 is positive at that point the stationary value is a minimum
     (c) if d^2y/dx^2 is zero the stationary value may be a maximum, minimum or point of inflexion and the curve should be sketched to examine it.

The condition (1) is frequently referred to as the *necessary* condition for a stationary value. That is, unless this is satisfied neither a maximum, nor minimum nor point of inflexion can occur. This condition is not, however, *sufficient* for a maximum, since it is also satisfied at a minimum or point of inflexion. Conditions (1) and (2) together make up the *necessary and sufficient* conditions for a maximum, minimum or point of inflexion since they include all the possibilities. If a stationary value occurs, it satisfies these conditions. If a value of x occurs which satisfies these conditions it is a stationary value.

5.8   Cost and revenue analysis

The differential calculus enables us to determine the stationary values of a function and to decide whether these values are maxima or minima. This important result is of frequent use in economics as the following example shows:

Example

Let the total cost function be

$$C = 100 + 2q + \tfrac{1}{10}q^2$$

and let the demand function be

$$p = 20 - \tfrac{1}{5}q$$

Then the total revenue $TR = pq$. Substituting the above expression for p in the revenue function gives

$$TR = pq = (20 - \tfrac{1}{5}q)q$$
$$= 20q - \tfrac{1}{5}q^2$$

This is a quadratic equation and it can be sketched roughly by determining a few points on the curve.

1. $TR = 0$ when $20q - \tfrac{1}{5}q^2 = 0$

   that is, when $q(20 - \tfrac{1}{5}q) = 0$

$$q = 0 \text{ or } 100$$

2. $$\frac{d(TR)}{dq} = 20 - \tfrac{2}{5}q$$

   This equals zero when

$$\tfrac{2}{5}q = 20, \qquad q = 50$$

   and at this point

$$TR = 20 \times 50 - \tfrac{1}{5} \times 50 \times 50 = 500$$

To check whether this is a maximum or a minimum it is necessary to obtain the second-order derivative

$$\frac{d^2(TR)}{dq^2} = -\tfrac{2}{5}$$

This is negative, and therefore the maximum total revenue occurs at an output of 50 units.

The revenue function is shown in Fig 5.11, along with the cost function. The vertical distance between the two curves indicates the net revenue, NR, or profit, defined by

net revenue = total revenue − total costs

or $$NR = TR - C$$

This is positive over the range $q = q_1$ to $q = q_2$. It is not, however, necessarily at its maximum at the output where total revenue is at its maximum, because costs are increasing with output. To determine the output where net revenue is a maximum use can be made of the differential calculus:

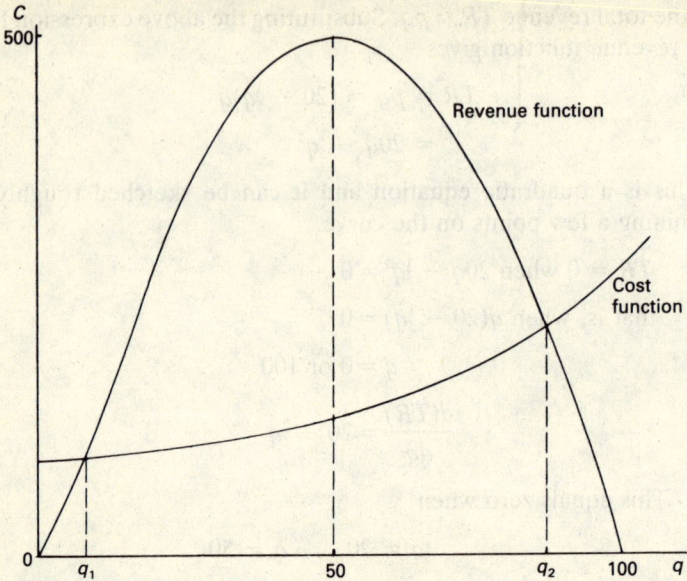

Fig. 5.11

$$NR = TR - C$$
$$= (20q - \tfrac{1}{5}q^2) - (100 + 2q + \tfrac{1}{10}q^2)$$
$$= 18q - \tfrac{3}{10}q^2 - 100$$

$$\therefore \quad \frac{d(NR)}{dq} = 18 - \tfrac{3}{5}q$$

This is zero when

$$\tfrac{3}{5}q = 18 \qquad \text{or} \qquad q = 30$$

At this output

$$NR = 18 \times 30 - \tfrac{3}{10} \times 30 \times 30 - 100 = 170$$

To check whether this is a maximum or a minimum we differentiate a second time with the result

$$\frac{d^2(NR)}{dq^2} = -\tfrac{3}{5}$$

This is negative and, therefore, this net revenue is a maximum. The maximum total revenue occurs when $q = 50$. At this output

$$NR = 18 \times 50 - \tfrac{3}{10} \times 50 \times 50 - 100 = 50$$

This is considerably less than at the reduced output of 30. Finally,

$$\frac{d(TR)}{dq} = 20 - \tfrac{2}{5}q \quad \text{and} \quad \frac{dC}{dq} = 2 + \tfrac{1}{5}q$$

Therefore at an ouput of 30

$$\text{Marginal revenue } = \frac{d(TR)}{dq} = 20 - \tfrac{2}{5} \cdot 30 = 8$$

$$\text{Marginal cost } = \frac{dC}{dq} = 2 + \tfrac{1}{5} \cdot 30 = 8$$

This confirms that maximum net revenue is obtained when

$$\text{marginal revenue} = \text{marginal cost}$$

and $\qquad$ for $q < 30, \qquad MR > MC$

$\qquad$ for $q > 30, \qquad MR < MC$

5.9 Exercises

1. Determine the stationary values of the following functions:

    (a) $y = 3x^2 - 120x + 30$

    (b) $y = 16 - 8x - x^2$

    (c) $y = x^4$

    (d) $y = (x - 5)^3$

2. The total cost of producing an output q is given by

$$C = 500 + 4q + \tfrac{1}{2}q^2$$

Obtain the value of q which minimises the average cost.

3. If the demand function for a monopolist's product is $q + 2p = 10$, determine the price and output required to maximise the total revenue. What is the elasticity of demand at this price?

4. The total cost function for a product is $TC = q^3 + 20q^2 + 20$ and the demand function is $q + p = 240$. Find the price and level of output required to maximise a monopolist's net revenue.

5. Compare the levels of output which (a) maximise total revenue (b) minimise average costs and (c) maximise profit for a monopolist faced with a demand function $2q + 4p = 200$ and the total cost function $TC = 256 + 2q + 2q^2$. Sketch the total revenue, average cost and profit functions.

5.10　Profit maximisation in several markets

A common problem in economics is where a monopolist trades in more than one market. For example, a firm may face different demand curves in the home and overseas markets, or a company such as a national railway network may be able to charge different prices to different groups of customers (e.g. season-ticket holders, students, pensioners, groups, same-day returns). In these circumstances, two types of profit maximisation can occur: with price discrimination and without it. If the firm has all the relevant information it will be able to decide whether price discrimination is worthwhile.

For example, suppose that a national electricity supplier has two types of customer – domestic (d) and industrial (i) – and that the demand functions are

$$5Q_d + P_d = 1000$$

$$P_i + 2.5Q_i = 10000$$

Here the demand from industrial users is higher but declines more steeply as the price increases. The cost of producing electricity is the same for both groups of customers and is given by

$$TC = 100 + 20Q + 0.1Q^2 \qquad \text{where } Q = Q_d + Q_i$$

Initially we will assume that there is price discrimination and that profits are to be maximised. For the domestic market total revenue is

$$TR_d = P_dQ_d = (1000 - 5Q_d)Q_d$$

from using the demand function, where we have chosen to obtain TR_d in terms of Q_d because TC is in terms of Q. It is also possible to use the demand functions to eliminate the quantities and to operate on prices. The total cost is

$$TC = 100 + 20(Q_d + Q_i) + 0.1(Q_d + Q_i)^2$$

where Q_i is supplied to industrial customers. From the point of view of the market for domestic users, Q_i is exogenous and can be regarded as a constant. The profit function for the domestic market is

$$\pi_d = TR_d - TC$$
$$= (1000 - 5Q_d)Q_d - [100 + 20(Q_d + Q_i) + 0.1(Q_d + Q_i)^2]$$
$$= (980 - 0.2Q_i)Q_d - 5.1Q_d^2 - 100 - 20Q_i - 0.1Q_i^2$$

Profit is maximised when the derivative with respect to Q_d is zero, treating Q_i as a constant. Therefore,

$$\frac{d\pi_d}{dQ_d} = 980 - 0.2Q_i - 10.2Q_d$$

and setting this to zero gives

$$Q_d = 96.0784 - 0.0196Q_i$$

That is, the value of Q in the domestic market which maximises profits depends on the value of Q in the industrial market. This is to be expected when the cost function is non-linear since it is the total of Q_d and Q_i which determines TC. We note that the second-order derivative equals -10.2 and so the stationary value is a maximum.

In the industrial market the total revenue is

$$TR_i = P_iQ_i = (10000 - 2.5Q_i)Q_i$$

and profits are given by

$$\pi_i = TR_i - TC$$
$$= (10000 - 2.5Q_i)Q_i - [100 + 20(Q_d + Q_i) + 0.1(Q_d + Q_i)^2]$$
$$= (9980 - 0.2Q_d)Q_i - 2.6Q_i^2 - 100 - 20Q_d - 0.1Q_d^2$$

and

$$\frac{d\pi_i}{dQ_i} = 9980 - 0.2Q_d - 5.2\,Q_i$$

Setting this to zero gives

$$Q_i = 1919.2307 - 0.0385Q_d$$

Again the second-order derivative indicates a maximum value, and the value of Q_i depends on the value of Q_d. We now have two simultaneous equations resulting from maximising profits in each market:

$$Q_d = 96.0784 - 0.0196Q_i$$
$$Q_i = 1919.2307 - 0.0385Q_d$$

and solving these gives $Q_d = 58.5056$ and $Q_i = 1916.9782$. The prices are $P_d = 707.474$ and $P_i = 5207.556$ from the demand functions. The value of Q is 1975.48 giving $TC = 429863.29$. The total revenues are $TR_d = 41391.19$ and $TR_i = 9982771.3$ so that the overall profit is

$$TR_d + TR_i - TC = 9594299$$

Next we assume that there is no discrimination so that prices must be the same in both markets. Let P be the common price. The demand functions are

$$5Q_d + P = 1000 \quad \text{or } Q_d = 200 \ - 0.2P$$

$$P + 2.5Q_i = 10000 \text{ or } Q_i = 4000 - 0.4P$$

and so the total demand is

$$Q = Q_d + Q_i = 4200 - 0.6P$$

and the total revenue function is

$$TR = PQ = (7000 - 1.667Q)Q$$

Profit is

$$\pi = TR - TC = (7000 - 1.667Q)Q - (100 + 20Q + 0.1Q^2)$$
$$= 6980Q - 1.767Q^2 - 100$$

so that

$$\frac{d\pi}{dQ} = 6980 - 3.534Q$$

and setting this to zero gives $Q = 1975.1$ and $\pi = 6892996$ which is a maximum. Notice that the value of Q is approximately the same with and without price discrimination and that profits are higher with price discrimination.

5.11 Maximising tax revenue

In section 1.9 the effects of taxes on the supply of goods was discussed. We now consider the problem faced by the government of choosing the level of a tax so as to maximise revenue. Initially the case of a flat-rate tax of t per unit is considered. In the absence of a tax the demand and supply functions are

$$\text{demand: } q_d = f(p)$$
$$\text{supply: } q_s = g(p)$$

while with the flat-rate tax they are

$$\text{demand: } q_d = f(p)$$
$$\text{supply: } q_s = g(p')$$

where
$$p' = p - t$$

is the price received by the supplier. The total tax revenue is

$$T = tq$$

where q is the equilibrium quantity and T is to be maximised with respect to q.

For example, let the demand and supply functions be

$$\text{demand: } q_d = 194 - 3p$$
$$\text{supply: } q_s = 2p' - 6$$

and
$$p' = p - t$$

The equilibrium is when $q_s = q_d$ so that

$$194 - 3p = 2(p - t) - 6$$

or
$$200 + 2t = 5p$$

Since T is expressed in terms of q the value of p can be substituted into the demand equation to give the equilibrium q as

$$q = 194 - 3(200 + 2t)/5 = 74 - 1.2t$$

and hence
$$T = tq = 74t - 1.2t^2$$

To maximise T, $\dfrac{dT}{dt} = 74 - 2.4t$

and setting this to zero, $t = 30.83$. For this value, $q = 37$, $p = 52.33$ and the price received by the supplier is $p' = 21.5$. The second-order derivative is -2.4 and so the stationary point is a maximum. The maximum tax revenue is $T = 1140.7$. It is easily seen that if $t = 0$ the equilibrium is $p = 40$ and $q = 74$, showing that the flat-rate tax has a big effect on the value of q.

For a percentage tax the approach is similar except that a tax of $100r\%$ changes the supply curve in the example to

$$\text{supply: } q_s = 2p^r - 6$$

where $$p^r = p(1 - r)$$

is the price received by the supplier. The equilibrium is

when $$q_d = 194 - 3p = 2p(1 - r) - 6$$

or $$p = 200/(5 - 2r)$$

The corresponding q is $q = 194 - 600/(5 - 2r)$
Tax revenue is given by

$$T = rpq = \frac{200r}{(5 - 2r)}\left[194 - \frac{600}{(5 - 2r)}\right]$$

$$= \frac{74000r - 77600r^2}{(5 - 2r)^2}$$

To maximise T, we require the value of r for which dT/dr is zero. Treating the expression for T as a quotient,

$$\frac{dT}{dr} = \frac{(5 - 2r)^2(74000 - 155200r) - (74000r - 77600r^2)(-4)(5-2r)}{(5 - 2r)^4}$$

Setting the numerator to zero and simplifying,

$$(5 - 2r)(185 - 388r) + (740r - 776r^2) = 0$$

and the solution is $r = 0.5892$. That this is the maximum can be shown by checking the second-order derivative. The maximum value of T is 1140.9 which occurs when $q = 37$ and $p = 52.33$. The price received by the supplier is $p^r = 21.5$.

The results with these two types of taxes, the flat-rate and the percentage tax, are the same. This is to be expected since for any percentage tax there is an equivalent flat-rate tax. The tax per unit is rp for the percentage tax and t for the flat-rate tax, and the tax revenue functions which are maximised are $T = rpq$ and $T = tq$ respectively.

5.12 Inventory models

One of the costs of a manufacturing business is that of holding stocks or inventories. For example, suppose a firm has an annual demand for N units of a product and that the demand is evenly spread throughout the year. One way of meeting this demand would be to produce N units at the beginning of the year and use them up during the year. This would result in only one lot of set-up costs, but relatively high storage or inventory costs. Another way would be to produce $N/12$ units

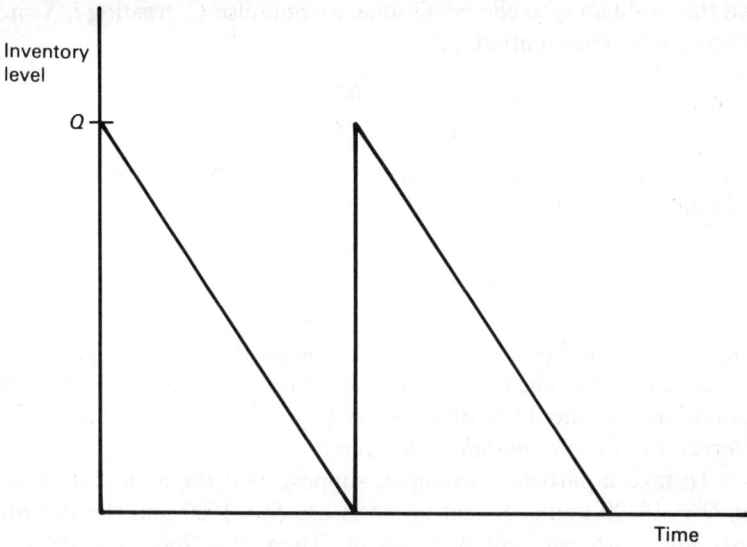

Fig. 5.12

each month. This would reduce inventory costs, since the maximum inventory is now $N/12$ instead of N, but would result in twelve times as high set-up costs.

The problem is to decide how to minimise the total annual costs and so balance the benefits of long production runs against the costs of holding inventories. For simplicity, we will assume that production is instantaneous and that no shortages are allowed. Let the set-up costs be S per batch and Q be the number of items produced in each batch. The total annual production is N so the number of batches per year is N/Q and the annual cost of setting up production is SN/Q. Since production is assumed to be instantaneous, the number of units in stock immediately after production has ceased is Q, and the stock then runs down until it reaches zero and the next lot of production starts. This implies that the average level of inventory is $Q/2$ (see Fig. 5.12) and if the annual cost of having an item in inventory is £i per unit then the total annual inventory costs are $iQ/2$. The total of the set-up costs and inventory costs is

$$C = \frac{SN}{Q} + \frac{iQ}{2}$$

and the problem is to choose Q so as to minimise C, treating i, S and N as constants. Differentiating,

$$\frac{dC}{dQ} = -\frac{SN}{Q^2} + \frac{i}{2}$$

and for a minimum this is zero so that $Q = \sqrt{(2SN/i)}$. To check that this is a minimum,

$$\frac{d^2C}{dQ^2} = \frac{2SN}{Q^3}$$

and since S, N and Q are all expected to be positive we have a minimum. The result is that each batch of production should be $Q = \sqrt{(2SN/i)}$ items and these should be produced N/Q times a year. This value of Q is referred to as the *economic order quantity*.

To take a particular example, suppose that the annual demand is for $N = 10000$ units, the set-up costs are $S = 1000$ and the inventory costs are $i = 5$ per unit per annum. Then $Q = 2000$ and there are $N/Q = 5$ production runs per year, with a total cost of $C = 10000$. Notice that when $Q = 2000$ the total inventory cost is 5000 and so is the total set-up cost. Next we will see what happens when the annual demand changes. Suppose the new value is $N = 2500$, a quarter of the previous value, and the costs are unchanged. Then

$$Q = \sqrt{(2.1000.2500/5)} = 1000$$

which is half of the previous value. Therefore the value of the economic order quantity, Q, is not very sensitive to changes in demand. From the formula for Q it can be seen that the value of Q is proportional to the square root of S and N, and inversely proportional to the square root of i. We will return to a discussion of sensitivity analysis in section 5.14 below.

The various assumptions made above can be relaxed. For example, if the assumption of instantaneous production is dropped and instead it is assumed that production is at the rate of R units per year, the situation is as in Fig. 5.13. If production occurs for the period t_1 (measured as a fraction of a year) then the number of units produced is Q and the demand during this period is t_1N. Notice that $t_1 = Q/R$. The resulting maximum inventory at the end of t_1 is $Q - t_1N$ or $Q(1 - N/R)$. This now declines at a rate N for a period t_2 until the inventory becomes zero and production starts again. The average level of the inventory is again half

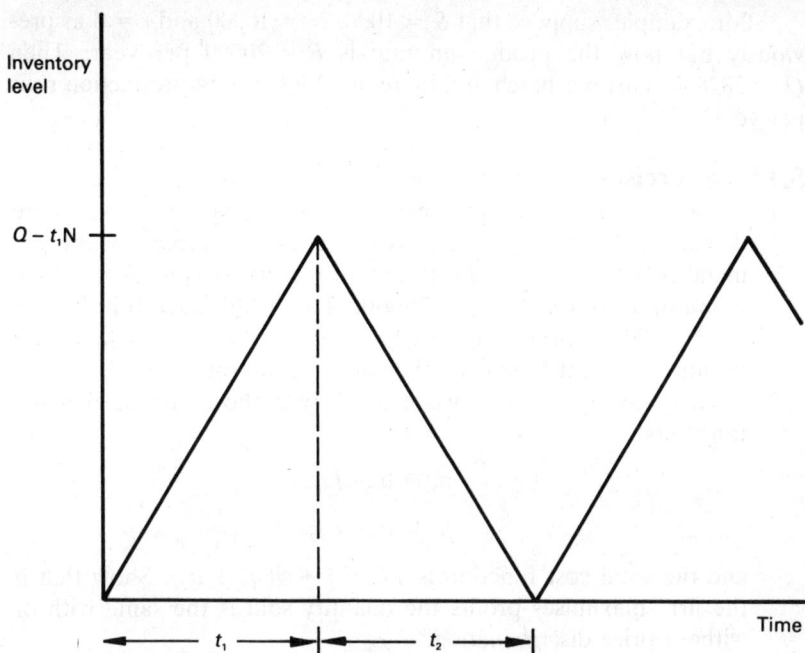

Fig. 5.13

of the maximum (because additions and withdrawals from the inventory take place at a steady rate) so that the total annual inventory cost is

$$IC = 0.5iQ(1 - N/R)$$

The total annual cost, C, is IC plus the set-up cost and so

$$C = 0.5iQ(1 - N/R) + SN/Q$$

This is to be minimised, treating everything except Q as a constant.

$$\frac{dC}{dQ} = 0.5i(1 - N/R) - SN/Q^2$$

Setting this to zero and solving,

$$Q^2 = \frac{2SNR}{i(R - N)}$$

It is easily shown that the second-order condition for a minimum is satisfied.

For example, suppose that $S = 1000$, $N = 10000$ and $i = 5$ as previously but now the production rate is $R = 20000$ per year. Then $Q = 2828.4$ items per batch and there are $N/Q = 3.54$ production runs per year.

5.13   Exercises

1. A car firm sells in three markets: to its own workers, where demand is $0.1P_1 + Q_1 = 50$, to the domestic market, where demand is $0.2P_2 + Q_2 = 20000$, and to the overseas market, where demand is $0.4P_3 + Q_3 = 25000$. The total cost function is $TC = 1000 + 15(Q_1 + Q_2 + Q_3)$. The firm wishes to maximise profits. Should it follow a policy of price discrimination?

2. A company operates in two markets with the following demand functions:

$$p_1 = a - bq_1$$

$$p_2 = c - eq_2$$

   and the total cost function is $TC = f + g(q_1 + q_2)$. Show that if the firm maximises profits the quantity sold is the same with or without price discrimination.

3. The demand for a product is given by $p + 2q = 250$ and the supply by $p - 4q = 100$.
   (a) If a flat-rate tax is imposed on each unit sold what is the maximum possible tax revenue?
   (b) Show that if a percentage tax is imposed on each unit sold the same maximum possible tax revenue is obtained.

4. If the market for a product is given by

$$\text{demand: } p = 110 - 3q^2$$

$$\text{supply: } p = 10 + q^2$$

   determine the equilibrium values of p and q,
   (a) in the absence of any taxes
   (b) if a flat-rate tax, chosen to maximise tax revenue, is imposed on each item sold.

5. An electronics company is under contract to supply 20,000 radios per annum at a uniform rate to a chain-store. The annual inventory cost per radio is £5 and the set-up cost for a production run is £2,000. Assuming production is instantaneous and shortages are

not permitted, how many radios should be produced in each run so as to minimise the total annual cost?

6. If the circumstances in Question 5 change so that the rate of production is 50,000 per annum, what effect does this have on the size of the optimal production run?

7. A manufacturer of business stationery has agreed to supply, at a constant rate, 3,000 boxes of A4 headed notepaper each year to a local company. The cost of storing a box is £3 per year and the cost of setting up the printer is £500. Once set up, production is at a constant rate of 9,000 boxes per year. If no shortages are permitted, how many boxes should be produced in each run in order to minimise the total annual cost?

5.14 Differentials

The derivative dy/dx represents the rate of change of y with respect to x for very small changes in x. It is defined as the *ratio*

$$\frac{dy}{dx} = \lim_{\delta x \to 0}\left(\frac{\delta y}{\delta x}\right)$$

where δy and δx are small but finite changes in y and x. The relationship between these is shown in Fig. 5.14

In fact,
$$\frac{\delta y}{\delta x} = \frac{dy}{dx} + \epsilon$$

or
$$\delta y = \left(\frac{dy}{dx}\right)\delta x + \epsilon\,\delta x$$

where $\epsilon \to 0$ as Q moves close to P. When this occurs, $\delta y \to dy$ $\delta x \to dx$ and $\epsilon\,\delta x \to 0$ so that

$$dy = \left(\frac{dy}{dx}\right)dx$$

We define dy as the *differential* of y and dx the differential of x. It follows that

$$\frac{dy}{dx} = \frac{\text{differential of } y}{\text{differential of } x} = \lim_{\delta x \to 0}\left(\frac{\delta y}{\delta x}\right)$$

Using this notation it is possible to calculate the change which will occur

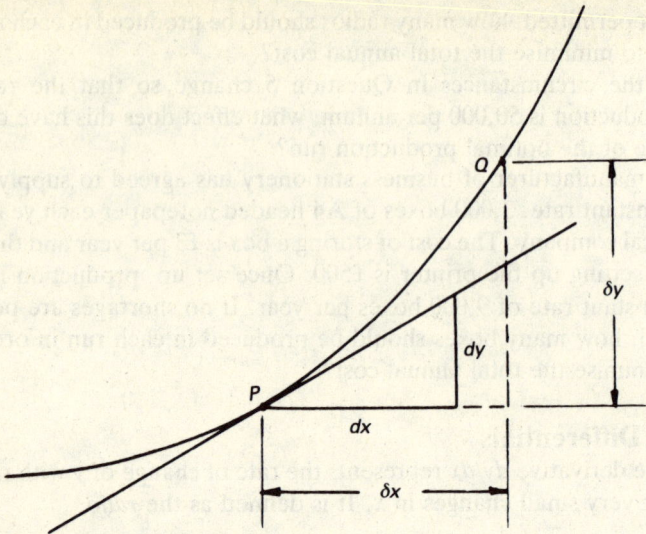

Fig. 5.14

in y for a small change in x when y is a function of x. For example if $y = 3x^2 + 2x + 1$, then $dy/dx = 6x + 2$ and $dy = (6x + 2)\ dx$ which means that the change in y which occurs due to a small change in x depends upon the value of x at which the change takes place.

If $x = 10$ then $dy = [(6 \times 10) + 2)]\ dx = 62\ dx$

If $x = 1$ then $dy = (6 + 2)\ dx = 8\ dx$

Therefore, the change in y resulting from a small change in x becomes larger as x increases in value.

Notice that the value given by the differential is an approximation which is close for small values of dx. Thus in the example, when $x = 1$, $y = 6$ and if the change in x is $dx = 0.1$, the approximate change in y is $dy = 0.8$ while the accurate value of y for $x = 1.1$ is 6.83 so that the actual change in y is 0.83.

Differentials can be used in what is known as *sensitivity analysis*, in which the importance of small changes in values of constants or variables is of interest. For example, in the simple inventory model discussed in section 5.12 above, the economic order quantity, Q, is given by

$$Q = \sqrt{\frac{2SN}{i}}$$

where S is the set-up cost, i is the inventory cost and N is the annual demand. Suppose that we have $S = 1000$, $N = 10000$, $i = 5$ so that $Q = 2000$ and we wish to see how sensitive the value of Q is to these initial values. To differentiate this expression it is convenient to take logarithms (to the base e) so that

$$\log Q = 0.5\log2 + 0.5\log S + 0.5\log N - 0.5\log i$$

and then

$$\frac{1}{Q}\frac{dQ}{dS} = \frac{0.5}{S} \quad \text{so } dQ = \frac{0.5Q}{S}dS$$

This gives the approximate change in Q resulting from a change (or error) of dS in S. For example, suppose that $dS = 100$, then $dQ = 100$ so that an error of 100 in the set-up cost changes the optimal batch size by 100, approx. Similarly,

$$dQ = \frac{0.5Q}{N}dN \quad \text{and} \quad dQ = -\frac{0.5Q}{i}di$$

and if $dN = 100$, $dQ = 10$ approx., while if $di = 1$, $dQ = -200$ approx. Here it is clear that errors in S and i can have a big effect on the value of Q.

5.15 Taylor's and MacLaurin's theorems

These are two theorems which have many important applications in advanced economic theory. Here they will be used to obtain some important series expansions. Taylor's theorem states that, for a general function f,

$$f(a+x) = f(a)+xf'(a) + \frac{x^2}{2!}f''(a) + \cdots + \frac{x^n}{n!}f^{(n)}(a) + \cdots$$

if the series converges, where

$f(a)$ is the value of the function $f(a + x)$ when $x = 0$,

$f'(a)$ is the value of the $f'(a + x)$ when $x = 0$,

$f^{(n)}(a)$ is the value of the nth derivative of $f(a + x)$ when $x = 0$

and provided the derivatives are finite and continuous. This theorem allows any general function to be approximated by a polynomial. If the terms in the expansion decrease quickly, the polynomial may be of low order, such as a quadratic or linear one.

To show that Taylor's theorem is valid suppose that

$$f(x) = 3 + 2x + x^2$$

which has

$$f'(x) = 2 + 2x$$

$$f''(x) = 2$$

$$f'''(x) = 0$$

and we let $a = 1$ so that

$$f(1 + x) = f(1) + xf'(1) + \frac{x^2}{2!} f''(1)$$

Further terms can be ignored since they are zero. By putting $x = 1$ we get $f(1) = 6$, $f'(1) = 4$, $f''(1) = 2$ and so

$$f(1 + x) = 6 + 4x + \frac{2x^2}{2} = 6 + 4x + x^2$$

which is the same as $f(x)$ with x replaced by $1 + x$.

As an illustration of the use of Taylor's theorem let us determine the series expansion of $\log(1 + x)$, where logarithms to the base e are used.

Let

$$f(a + x) = \log (1 + x)$$

Then

$$f(a) = \log (1 + 0) = 0$$

$$f'(a + x) = \frac{1}{1 + x} \qquad \text{and} \qquad f'(a) = 1$$

$$f''(a + x) = \frac{-1}{(1 + x)^2} \qquad \text{and} \qquad f''(a) = -1$$

$$f'''(a + x) = \frac{2!}{(1 + x)^3} \qquad \text{and} \qquad f'''(a) = 2!$$

$$f^{(4)}(a + x) = \frac{-3!}{(1 + x)^4} \qquad \text{and} \qquad f^{(4)}(a) = -3!$$

so that

$$\log (1 + x) = 0 + x - \frac{x^2}{2!} + \frac{x^3}{3!}(2!) + \frac{x^4}{4!} (-3!) + \cdots$$

$$= x - \frac{x^2}{2} + \frac{x^3}{3} - \frac{x^4}{4} + \frac{x^5}{5} + \cdots$$

which is an infinite series. It can be shown that this series converges provided $-1 < x < +1$.

MacLaurin's theorem can be regarded as the special case of Taylor's theorem which occurs when $a = 0$. It states

$$f(x) = f(0) + xf'(0) + \frac{x^2}{2!} f''(0) + \cdots + \frac{x^n}{n!} f^{(n)}(0) + \cdots$$

if the series converges. The derivatives are evaluated at $x = 0$.

Example 1

$$f(x) = \sin x \qquad f(0) = \sin 0 = 0$$
$$f'(x) = \cos x \qquad f'(0) = \cos 0 = 1$$
$$f''(x) = -\sin x \qquad f''(0) = -\sin 0 = 0$$
$$f'''(x) = -\cos x \qquad f'''(0) = -\cos 0 = -1$$

This process can be repeated indefinitely and an infinite series of terms is obtained of which all the odd terms are zero and the even terms are alternately $+1$ and -1. Applying MacLaurin's theorem, we obtain

$$f(x) = \sin x = 0 + x(1) + \frac{x^2}{2!} (0) + \frac{x^3}{3!} (-1) + \cdots$$

$$= x - \frac{x^3}{3!} + \frac{x^5}{5!} - \frac{x^7}{7!} + \cdots$$

Sin x is known as an odd function of x because its expansion contains only odd powers of x.

MacLaurin's theorem can also be used to expand the functions $\cos x$ and e^x

Example 2

$$f(x) = \cos x \qquad f(0) = \cos 0 = 1$$
$$f'(x) = -\sin x \qquad f'(0) = -\sin 0 = 0$$
$$f''(x) = -\cos x \qquad f''(0) = -\cos 0 = -1$$
$$f'''(x) = \sin x \qquad f'''(0) = \sin 0 = 0$$

Extending this process further and substituting the results in Mac-Laurin's expansion gives

$$\cos x = 1 - \frac{x^2}{2!} + \frac{x^4}{4!} - \frac{x^6}{6!} + \cdots$$

Cos x is known as an even function of x because its expansion contains only even powers of x.

Example 3

$$f(x) = e^x \qquad f(0) = e^0 = 1$$
$$f'(x) = e^x \qquad f'(0) = e^0 = 1$$
$$f''(x) = e^x \qquad f''(0) = e^0 = 1$$
$$f'''(x) = e^x \qquad f'''(0) = e^0 = 1$$

which can be continued to give

$$e^x = 1 + x + \frac{x^2}{2!} + \frac{x^3}{3!} + \cdots$$

5.16 Euler relations

These arise from the three series which were obtained using Mac-Laurin's theorem. Since

$$e^x = 1 + x + \frac{x^2}{2!} + \frac{x^3}{3!} + \frac{x^4}{4!} + \cdots$$

if the index is multiplied by the imaginary number i (see section 3.3) and the function expanded as an infinite series, we obtain

$$e^{ix} = 1 + (ix) + \frac{(ix)^2}{2!} + \frac{(ix)^3}{3!} + \frac{(ix)^4}{4!} + \cdots$$

and remembering that $i = \sqrt{(-1)}$, $i^2 = -1$, $i^3 = -i$, $i^4 = 1$, etc, we have

$$e^{ix} = 1 + ix - \frac{x^2}{2!} - \frac{ix^3}{3!} + \frac{x^4}{4!} + \cdots$$

This can be separated into real and imaginary parts as follows:

$$e^{ix} = \left(1 - \frac{x^2}{2!} + \frac{x^4}{4!} - \cdots\right) + i\left(x - \frac{x^3}{3!} + \frac{x^5}{5!} - \cdots\right)$$

$$= \cos x + i \sin x.$$

By similar reasoning it can be shown that

$$e^{-ix} = \cos x - i \sin x.$$

These relationships are important and will be found extremely useful in the treatment of certain differential equations (see section 9.5).

5.17 Newton's method

It is possible to use the differential calculus to determine the real roots of any polynomial in x. This is particularly useful in the case of cubic and higher-order equations when the roots are not integer values.

For example let the cubic equation be

$$f(x) = x^3 + 2x^2 + x - 5 = 0$$

We can determine a very approximate value for a root to this equation by substituting various integer values for the variable. For example,

$$\text{let } x = 0 \quad \text{then} \quad f(x) = 0 + 0 + 0 - 5 = -5$$
$$x = 1 \quad\quad\quad f(x) = 1 + 2 + 1 - 5 = -1$$
$$x = 2 \quad\quad\quad f(x) = 8 + 8 + 2 - 5 = 13$$

With this information we can see that the function changes sign between the values $x = 1$ and $x = 2$. It follows that at least one root of the cubic equation lies between these two values. One of these values is taken as a starting point from which it is possible to proceed in a series of steps to the actual solution. The procedure can be stated, in general, as follows.

Let a be one of the two consecutive integer values between which the function changes sign. Then the true root of the equation is $a + h$, where h is an unknown whose absolute value lies between 0 and 1. If a is the lower of the two integers then h is positive whereas if a is the higher integer h is negative.

Taylor's theorem states

$$f(a + h) = f(a) + hf'(a) + \frac{h^2}{2!} f''(a) + \cdots$$

If h is small then approximately

$$f(a + h) = f(a) + hf'(a)$$

Now for this to be a root, $f(a + h) = 0$ and hence

$$h_1 = \frac{-f(a)}{f'(a)} \tag{1}$$

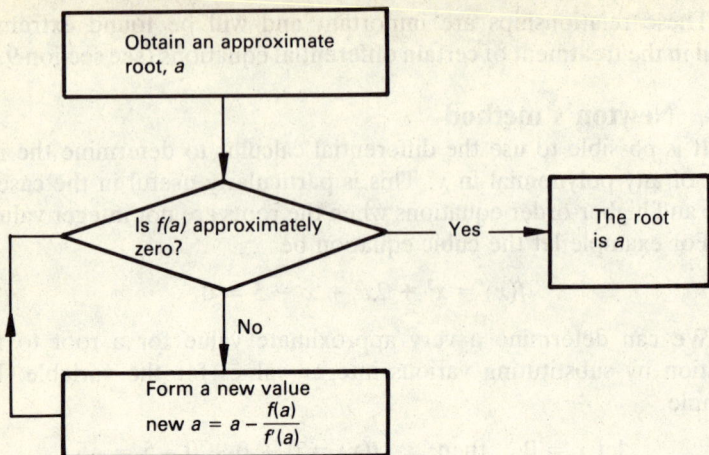

Fig. 5.15

is close to h. Thus a first approximation to the root $a + h$ is given by $a + h_1$ where a is a known value and h_1 is $-f(a)/f'(a)$.

This suggests an iterative (or repeated) strategy for obtaining a root of an equation:

1. Obtain an approximate root, a.
2. Evaluate $f(a)$ and if $f(a)$ is close enough to zero to be acceptable, stop the procedure. Otherwise continue.
3. Obtain a better approximation to the root by using

$$a + h_1 = a - \frac{f(a)}{f'(a)}$$ and go back to step 2.

This procedure is summarised in the flow diagram (see Fig. 5.15) where, starting at the top with the first step of obtaining an approximate root, we follow the arrow to the second step, represented by the diamond, indicating that a decision is needed. If the answer to 'is $f(a) = 0$?' is yes the process stops. If the answer is no, we move to the lowest box (for step 3) and form a new approximation, the new value of a, and move to step 2.

In the example we are considering

$$f(x) = x^3 + 2x^2 + x - 5 \quad \text{and} \quad a = 1$$
$$\therefore \quad f(a) = f(1) = 1 + 2 + 1 - 5 = -1$$

Also $f'(x) = \dfrac{d}{dx}\{f(x)\} = 3x^2 + 4x + 1$

$\therefore\ f'(a) = f(1) = 3 + 4 + 1 = 8.$

Substituting these values in (1)

$$h_1 = \frac{-f(1)}{f'(1)} = \frac{-(-1)}{8} = \frac{1}{8} = 0.125$$

At this stage in the procedure a better approximation to the root of the cubic equation is given by $(a + h_1)$, that is $(1 + 0.125) = 1.125$. Let us call this value a_1). Then we can approach even closer to the true root by letting this be denoted by $(a_1 + h_2)$ where

$$f(a_1 + h_2) = f(a_1) + h_2 f'(a_1) = 0$$

or $$h_2 = \frac{-f(a_1)}{f'(a_1)}$$

As before $f(x) = x^3 + 2x^2 + x - 5$

$\therefore\ f(a_1) = f(1.125) = (1.125)^3 + 2(1.125)^2 + (1.125) - 5 = 0.0801$

and $f'(a_1) = f'(1.125) = 3(1.125)^2 + 4(1.125) + 1 = 9.2969$

$$\therefore\ h_2 = \frac{-f(1.125)}{f'(1.125)} = -\frac{0.0801}{9.2969} = -0.0086$$

A closer approximation to the root of the cubic equation is given by $(a_1 + h_2) = (1.125 - 0.0086) = 1.1164$, for which $f(1.1164) = 0.0005$.

This procedure can be repeated an indefinite number of times and successive estimates of the root of the equation will oscillate about the true value with ever decreasing amplitude. It can be carried out on a computer because of its repetitive stepwise approach. The roots of any polynomial in x for which the first and second derivatives exist (i.e. for any continuous function of x) can be obtained by this method.

As another example let

$$f(x) = x\log x - 4 = 0$$

To find an approximate root, since logarithms are not defined for negative numbers, we try positive values of x. When $x = 1$, $f(1) = -4$, for $x = 2$, $f(2) = -2.6$ and for $x = 3$, $f(3) = -0.70$. Therefore there is a root near $x = 3$ and this can be used as the starting value for Newton's method. Since

$$f(x) = x\log x - 4$$
$$f'(x) = \log x + x/x = \log x + 1$$

When $a = 3$, $f(a) = -0.70$

$$f'(a) = \log 3 + 1 = 2.0986$$

and so
$$h_1 = \frac{-f(a)}{f'(a)} = \frac{0.70}{2.0986} = 0.3336$$

The new approximation to the root is $3 + h_1 = 3.3336$

and
$$f(3.3336) = 0.0138$$
$$f'(3.3336) = 2.2041$$

giving $h_2 = \dfrac{-f(a)}{f'(a)} = \dfrac{-0.0138}{2.2041} = -0.00626.$

The new approximation to the root is $3.3336 - 0.00626 = 3.32734$ for which $f(3.32734) = 0.00004$. This is close to zero and so we will take 3.32734 as the root. A closer approximation could be found by repeating the process.

In our discussion of Newton's method we have used the first two terms of Taylor's theorem and assumed that the other terms are zero. Another approach is to include the third term so that, approximately,

$$f(a + h) = f(a) + hf'(a) + \frac{h^2}{2} f''(a)$$

where a is the first approximate root and h is the adjustment to improve it. Setting this to zero and rearranging gives the solution for h, h_1, as

$$h_1 = - \frac{f(a)}{f'(a) + 0.5h_1 f''(a)} \qquad (2)$$

so that the new approximate root is $(a + h_1)$. Notice that comparing (2) with (1), there is an extra term in the denominator, and this includes h_1. Since h_1 is unknown we can take the approximate value given by (1), and substituting this into (2) in the denominator the following formula results

$$\frac{1}{h_1} = \frac{-f'(a)}{f(a)} + \frac{f''(a)}{2f'(a)} \qquad (3)$$

To demonstrate the effect of taking account of the third term, in the previous example

$$f(x) = x\log x - 4$$
$$f'(x) = \log x + 1$$
$$f''(x) = 1/x$$

and so taking $a = 3$ as the first approximate root,

$$f(3) = -0.70$$
$$f'(3) = 2.0986$$
$$f''(3) = 0.3333$$

and so $\dfrac{1}{h_1} = \dfrac{-2.0986}{-0.70} + \dfrac{0.3333}{2(2.0986)} = 3.0774$

Therefore, from using (3), $h_1 = 0.3249$, the new approximation to the root is 3.3249 and $f(3.3249) = -0.0053$ compared with $f(3.3336) = 0.0138$ from using (1). Using (3) has resulted in a better approximate root after one iteration. Whether the extra complication of using (3) is worthwhile depends on how easy it is to evaluate $f''(x)$.

5.18 Plotting graphs

The use of differential calculus to determine the stationary values of a function can make the plotting of graphs easier. For any function the important characteristics are first, where the graph cuts axes and secondly, where the stationary values are and whether they are local maxima, local minima or points of inflexion. If y is a function of x, the graph cuts the x-axis when $y = 0$ and putting this value into the equation results in an equation in the one unknown, x. If this equation is a quadratic it can be solved as shown in Chapter 3. For higher-order polynomials and other functions finding the solutions can be difficult and Newton's method may be needed.

For example, to sketch the graph of

$$y = 2x^2 + 3x - 14$$

the points where it cuts the x-axis are when $y = 0$ and, solving the quadratic, $x = 2$ and $x = -3.5$. It cuts the y-axis when $x = 0$ and $y = -14$. The stationary values are when the derivative is zero or when

$$\frac{dy}{dx} = 4x + 3 = 0 \text{ so } x = -0.75$$

The second-order derivative is 4 and so $x = -0.75$ is a local minimum. There are no other stationary values and so all that is needed is a few points to give the general shape of the graph. The particular values already found are used (i.e. $x = 2, -3.5, 0, -0.75$) plus a few other values to give a wider picture. Here we include two extra values, for $x = -5$ and $x = 5$:

x	-5	-3.5	-0.75	0	2	5
y	21	0	-15.125	-14	0	51

The graph is plotted in Fig. 5.16.

As a further example consider plotting the graph of

$$y = 2x^3 + 0.2x^2 - 19.3x + 15.75$$

When $x = 0$, $y = 15.75$. When $y = 0$,

$$f(x) = 2x^3 + 0.2x^2 - 19.3x + 15.75 = 0$$

To solve this we try to find an approximate root. When $x = 1$, $f(x) = -1.35$ so that the change of sign between $f(0) = 15.75$ and $f(1)$ indicates there is at least one root between $x = 0$ and $x = 1$, and the values of $f(x)$ indicate the root is near $x = 1$. Taking $x = 1$ as an approximate root and using Newton's method,

$$f'(x) = 6x^2 + 0.4x - 19.3$$

and

$$f'(1) = -12.9$$

so

$$h = -\frac{f(1)}{f'(1)} = -\frac{1.35}{12.9} = -0.1047$$

The new approximate root is $1 - 0.1047 = 0.8953$ and $f(0.8953) = 0.0663$. Using Newton's method again, $f'(0.8953) = -14.1325$ and so

$$h = \frac{-0.0663}{-14.1325} = 0.0046$$

The new approximate root is 0.8999 and $f(0.8999) = 0.0014$. Now $f'(0.8999) = -14.0811$ and so $h = 0.0001$. The new approximate root is

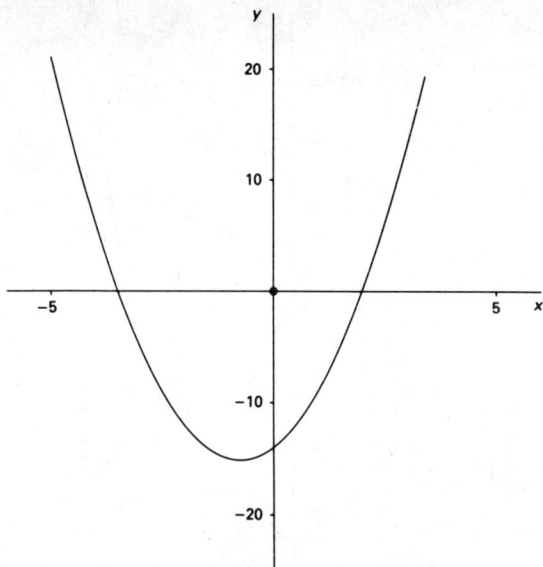

Fig. 5.16

0.9 and $f(0.9) = 0.0$. The factor $(x - 0.9)$ can now be removed from the cubic equation by long division:

$$
\begin{array}{r}
2x^2 + 2x - 17.5 \\
x - 0.9\overline{\smash{\big)}\, 2x^3 + 0.2x^2 - 19.3x + 15.75} \\
\underline{2x^3 - 1.8x^2} \\
2.0x^2 - 19.3x \\
\underline{2.0x^2 - 1.8x} \\
-17.5x + 15.75 \\
\underline{-17.5x + 15.75} \\
0
\end{array}
$$

Therefore, $f(x) = (x - 0.9)(2x^2 + 2x - 17.5) = 0$
The roots of the quadratic are 2.5 and −3.5.

The stationary points are when the derivative is zero. Here,

$$\frac{dy}{dx} = 6x^2 + 0.4x - 19.3$$

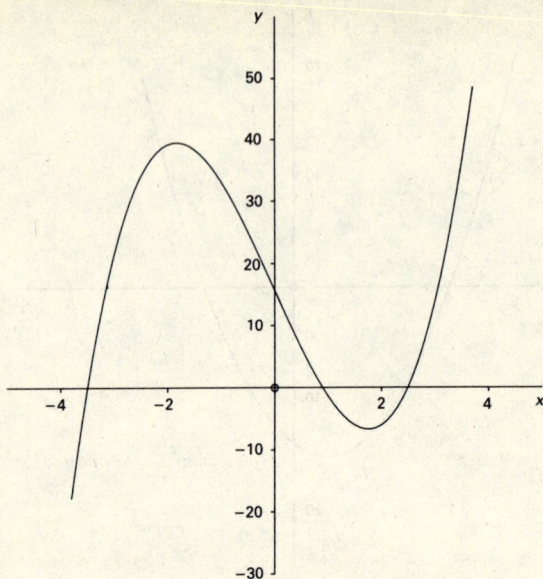

Fig. 5.17

and setting this to zero gives $x = 1.76$ and $x = -1.83$. The second-order derivative is

$$\frac{d^2y}{dx^2} = 12x + 0.4$$

and so $x = 1.76$ is a local minimum and $x = -1.83$ is a local maximum. The table of values is:

x	-3.5	-1.83	0	0.9	1.76	2.5
y	0	39.48	15.75	0	-6.69	0

The graph is shown in Fig. 5.17.

5.19 Exercises

1. Use Taylor's theorem to expand $(1 + x)^4$ and hence show that $1.5^4 = 5.0625$.
2. Use MacLaurin's theorem to expand $(1 + x)^{-1}$ as a power series.
3. Use Newton's method to obtain a value of x which satisfies the

following equations using the given approximate root as a starting point.

(a) $x^2 + 4.55x - 8.70 = 0$

Approximate root $x = 1.5$

(b) $x^3 - 2.34x^2 + 2x - 4.68 = 0$

Approximate root $x = 2.25$

(c) $2x^3 - 8.2x^2 - 3x + 12.3 = 0$

Approximate root $x = 4$

4. What rate of interest makes the present values of the following two projects equal?

Return at end of year	1	2	3	4
Project A	70	80	100	100
Project B	60	90	80	130

(*Hint*: If i is the rate of interest and $R = 1 + i$ then the cubic equation in R can be solved using $R = 1$ as an approximate root.)

5. Plot the graphs of the following functions:

(a) $y = x^2 + x + 12$

(b) $y = x^3 - x^2$

(c) $y = x^3 - 3x^2 - 24x + 26$

(d) $y = x^5$

5.20 Revision exercises for Chapter 5 (without answers)

1. Differentiate

(a) $y = 6x^4 - 3x^2 + (9/x^3)$

(b) $y = (3x - 4)(5x^2 - 2x + 20)$

(c) $y = (2x^3 + x^{-1} + 7)(4x^2 - x^3)$

(d) $y = (x^4 + 3x)/(7x^2 + 4x + 4)$

(e) $y = (3x^2 - x^3 + 5)^3$

(f) $y = \log (4x^5 + 3x^3 - 2x)$

(g) $y = 4e^{2x} - 3e^{-3x}$

(h) $y = e^{4x} \log (x^2 - 6x^4)$

(i) $y = e^{3x} \sin 2x - e^{-3x} \cos 2x$.

2. For the following demand equations, where p = price, q = quantity, determine the marginal revenue functions and the elasticity of demand when $p = 5$:

(a) $3p = 200 - 5q$

(b) $p^2 = 350 - 13q$

(c) $p = 500 - 3e^{2q}$.

3. Find the stationary values of

(a) $y = 4x^2 - 16x + 25$

(b) $y = (2x - 5)^3$

(c) $y = x^6$.

4. A monopolist has a demand function $p = 400 - 5q$ and the total cost function is $TC = 50 + 6q + q^2$. Determine the level of output (q) which gives
(a) the maximum value of total revenue
(b) the minimum value of average costs
(c) the maximum value of profit.
Compare the values of profit for each of these.

5. A whisky producer can sell on the home market, for which the demand function is $p^h = 1000 - 4q^h$, and on the overseas market, for which the demand function is $2p^o = 1500 - 9q^o$. The total cost function is $TC = 300 + 4q^h + 5q^o$. Should the producer charge different prices in the two markets?

6. The demand function for a product is $p = a - bq^d$ and the supply function is $p = c + eq^s$, where a, b, c, e are all positive. Show that imposing a flat-rate tax of t per unit sold reduces the equilibrium quantity and also reduces the equilibrium price. What value of t maximises the tax revenue?

7. A company produces components on a machine with set-up costs of £500 and the cost of storing the components is £4 per year. Production is at the constant rate of 12,000 per year. The only customer purchases 6,000 components at a uniform rate each

year. How many components should be produced in each run in order to minimise the total annual cost?

8. Use MacLaurin's theorem to expand $(1 - x)^{-1}$ as a power series.

9. Use Newton's method to find a root of $4x^4 - 17x^2 + 0.2 = 0$ starting with $x = 0$ as an approximate root.

10. Plot the graphs of

   (a) $y = 2x^3 - 6x^2 + 15$

   (b) $y = x^4 - 8x^2 - 5$

   (c) $y = x^3 + 3x^2 + 6x - 2$.

Chapter 6
Integral Calculus

6.1 Introduction

If x is the level of output and TC is the total cost, then given a cost curve

$$TC = 3x^2 + 20x + 13$$

by differentiation the marginal cost curve is

$$MC = \frac{dTC}{dx} = 6x + 20$$

That is, by the process of differentiation the total cost function becomes the marginal cost function. If we are now told that a marginal cost function is

$$MC = 6x + 20$$

and asked to determine the total cost function we must reverse the process of differentiation, that is, we must *integrate* the marginal cost equation. From our knowledge of differential calculus we might guess that

$$TC = 3x^2 + 20x$$

and we should also realise that there may have been a constant term in the TC function which would have disappeared on differentiation. However, we cannot determine the value of this constant, if indeed there was one, from the marginal cost function alone. For example

$$TC = 3x^2 + 20x, \quad TC = 3x^2 + 20x + 13, \quad TC = 3x^2 + 20x + 50$$

each have the same marginal cost function

$$MC = 6x + 20$$

We therefore write the total cost function associated with this marginal cost function as

224

$$TC = 3x^2 + 20x + C$$

where C is an arbitrary constant, *the constant of integration*, whose value can be determined only if additional information is available. This additional information may be in the form of an *initial condition*, which corresponds to the starting value of x, such as when $x = 0$ $TC = 20$, resulting in $C = 20$, or a *boundary condition*, which states the maximum value of x, such as when $x = 100$ $TC = 33,000$, resulting in $C = 1,000$, or a particular value such as that $TC = 100$ when $x = 3$ so that $C = 13$.

The notation commonly adopted is

$$\int (6x + 20) \, dx = 3x^2 + 20x + C$$

where the $\int$ is the *integral sign*, the process is known as *integration* and the result is called the *indefinite integral* of $6x + 20$. The term dx is included to show that x is the variable being considered.

Other examples of applications in economics and business of integration as the reverse process to differentiation follow directly from Chapter 5. Thus, for example, given a marginal-revenue function and an initial condition, the total-revenue function can be found. Also, if the expression for the marginal propensity to consume is known, then by integrating, the corresponding consumption function can be found and it will include a constant of integration.

By making use of the fact that integration can be thought of as the reverse of differentiation it is possible to find, by trial and error, the integral of many common functions. However, this may not always be easy because a function may be altered in appearance after differentiation by grouping together like terms or by cancelling out of common factors.

It is because of this that integration is generally more difficult than differentiation and requires a good deal of practice and a bit of intuition to become proficient. We have at our disposal, however, standard methods for finding the integral of most common functions, and a number of examples are considered here to illustrate these methods. We defer some more difficult cases to Chapter 9 where differential equations are considered.

6.2 Exercises

1. Derive the total cost function for processes in which the
   (a) marginal cost is £1 per unit and the fixed cost is £600
   (b) marginal cost is £1 per unit and the cost of production of 250 units is £300

(c) marginal cost is given by

$$MC = 2x + 3$$

and the total cost is 100 when $x = 5$.

2. For the consumption function $C = a + bY + cY^2$ determine the marginal propensity to consume (*mpc*) function. Hence, if *mpc* $= 0.1 + 0.02Y$ and when $Y = 0$, $C = 0$, determine the values of a, b and c.

3. If capital formation occurs continuously as time passes then the rate of change of capital stock, with respect to time, is net investment. Suppose that net investment is given by

$$I(t) = \frac{dK}{dt} = a + 2bt + 3ct^2$$

Determine the equation for the capital stock if when $t = 0$, $K = 0$.

6.3   Techniques of integration

In this section the integrals of the functions discussed in Chapter 5 are to be determined mainly by reversing the process of differentiation. More difficult forms are deferred until section 6.14.

(a) If $y = ax^n$, $dy/dx = anx^{n-1}$
Hence,

$$\int ax^n \, dx = \frac{ax^{n+1}}{n + 1} + C$$

which is true except in the case of $n = -1$ where the denominator would become zero. This exception is considered under (b) below.

Examples

1. $a = 1$, $n = 3$, $\displaystyle\int x^3 \, dx = \frac{x^4}{4} + C$

2. $a = 2$, $n = 0.5$, $\displaystyle\int 2x^{0.5} \, dx = \frac{2x^{1.5}}{1.5} + C$

3. $a = 4$, $n = -2$, $\displaystyle\int 4x^{-2} \, dx = \frac{4x^{-1}}{-1} + C$

(b) If $y = \log x$, $dy/dx = 1/x$

Hence,

$$\int \frac{1}{x} \, dx = \log x + C$$

which is the special case referred to above. More generally, if

$$y = \log u, \qquad \frac{dy}{dx} = \frac{1}{u} \frac{du}{dx}$$

Hence,

$$\int \frac{1}{u} \frac{du}{dx} = \log u + C$$

Thus if the numerator is the derivative of the denominator the integral is the logarithm of the denominator.

Examples

1. $u = x$, $\qquad \dfrac{du}{dx} = 1$ and $\displaystyle\int \frac{1}{x} \, dx = \log x + C$

2. $u = 2x + 3$, $\qquad \dfrac{du}{dx} = 2$ and $\displaystyle\int \frac{2 \, dx}{2x + 3} = \log (2x + 3) + C$

3. $u = x^3 - 4x + 1$, $\qquad \dfrac{du}{dx} = 3x^2 - 4$

$$\int \frac{3x^2 - 4}{x^3 - 4x + 1} \, dx = \log (x^3 - 4x + 1) + C$$

4. In some cases the expressions to be integrated need to be modified slightly to obtain the standard form. For example,

$$\int \frac{x^2 + 2x}{x^3 + 3x^2 + 5} \, dx$$

Here, $u = x^3 + 3x^2 + 5$ and

$$\frac{du}{dx} = 3x^2 + 6x = 3(x^2 + 2x)$$

Therefore write

$$\frac{x^2 + 2x}{x^3 + 3x^2 + 5} = \frac{1}{3} \left(\frac{3x^2 + 6x}{x^3 + 3x^2 + 5} \right)$$

Hence

$$\int \frac{x^2 + 2x}{x^3 + 3x^2 + 5} \, dx = \frac{1}{3} \int \frac{3x^2 + 6x}{x^3 + 3x^2 + 5} \, dx = \frac{1}{3} \log (x^3 + 3x^2 + 5) + C$$

(c) If $y = ae^{bx}$, $dy/dx = abe^{bx}$

Hence

$$\int ae^{bx} \, dx = \frac{ab^{bx}}{b} + C$$

Examples

1. $a = 3$, $b = 2$, $\displaystyle \int 3e^{2x} \, dx = \frac{3e^{2x}}{2} + C$

2. $a = 4$, $b = -1$, $\displaystyle \int 4e^{-x} \, dx = -4e^{-x} + C$

(d) If $y = \sin x$, $dy/dx = \cos x$

Hence $\displaystyle \int \cos x \, dx = \sin x + C$

Similarly $\displaystyle \int \sin x \, dx = -\cos x + C$

The more general versions of these and also the integrals of other trigonometric functions are deferred to Appendix A, Section A.6.

6.4   Exercises

1. Obtain the total cost functions from the following marginal cost functions:

   (a) $MC = 2x + 3$ with $TC = 50$ when $x = 5$

   (b) $MC = x^2 + 2x + 4$ with $TC = 450$ when $x = 9$.

2. Integrate the following with respect to x:

   (a) $3x - 4$ (b) $2x^2 - 4x + 3$

   (c) $3x^5 - \dfrac{4}{x^2} + 2x$ (d) $\dfrac{2}{x^3} - \dfrac{4}{x}$

   (e) $\dfrac{3}{3x - 5}$ (f) $\dfrac{4x + 1}{2x^2 + x - 4}$

   (g) $\dfrac{x^3 - x}{x^4 - 2x^2 + 2}$ (h) $\dfrac{4x + 6}{x^2 + 3x + 1}$

(i) $2e^{3x} + e^{-x}$ (j) $2e^x - 4e^{3x}$

(k) $2 \sin x - 4 \cos x$

3. If marginal revenue is given by $MR = a - 2bq$ where a and b are positive constants, determine the total revenue curve if $TR = 0$ when $q = 0$. Hence determine the demand function.
4. The elasticity of demand for a certain product is $e^d = -0.3p/q$. What is the demand curve if when $p = 10$, $q = 100$?
5. If the marginal propensity to consume is $mpc = 0.5 + Y^{-1} - Y^{-2}$ and when $Y = 1$, $C = 1$, what is the consumption function?

6.5   The calculation of areas

The area under a continuous function can generally be determined by using the integral calculus.

Let AB be a section of the curve $y = f(x)$ between the values $x = X_1$, and $x = X_2$, and let Z be ABX_2X_1, the area which is required (Fig. 6.1). In this illustration $f(x)$ is taken to be steadily increasing: the treatment in other cases is similar.

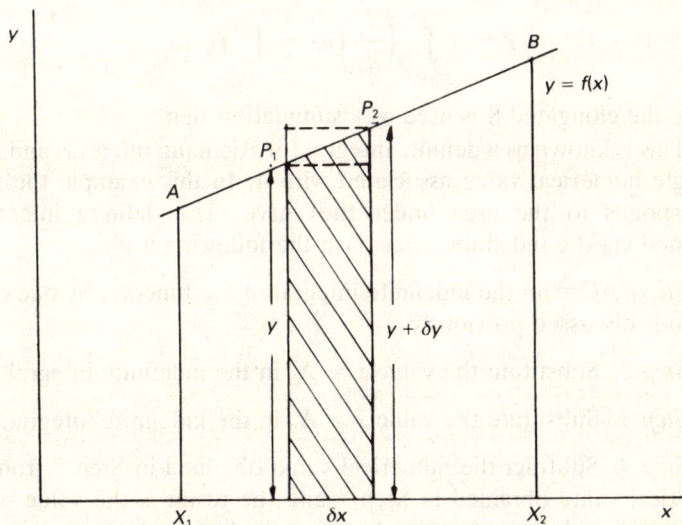

Fig. 6.1

Let us consider two points P_1 and P_2 on the curve which are a small distance apart and from which perpendiculars are dropped to the x-axis. If the heights of these perpendiculars are y and $y + \delta y$ respectively and their horizontal distance apart is δx, then it is possible to estimate the area between these two lines, the curve, and the x-axis. This area, which we call δz, is shaded in the diagram and it must be greater than the area of the rectangle $y\,\delta x$ and less than the rectangle $(y + \delta y)\,\delta x$

that is
$$y\,\delta x < \delta z < (y + \delta y)\delta x$$

or
$$y < \frac{\delta z}{\delta x} < y + \delta y$$

When $\delta x \to 0$, $\delta y \to 0$ and $\delta z/\delta x \to dz/dx$, so that

$$\frac{dz}{dx} = y = f(x) \tag{1}$$

and the rate of change of the area depends on the value of x. The total area ABX_2X_1 is made up of an unfinitely large number of these small areas such as dz. Its value is therefore given by the sum of all such areas which are located between the values $x = X_1$, and $x = X_2$. This is usually written as

$$Z = \int_{x=X_1}^{x=X_2} \left(\frac{dz}{dx}\right) dx = \int_{X_1}^{X_2} f(x)\,dx \tag{2}$$

where the elongated S is used as a summation sign.

This is known as a definite integral (or Riemann integral) and it has a single numerical value associated with it. In this example the value corresponds to the area under the curve. The definite integral is obtained via the indefinite integral in the following way:

Step 1. Obtain the indefinite integral of the function by one of the methods discussed previously.

Step 2. Substitute the value $x = X_1$ in the indefinite integral.

Step 3. Substitute the value $x = X_2$ in the indefinite integral.

Step 4. Subtract the numerical value obtained in Step 2 from the numerical value obtained in Step 3 and the result is the value of the definite integral of the function between the limits $x = X_1$ and $x = X_2$.

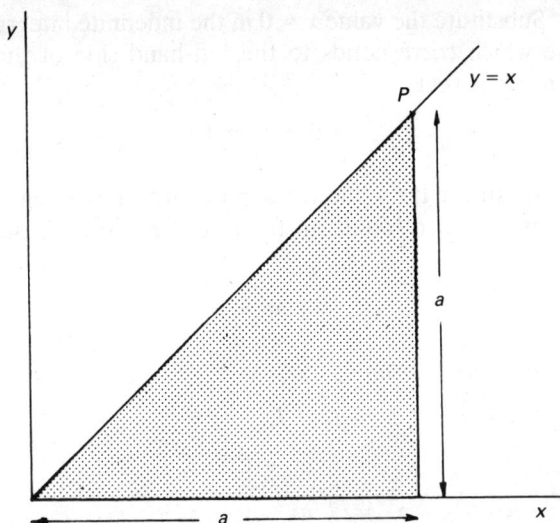

Fig. 6.2

A derivation of these rules is rather difficult but a justification is provided by the following simple example.

Example 1

Let us consider the linear function $y = x$. This is a straight line which passes through the origin and bisects the angle between the x- and the y-axis (Fig. 6.2).

Any point on this line has the same value for y as it does for x, and P, with co-ordinates (a, a), can be considered as a typical point.

The area which is bounded by the line $y = x$, the ordinate $x = a$ and the x-axis is triangular in shape and is shaded in the diagram. Its area is calculated very easily by simple geometry and is equal to $\frac{1}{2} a^2$. Let us compare this with the result that is obtained via the integral calculus.

Step 1. Determine the indefinite integral of the function with respect to x.

$$
\begin{aligned}
I &= \int y \, dx \\
&= \int x \, dx \text{ (by substitution } y = x) \\
&= \tfrac{1}{2} x^2 + C
\end{aligned}
$$

Step 2. Substitute the value $x = 0$ in the indefinite integral ($x = 0$ is the ordinate which corresponds to the left-hand side of the required area). The result is that

$$I_1 = 0 + C = C$$

Step 3. Substitute the value $x = a$ in the indefinite integral ($x = a$ is the ordinate which corresponds to the right-hand side of the required area). Then

$$I_2 = \tfrac{1}{2} a^2 + C$$

Step 4. Subtract I_1 from I_2

$$Z = I_2 - I_1$$
$$= \tfrac{1}{2} a^2 + C - C$$
$$= \tfrac{1}{2} a^2.$$

This is an identical result to that obtained by simple geometry and it is obvious now that the definite integral will never contain a constant of integration. This is always eliminated during Step 4 of the above method.

It is not really necessary to use the integral calculus to determine the areas in simple cases such as those where the answer is obvious from the geometry of the system. It can be of great use, however, in many other cases, particularly where the function is non-linear as the following example shows.

Example 2

For a company using a process where the marginal cost of production is a function of the level of output, decreasing at first and then increasing when the output is above a certain level, the following might apply:

$$MC = 10 - q + \tfrac{1}{20} q^2, \quad 0 < q \leqslant 20$$

This can be presented graphically as shown in Fig. 6.3.

The area under the marginal cost curve between the values $q = 10$ and $q = 20$ is equal to the extra cost which must be incurred when production is increased from 10 units to 20 units of output.

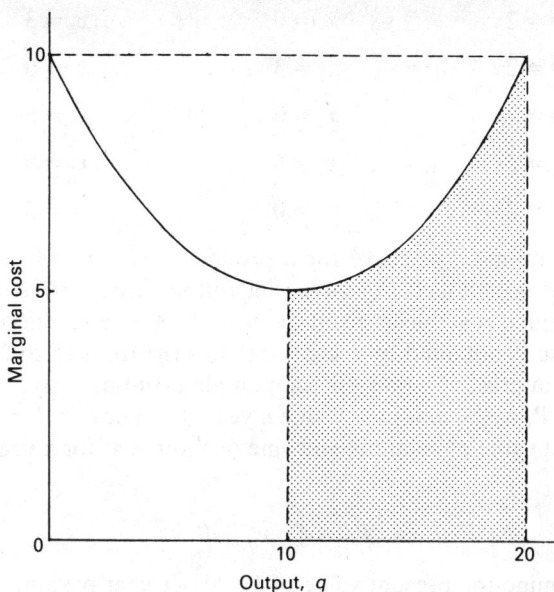

Fig. 6.3

$$\text{Extra cost} = \int\limits_{q=10}^{q=20} (MC)\, dq$$

$$= \int\limits_{10}^{20} (10 - q + \tfrac{1}{20}q^2)\, dq$$

$$= [10q - \tfrac{1}{2}q^2 + \tfrac{1}{60}q^3]_{10}^{20}$$

$$= \{10(20) - \tfrac{1}{2}(20)^2 + \tfrac{1}{60}(20)^3\} - \{10(10) - \tfrac{1}{2}(10)^2 + \tfrac{1}{60}(10)^3\}$$

$$= 200 - 200 + \frac{400}{3} - 100 + 50 - \frac{50}{3} = \frac{200}{3}$$

6.6 Exercises

1. Calculate the area between the axis of x, the straight line given by the following equations and the two values of x.

(a) $y = 2x$ $x_1 = 0$ $x_2 = 3$

(b) $y = 2x$ $x_1 = 3$ $x_2 = 6$

(c) $y = 2x$ $x_1 = 0$ $x_2 = 6$

(d) $y = 5x$ $x_1 = 6$ $x_2 = 8$

(e) $y = 2 + 3x$ $x_1 = 0$ $x_2 = 3$.

2. The marginal cost curve for a product is $MC = 100 - 4q + q^2$. What is the total cost of increasing output from $q = 20$ to $q = 30$?

3. If marginal revenue is given by $MR = 200 - 6q$, what extra total revenue is obtained by increasing sales (q) from 15 to 20?

4. When interest at a rate r is compounded continuously, the present value, P, of the amount, S, due n years from now is $P = Se^{-rn}$. The present value of an annual income of S per year for n years is given by

$$P_n = \int_0^n Se^{-rt}\, dt$$

Determine the present value of 1,000 per year payable 10 times if the interest rate is 5% compounded continuously.

6.7   Absolute area

When the graph of $f(x)$ cuts the x-axis, areas below the x-axis are taken to be negative. For some applications this is the correct interpretation. For others, however, the *absolute area* under the curve is required, i.e. the deviations of the curve from the x-axis are to be summed. When this happens, the direct evaluation of the integral gives an incorrect solution. This can be seen by reference to the function $y = x$ and by considering the area shaded in Fig. 6.4.

The shaded area between the ordinates $x = -a$ and $x = +a$ is obviously equal to $\frac{1}{2}a^2 + \frac{1}{2}a^2$, that is, a^2. Using the integral calculus as before would lead to

$$Z = \int_{x=-a}^{x=+a} y\, dx = \int_{x=-a}^{x=+a} x\, dx = \left[\tfrac{1}{2}x^2 \right]_{-a}^{+a} = \tfrac{1}{2}a^2 - \tfrac{1}{2}a^2 = 0$$

This is true if we consider the area above the x-axis to be positive and the area below the x-axis to be negative. But if the total shaded area is required irrespective of whether it is above or below the axis, then the two sections must be calculated separately and their individual values added together. The dividing point is that at which the function crosses

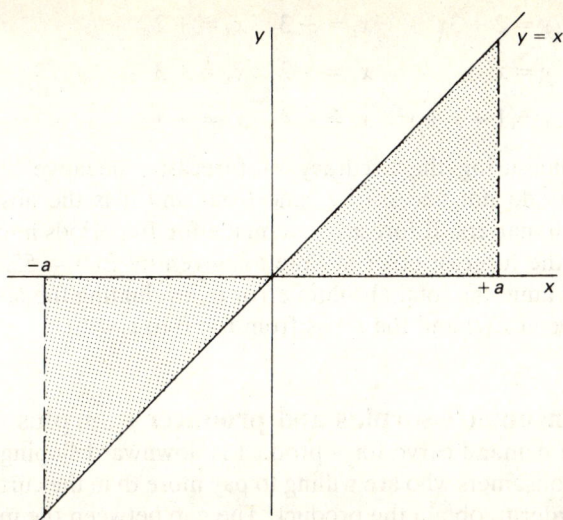

Fig. 6.4

the x-axis. (Functions of higher order than the simple linear one may cross the axis in a number of points and in such cases it may be necessary to consider all the areas separately.)

In the example with the linear function $y = x$ the dividing point is at $x = 0$ and the area is found as follows:

$$Z_1 = \int_{x=-a}^{x=0} x \, dx = \left[\tfrac{1}{2} x^2 \right]_{x=-a}^{x=0} = -\tfrac{1}{2} a^2$$

$$Z_2 = \int_{x=0}^{x=a} x \, dx = \left[\tfrac{1}{2} x^2 \right]_{x=0}^{x=a} = \tfrac{1}{2} a^2$$

The area is the sum of the absolute values of Z_1 and Z_2, i.e. the sum of their numerical values when the negative sign associated with Z_1 is ignored:

$$Z = \tfrac{1}{2} a^2 + \tfrac{1}{2} a^2 = a^2$$

6.8 Exercises

1. Sketch the following curves and calculate the absolute area between the curves, the axis of x and the given values x_1 and x_2.

   (a) $y = 2x$ $\qquad\qquad x_1 = -1 \quad x_2 = +1$

(b) $y = 2 + 3x$ $x_1 = -3$ $x_2 = +2$

(c) $y = x^2$ $x_1 = -2$ $x_2 = +3$

(d) $y = 1 + x + x^2$ $x_1 = -4$ $x_2 = +1$

2. In measuring the accuracy of forecasts, negative and positive errors do not necessarily cancel out and it is the absolute error which matters. If forecasts are made for 10 periods into the future and the forecast error at time t is given by $E(t) = 55 - 16t + t^2$, determine the total absolute error by evaluating the absolute area between $E(t)$ and the t-axis from $t = 0$ to $t = 10$.

6.9 Consumer's surplus and producer's surplus

If the demand curve for a product is downward sloping, there will be some consumers who are willing to pay more than the current market price in order to obtain the product. The gap between the market price and the price the consumer is willing to pay is a measure of the consumer's surplus satisfaction and is known as *consumer's surplus*. Since the demand function is continuous the total consumer's surplus is measured by the area ABP_1, shown in Fig. 6.5, where P_1 and X_1 are the market price and quantity. This area can be evaluated by integration as the area ABX_1O minus the area P_1BX_1O.

For example, if the demand function is

$$p = 45 - 2x - x^2$$

and the market price is $P_1 = 10$ and the quantity is $X_1 = 5$, then the area ABX_1O is given by

$$I = \int_{x=0}^{x=5} p\, dx = \int_{x=0}^{x=5} (45 - 2x - x^2)\, dx$$

$$= \left[45x - x^2 - \frac{x^3}{3} \right]_0^5$$

$$= \left(225 - 25 - \frac{125}{3} \right) - (0) = 158.33$$

The area P_1BX_1O is $P_1X_1 = 50$ and therefore the consumer's surplus is measured as $158.33 - 50 = 108.33$.

The concept of producer's surplus is rather similar to that of consumer's surplus but relates to the supply curve.

Since the supply curve is assumed to be upward sloping, for any

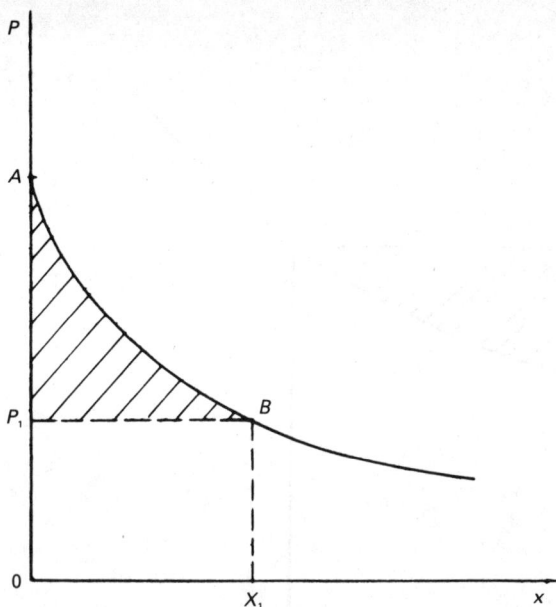

Fig. 6.5

given market price, P_1, say, some producers would have been willing to supply a lower quantity at a lower price. The vertical gap between the market price, P_1, and the supply curve (see Fig. 6.6) is a measure of the producer's extra satisfaction and is known as *producer's surplus*. It is evaluated by finding the total area which is above the supply curve and below the horizontal line $P = P_1$. For the price P_1 and quantity X_1 the shaded area is given by

$$I = P_1 X_1 - \int_0^{x_1} P \, dx$$

For example, suppose that the supply curve is

$$p = 20 + 2x + 0.3x^2$$

and the producer's surplus is required for a price $p = 70$ and quantity $x = 10$, then

$$I = 700 - \int_0^{10} (20 + 2x + 0.3x^2) \, dx$$

$$= 700 - [20x + x^2 + 0.1x^3]_0^{10}$$

$$= 700 - (200 + 100 + 100) = 300$$

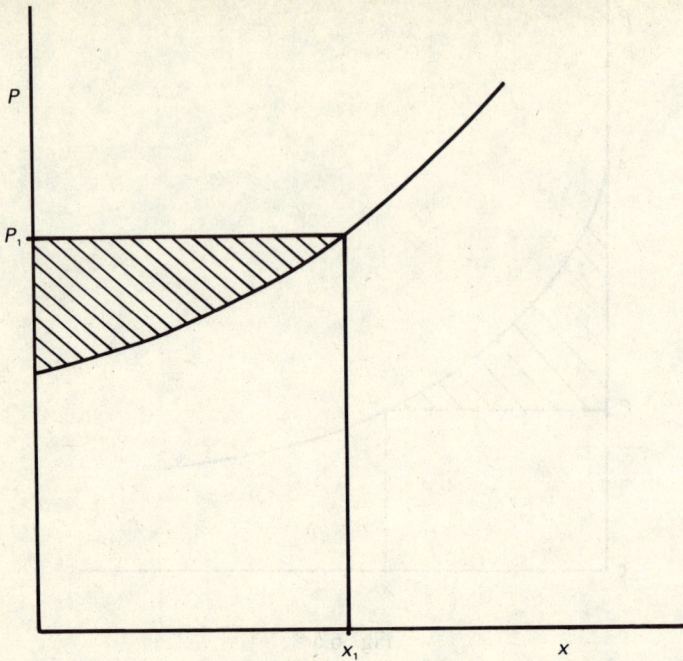

Fig. 6.6

6.10   Areas between curves

We now consider two problems where the area of interest is in between two curves. The first concerns profit, given by the area between a marginal revenue function and a marginal cost function. The second relates to depreciating assets and the decision to scrap a capital good.

In discussing profit (or net revenue) maximisation in section 5.8 the starting point was the net revenue function, defined as being total revenue minus total costs. Net revenue was maximised when marginal revenue equalled marginal cost and, given the maximising value of output, the maximum value of net revenue could be obtained. Here we start with the marginal cost (MC) and marginal revenue (MR) functions. The difference between these can be defined as being the marginal profit (MP) function so that

$$MP = MR - MC$$

If the level of output is q_1 then the total value of profits is

$$\text{profit} = \int_0^{q_1} (MR - MC) \, dq$$

For example, suppose that the marginal revenue function is

$$MR = 50 - 2q$$

and the marginal cost function is

$$MC = 10 + q$$

For an output of $q = 10$ the value of profit is

$$\text{Profit} = \int_0^{10} (50 - 2q - 10 - q)\, dq$$

$$= \int_0^{10} (40 - 3q)\, dq = [40q - 1.5q^2]_0^{10}$$

$$= (400 - 150) - (0) = 250$$

The situation is represented graphically as the shaded area in Fig. 6.7. Clearly, at $q = 10$ $MR > MC$ and so it pays to increase output. Profit is maximised when $MR = MC$ and here this is when $q = 13.33$, giving a profit of 266.67.

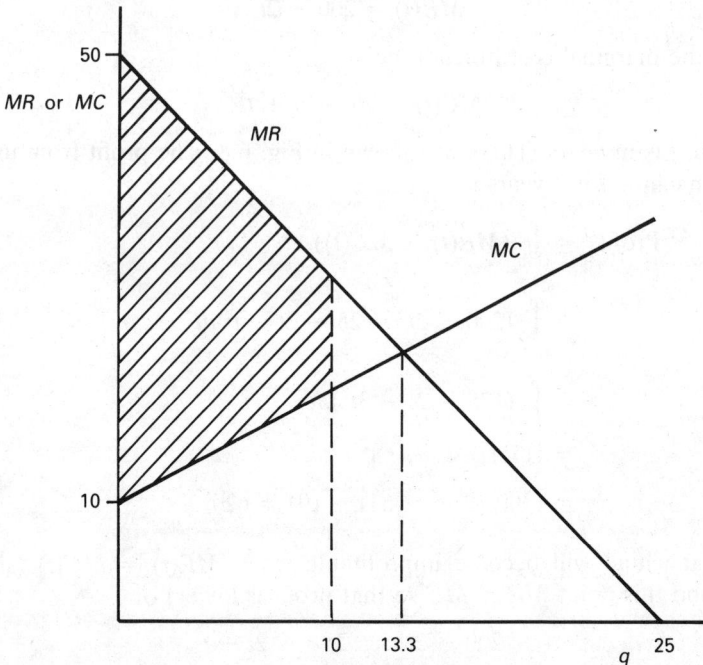

Fig. 6.7

Before leaving this approach to measuring profit it is worth pointing out that the use of the *MR* and *MC* curves means that the fixed costs of production (if any) are ignored. This is because in using marginal analysis we are concerned with short-run behaviour and therefore fixed costs, which are defined as those costs which cannot be changed easily, are not relevant.

Turning now to the second application of areas between two curves, we consider problems with depreciating assets and the decision of when to scrap a capital good. Suppose that a company instals a machine which dispenses hot drinks. The revenue from this is $MR(t)$ which is a function of time. Notice that this is the marginal revenue at time t and not the total revenue from time 0 to t. As the machine gets older it is more prone to being out of order so that revenue falls as time passes. Also, the cost of servicing the machine, $MC(t)$, rises as time passes, since parts wear out and need to be replaced. Initially we will assume that the machine has no scrap value. The profit from having the machine is the difference between $MR(t)$ and $MC(t)$ and is the area between the two curves. Let the marginal revenue function be

$$MR(t) = 200 - 2t^2$$

and the marginal cost function be

$$MC(t) = 25 + 2t + t^2,$$

where t is in years. These are shown in Fig. 6.8. The profit from using the machine for 4 years is

$$\begin{aligned}
\text{Profit} &= \int_0^4 (MR(t) - MC(t))\, dt \\
&= \int_0^4 (200 - 2t^2 - 25 - 2t - t^2)\, dt \\
&= \int_0^4 (175 - 2t - 3t^2)\, dt \\
&= [175t - t^2 - t^3]_0^4 \\
&= (700 - 16 - 64) - (0) = 620
\end{aligned}$$

The machine will become unprofitable when $MR(t) = MC(t)$ (since beyond this point $MR < MC$ so that profit is lower) or

$$200 - 2t^2 = 25 + 2t + t^2$$

or

$$0 = 3t^2 + 2t - 175$$

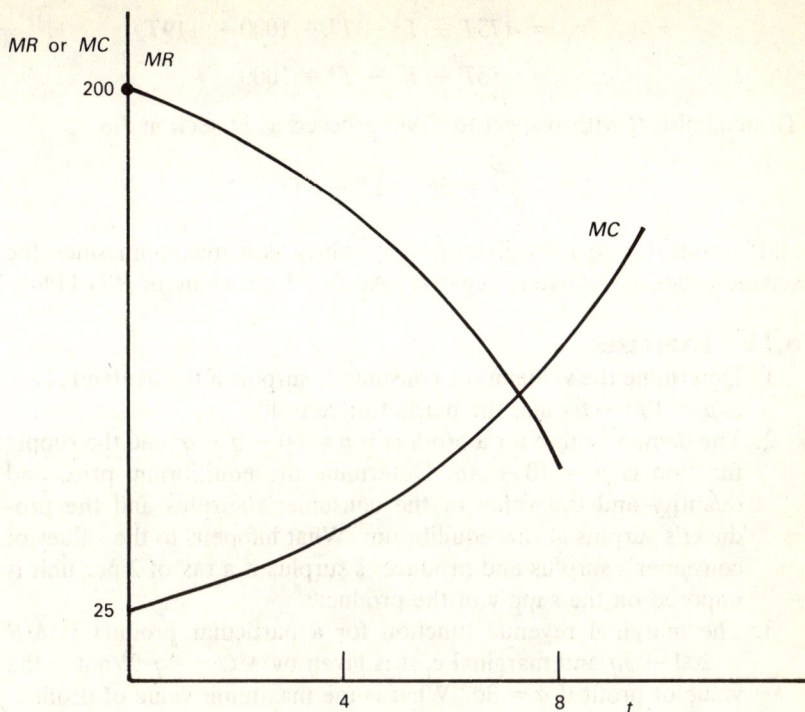

Fig. 6.8

and so $t = 7.31$ years. The maximum total profit is earned if the machine is scrapped after 7.31 years and this is

$$[175t - t^2 - t^3]_0^{7.31} = 835.20$$

If the machine were scrapped after 5 years the loss of revenue compared with scrapping after 7.31 years is the area from $t = 5$ to $t = 7.31$ or

$$[175t - t^2 - t^3]_5^{7.31} = 835.20 - 725 = 110.20$$

Now suppose that the machine has a scrap value given by

$$S(t) = 1000 - 119t$$

This changes the situation because the value of the project of installing the machine now has two components: the profit earned and the scrap value. If the machine is scrapped at time T the total revenue is

$$R(T) = \int_0^T (MR(t) - MC(t))dt + S(T)$$

$$= 175T - T^2 - T^3 + 1000 - 119T$$
$$= 56T - T^2 - T^3 + 1000$$

To maximise R with respect to T we proceed as in section 5.8:

$$\frac{dR}{dT} = 56 - 2T - 3T^2$$

and setting this to zero gives $T = 4$, which is a maximum since the second-order derivative is negative. At $T = 4$ the value of R is 1144.

6.11 Exercises

1. Determine the value of the consumer's surplus if the demand curve is $p = 100 - 6q$ and the market price is 40.
2. The demand curve for a product is $p = 60 - q - q^2$ and the supply function is $p = 10 + 4q$. Determine the equilibrium price and quantity and the value of the consumer's surplus and the producer's surplus at this equilibrium. What happens to the values of consumer's surplus and producer's surplus if a tax of 2 per unit is imposed on the supply of the product?
3. The marginal revenue function for a particular product is $MR = 200 - 3q$ and marginal cost is given by $MC = 2q$. What is the value of profit if $q = 30$? What is the maximum value of profit?
4. A marketing manager expects that, as a result of an advertising campaign which is about to start in month 0, sales in month t will be to the value of $S(t) = 25 + 30t^2 - 4t^3$ and the cost of the campaign in month t will be $C(t) = 15 + 5t$. What is the total net revenue expected to be in the first 4 months? The campaign is planned to last 6 months. Is it worthwhile extending it to 8 months?
5. A new machine is installed which is expected to produce a marginal revenue at time t of $MR = 600 - 10t - 3t^2$ and the marginal cost function is $MC = 50 + 15t$. Determine the economic life of the machine and the expected total net revenue over its life. If the machine has a scrap value given by $S(t) = 5000 - 300t$, when should the machine be scrapped?

6.12 Numerical methods of integration

So far we have assumed that it is possible to evaluate the definite integral directly. However, in some cases it may prove to be difficult to find the integral of the function in question, while in other cases the function may not be known and the problem is one of finding the area

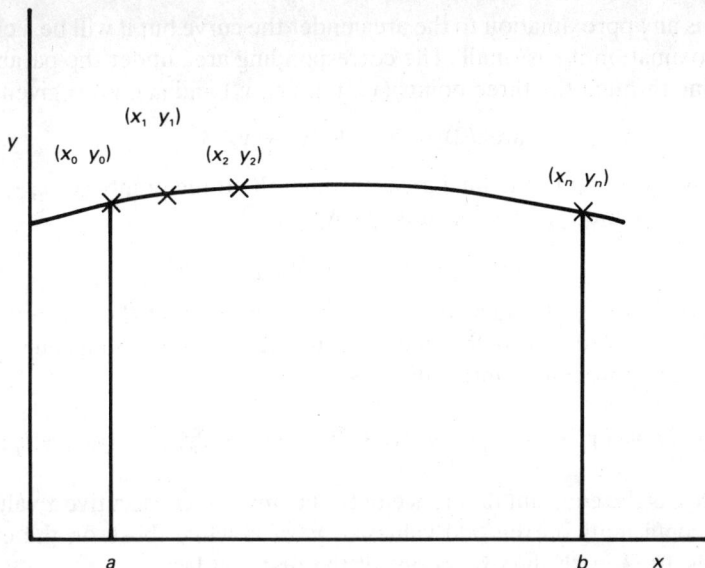

Fig. 6.9

under a curve defined by a table of values. We will consider two procedures which give approximate, numerical solutions to these problems. Each assumes that the function is smooth and has no discontinuities in the range of interest. The first is known as Simpson's rule, and relies on obtaining a table of values of the function, while the second uses Taylor's theorem which relies on being able to evaluate derivatives of the function.

Using Simpson's rule involves first splitting the area to be evaluated into an even number, n, of parts with equal width. The area under each pair of parts is then evaluated approximately by using a parabolic arc.

Suppose that the problem is to obtain the area under $f(x)$ between $x = a$ and $x = b$, and that the distance between a and b is split into n parts, each of width $c = (b - a)/n$. Here n is an even number. Referring to Fig. 6.9, where for simplicity it is assumed that the whole of the area of interest is above the x-axis, the consecutive x-values are x_0, $x_1, x_2, \ldots, x_n$, with the first corresponding to a and the last to b. The area under the parabola passing through the three points (x_0, y_0), (x_1, y_1) and (x_2, y_2) is given by

$$\text{area}(1) = c(y_0 + 4y_1 + y_2)/3$$

This is an approximation to the area under the curve but it will be a close approximation if c is small. The corresponding area under the parabola passing through the three points (x_2, y_2), (x_3, y_3) and (x_4, y_4) is given by

$$\text{area}(2) = c(y_2 + 4y_3 + y_4)/3,$$

and similarly, up to the final area, through the three points (x_{n-2}, y_{n-2}), (x_{n-1}, y_{n-1}) and (x_n, y_n) which is given by

$$\text{area}(n/2) = c(y_{n-2} + 4y_{n-1} + y_n)/3$$

The total area is the sum of all these and, noting that the last term in each area expression is the first term in the next one, Simpson's rule gives the approximate integration as

$$\int_a^b f(x)dx \approx c(y_0 + 4y_1 + 2y_2 + 4y_3 + 2y_4 + \cdots + 2y_{n-2} + 4y_{n-1} + y_n)/3,$$

where c is the constant difference between any two consecutive x values. The coefficients on the odd values of y are 4 while those on the even values are 2, with the exceptions of the first and last.

To illustrate Simpson's rule we will use it to evaluate approximately the area under the curve $y = x^{-1}$ between the values $x = 1$ and $x = 4$ and compare it with the exact value.

First the values of n and c need to be chosen. As stated above, n should be even, and the larger the value the more accurate the approximation will be. Since the area between $x = 1$ and $x = 4$ is required, and $4 - 1 = 3$, which is not even, we might choose $n = 6$ with $c = 0.5$. Next we evaluate $y = x^{-1}$ for $x = 1, 1.5, 2, \ldots, 3.5, 4$. This gives the following set of values, where we work to four places after the decimal point and include the value of the subscript, i on y:

i	0	1	2	3	4	5	6
x	1	1.5	2	2.5	3	3.5	4
y	1	0.6667	0.5000	0.4000	0.3333	0.2857	0.2500

The approximate area is therefore,

$$\int_1^4 x^{-1}dx \approx c(y_0 + 4y_1 + 2y_2 + 4y_3 + 2y_4 + 4y_5 + y_6)/3$$

$$\approx 0.5(1 + 2.6668 + 1 + 1.6 + 0.6666 + 1.1428 + 0.2500)/3$$

$$\approx 0.5(8.3262)/3 = 1.3877$$

Direct integration gives the exact value as

$$\int_1^4 x^{-1}dx = [\log x]_1^4 = \log 4 - \log 1 = 1.3863$$

The approximate value is close to the accurate value here, and a more accurate approximation could have been obtained by taking a smaller value of c.

As a second example suppose that we have the following data on marginal cost (MC) and level of output (q) and we require the total cost of increasing output from $q = 10$ to $q = 50$,

q	10	20	30	40	50
MC	15	16	20	30	50

Here $n = 4$ (since the area is split into 4 parts) and $c = 10$ so Simpson's rule gives the approximate area as

$$\int_{10}^{50} MCdq \approx c(MC_0 + 4MC_1 + 2MC_2 + 4MC_3 + MC_4)/3$$

$$\approx 10(15 + 64 + 40 + 120 + 50)/3$$

$$\approx 10(289)/3 = 963.3$$

In this example it is not possible to evaluate the integral directly since the functional form is not given.

The second method of numerical integration which we consider uses Taylor's theorem (see section 5.15). This states that

$$f(a + x) = f(a) + xf'(a) + \frac{x^2}{2!} f''(a) + \cdots + \frac{x^n}{n!} f^{(n)}(a) + \cdots$$

where the primes indicate derivatives. It is assumed that this series converges and if it does not, this method should not be used. Now since we wish to integrate $f(x)$ rather than $f(a + x)$ it is convenient to let $z = a + x$ so that $x = z - a$ and substituting these into the series expansion gives

$$f(z) = f(a) + (z - a)f'(a) + \frac{(z - a)^2}{2!}f''(a) + \cdots + \frac{(z - a)^n}{n!}f^{(n)}(a) + \cdots$$

and replacing z by x results in the value of $f(x)$ as

$$f(x) = f(a) + (x - a)f'(a) + \frac{(x - a)^2}{2!}f''(a) + \cdots + \frac{(x - a)^n}{n!}f^{(n)}(a) + \cdots$$

To integrate the left hand side with respect to x we just integrate the right hand side, noting that we have a polynomial in x, which is simple to integrate. The terms $f(a)$, $f'(a)$, $f''(a)$, . . . are all constants and, because differentiation is generally straightforward, are easily obtained. Of course, if the function is a polynomial then the Taylor's series expansion will give the same polynomial and so no benefit arises from using this method. The value of a, around which the series expansion is taken, is usually chosen to be in the middle of the interval over which integration is taking place. The general statement is

$$\int_b^c f(x)dx = \int_b^c \{f(a) + (x - a)f'(a) + \frac{(x - a)^2}{2!}f''(a) + \cdots$$

$$+ \frac{(x - a)^n}{n!}f^{(n)}(a) + \cdots\}dx$$

$$= [f(a)x + \frac{(x - a)^2}{2!}f'(a) + \frac{(x - a)^3}{3!}f''(a) + \cdots]_b^c$$

To illustrate the use of Taylor's theorem we will take the first example from p. 244 above. We require the area under the curve $y = x^{-1}$ between $x = 1$ and $x = 4$. Taking the first three derivatives we have

$$f(x) = x^{-1}, f'(x) = -x^{-2}, f''(x) = 2x^{-3}, f'''(x) = -6x^{-4}$$

Choosing $a = 2.5$, being halfway between the limits of 1 and 4,

$$f(2.5) = 0.4, f'(2.5) = -0.16, f''(2.5) = 0.128, f'''(2.5) = -0.1536$$

and so the integral of Taylor's expansion gives

$$\int_1^4 f(x)dx = [f(a)x + \frac{(x - a)^2}{2!}f'(a) + \frac{(x - a)^3}{3!}f''(a) + \cdots]_1^4$$

$$= [0.4x + \frac{(x - 2.5)^2}{2!}(-0.16) + \frac{(x - 2.5)^3}{3!}(0.128) + \cdots]_1^4$$

$$= (1.6 - 0.18 + 0.072 - \cdots) - (0.4 - 0.18 - 0.072 - \cdots)$$

$$= (1.492) - (0.148) = 1.344$$

The accurate value is 1.3863 and so the Taylor's expansion value is reasonably close. Notice that the third terms in the expansion are relatively large. A closer approximation would result from taking account of extra terms in the expansion.

As a second example of the use of Taylor's theorem we evaluate the area between $y = (1 - x^2)^{-1}$ between $x = -0.5$ and $x = 0.5$. The first four derivatives are

$$f'(x) = 2x(1 - x^2)^{-2}, f''(x) = (2 + 4x^2 - 6x^4)(1 - x^2)^{-4}$$

$$f'''(x) = 24x(1 - x^4)(1 - x^2)^{-5}$$

$$f''''(x) = (24 + 216x^2 - 120x^4 - 120x^6)(1 - x^2)^{-6}$$

Since the limits are -0.5 and $+0.5$ we choose $a = 0$ so that

$$f(0) = 1, f'(0) = 0, f''(0) = 2, f'''(0) = 0, f''''(0) = 24$$

and the integral of the series expansion is

$$\int_{-0.5}^{0.5} f(x)dx = [f(a)x + \frac{(x - a)^2}{2!}f'(a) + \frac{(x - a)^3}{3!}f''(a) + \cdots]_{-0.5}^{0.5}$$

$$= [x + \frac{x^3}{3} + \frac{x^5}{5} \cdots]_{-0.5}^{0.5}$$

$$= (0.5 + 0.0417 + 0.00625 + \cdots)$$

$$- (-0.5 - 0.0417 - 0.00625 - \cdots)$$

$$\approx 0.54795 + 0.54795 = 1.0959$$

This is the approximate value of the area between $x = -0.5$ and $x = 0.5$, the x-axis and the curve $y = (1 - x^2)^{-1}$. The accurate value is, using the method of partial fractions explained in section 6.14,

$$\int_{-0.5}^{0.5} (1 - x^2)^{-1}dx = 0.5[\log\{(1 + x)/(1 - x)\}]_{-0.5}^{0.5}$$

$$= 0.5(\log 3 - \log 0.3333)$$

$$= 0.5(1.0986 - \{-1.0986\})$$

$$= 1.0986$$

Here the Taylor's theorem approximation is close to the accurate value and a more accurate approximation would be obtained by taking more terms in the series expansion.

6.13 Exercises

1. Evaluate the area between the x-axis, the curve $y = 4x^3$ and the values $x = 1$ and $x = 3$ by (a) exact integration and (b) Simpson's rule.

2. Use Simpson's rule to obtain the total cost of expanding output from $q = 10$ to $q = 14$ given the following marginal cost data:

q	10	11	12	13	14
MC	3	4	9	10	15

3. Obtain the approximate value of the integral $\int_0^8 x^4 dx$ using Simpson's rule with (a) $n = 2$ and (b) $n = 8$ and compare these with the exact value.

4. To see how important the choice of the value of 'a' is in integration by use of Taylor's series, use the working in the text (p. 244) to integrate $y = x^{-1}$ between $x = 1$ and $x = 4$ with (a) $a = 1$ and (b) $a = 3$, and then (c) compare with the results for $a = 2.5$ and the exact value.

5. Evaluate $\int_0^4 e^x dx$ by (a) exact integration (b) Simpson's rule with $n = 4$ and (c) Taylor's series using the first three derivatives.

6.14   More difficult integration

INTEGRATION BY PARTS

Integration by parts can be thought of as the reverse of differentiation of a product. We know that if u and v are functions of x then

$$\frac{d}{dx}(uv) = u\frac{dv}{dx} + v\frac{du}{dx}$$

Integrating both sides of this, we have

$$\int \frac{d}{dx}(uv)\,dx = \int u\frac{dv}{dx}dx + \int v\frac{du}{dx}dx$$

$$uv = \int u\,dv + \int v\,du$$

$$\therefore \int u\,dv = uv - \int v\,du$$

Thus to find the integral of a function such as xe^x we can let $x = u$ and $e^x\,dx = dv$.

Then $v = \int dv = \int e^x\,dx = e^x,$ and $du = dx$

$$\therefore \int xe^x\,dx = xe^x - \int e^x\,dx$$

$$= xe^x - e^x + C$$
$$= e^x(x - 1) + C$$

We can check this by differentiation:

$$\frac{d}{dx} [e^x(x - 1) + C] = e^x(x - 1) + e^x = xe^x$$

This method is of great importance in integration but care must be taken in deciding which part of the function should be represented by u and which by v. It is usually better to represent by dv that part which can be integrated most easily. If both parts are simple to integrate then represent by u that part which is reduced to a constant value by differentiation in the shortest possible time. There is, however, no hard and fast rule about this selection. The following examples have been chosen to show the method of selection in a few typical cases.

Example 1

$$\int x^2 e^x \, dx$$

Let $\qquad x^2 = u \qquad$ and $\qquad e^x \, dx = dv$

Then $\qquad \int x^2 e^x \, dx = x^2 e^x - \int e^x 2x \, dx$
$$= x^2 e^x - 2 \int e^x x \, dx$$

The second part of this must be integrated by the same method

Let $\qquad x = u \qquad$ and $\qquad e^x \, dx = dv$

Then $\qquad \int e^x x \, dx = \int x e^x \, dx$
$$= xe^x - \int e^x \, dx$$
$$= xe^x - e^x + C$$
$$= e^x(x - 1) + C$$

Then taking the two parts together, we have

$$\int x^2 e^x \, dx = x^2 e^x - 2e^x(x - 1) + C$$
$$= e^x(x^2 - 2x + 2) + C$$

It is not necessary to represent the terms by u and v in the order in which they appear in the function as the second part of Example 1 and Example 2 show.

Example 2

$$\int x \log_e x \, dx$$

Let $\qquad\qquad \log_e x = u \quad$ and $\quad x \, dx = dv$

Then $\qquad\qquad \dfrac{1}{x} \, dx = du \quad$ and $\quad \tfrac{1}{2}x^2 = v + C$

$$\therefore \int x \log_e x \, dx = \int (\log_e x) x \, dx$$

$$= (\log_e x)(\tfrac{1}{2}x^2) - \int (\tfrac{1}{2}x^2)\left(\frac{1}{x}\right) dx$$

$$= \tfrac{1}{2}x^2 \log_e x - \tfrac{1}{2} \int x \, dx$$

$$= \tfrac{1}{2}x^2 \log_e x - \tfrac{1}{4}x^2 + C$$

$$= \tfrac{1}{2}x^2(\log_e x - \tfrac{1}{2}) + C$$

It may be necessary to use more than two stages to complete the integration, but when this happens we should be reasonably certain that we are applying the appropriate technique. In general, this method reduces the part to be integrated to a simpler form at each stage.

THE METHOD OF PARTIAL FRACTIONS

In attempting to integrate some functions it is convenient to express the function as a series of partial fractions. For example, to integrate $(a^2 - x^2)^{-1}$ notice that

$$a^2 - x^2 = (a + x)(a - x)$$

and hence

$$\frac{1}{a^2 - x^2} = \frac{1}{(a + x)(a - x)} = \frac{A}{a + x} + \frac{B}{a - x}$$

where A and B are as yet unknown constants. Their values are determined as follows:

$$\frac{1}{a^2 - x^2} = \frac{A}{a + x} + \frac{B}{a - x} = \frac{A(a - x) + B(a + x)}{a^2 - x^2}$$

Since the denominators are equal then

$$1 = A(a - x) + B(a + x)$$

$$= (Aa + Ba) + x(B - A)$$

This can be true only if the coefficients of x in this equation are equal, so that

$$0 = B - A$$

and also the constant terms are equal so that

$$1 = Aa + Ba$$

From these two equations the values of A and B which satisfy the identity can be found and the result is

$$A = \frac{1}{2a}, \qquad B = \frac{1}{2a}$$

The original problem can now be restated as

$$\int \frac{1}{(a^2 - x^2)} dx = \int \left[\frac{1/2a}{(a + x)} + \frac{1/2a}{(a - x)} \right] dx$$

$$= \frac{1}{2a} \left[\int \frac{1}{(a + x)} dx + \int \frac{1}{(a - x)} dx \right]$$

$$= \frac{1}{2a} [\log_e (a + x) - \log_e (a - x)] + C$$

The negative sign in front of the second term is due to the fact that the denominator of the second fraction has a negative sign prefixing the variable x. Now the difference between the logarithms of two numbers can be written as the logarithm of the quotient of the numbers:

$$\int \frac{1}{(a^2 - x^2)} dx = \frac{1}{2a} \log_e \left(\frac{a + x}{a - x} \right) + C$$

The method of partial fractions can be extended to cover cases where the denominator is of higher degree than the second. It may be possible to split the denominator into several factors but this is likely to result in a factor which is a quadratic. In this case we let the numerator of the fraction containing this factor be linear in x. For example,

$$\int \frac{1}{(x + 2)(x^2 + 3x + 1)} dx \equiv \int \left[\frac{A}{(x + 2)} + \frac{Bx + C}{(x^2 + 3x + 1)} \right] dx$$

where $\qquad 1 \equiv A(x^2 + 3x + 1) + (Bx + C)(x + 2)$

$$\equiv Ax^2 + 3Ax + A + Bx^2 + 2Bx + Cx + 2C$$

$$\equiv x^2(A + B) + x(3A + 2B + C) + (A + 2C)$$

Then equating the coefficients of x^2, of x and of the constant term on each side of the identity we obtain three equations in three unknowns.

$$A + B = 0 \tag{1}$$

$$3A + 2B + C = 0 \tag{2}$$

$$A + 2C = 1 \tag{3}$$

Solving these equations gives

$$A = -1, \qquad B = 1, \qquad C = 1$$

$$\therefore \frac{1}{(x + 2)(x^2 + 3x + 1)} = \frac{-1}{x + 2} + \frac{x + 1}{x^2 + 3x + 1}$$

The first term on the right hand side is integrated easily:

$$\int \frac{-1}{x + 2} dx = -\log_e(x + 2)$$

The second term on the right hand side requires further consideration. It can be written as

$$\frac{x + 1}{x^2 + 3x + 1} = \frac{x + \frac{3}{2} - \frac{1}{2}}{x^2 + 3x + 1}$$

$$= \frac{\frac{1}{2}(2x + 3 - 1)}{x^2 + 3x + 1}$$

$$= \frac{1}{2}\left[\frac{2x + 3}{x^2 + 3x + 1} - \frac{1}{x^2 + 3x + 1} \right]$$

Again, the first term on the right hand side can be integrated easily because the numerator is the derivative of the denominator.

$$\int \frac{1}{2}\left(\frac{2x + 3}{x^2 + 3x + 1} \right) dx = \frac{1}{2} \log_e (x^2 + 3x + 1)$$

The only term remaining is

$$\int \frac{-1}{x^2 + 3x + 1} dx$$

Since $-(x^2 + 3x + 1) = (\frac{5}{4}) - (x + \frac{3}{2})^2$ and, from above,

$$\int\frac{1}{a^2 - x^2} dx = \frac{1}{2a}\log_e\left(\frac{a + x}{a - x}\right) + C$$

then $\quad \int\frac{1}{\frac{5}{4} - (x + \frac{3}{2})^2} dx = \frac{2}{2\sqrt{5}}\log_e\left(\frac{\sqrt{\frac{5}{4}} + x + \frac{3}{2}}{\sqrt{\frac{5}{4}} - x - \frac{3}{2}}\right) + C$

Collecting the three parts together gives

$$\int\frac{1}{(x + 2)(x^2 + 3x + 1)} dx = -\log_e(x + 2) + \tfrac{1}{2}\log_e(x^2 + 3x + 1)$$

$$+ \frac{1}{2\sqrt{5}}\log_e\left(\frac{\sqrt{\frac{5}{4}} + x + \frac{3}{2}}{\sqrt{\frac{5}{4}} - x - \frac{3}{2}}\right) + C$$

The result could be further simplified but this is left to the reader. The method may appear rather complicated but it serves to illustrate the way in which the integral of more complex functions can often be obtained by breaking down the function in successive stages into a series of terms for which the integral is known.

The major problem is to recognise the most suitable way in which to break the function down in order to arrive at a standard form. Care must then be taken in the application of such techniques as partial fractions to ensure that the function which is finally integrated is in fact identical in all respects with the original function.

INTEGRATION BY SUBSTITUTION

This can be illustrated by a simple example where the result can be recognised at a glance.

Example 3

$$\int\frac{2x}{x^2 + 2}dx$$

Let $\quad\quad\quad\quad\quad\quad\quad u = x^2 + 2$

Then $\quad\quad\quad\quad\quad\quad \frac{du}{dx} = 2x$

and we treat this as $\quad\quad du = 2x\,dx$

Substituting these values in the original expression, we have

$$\int \frac{2x}{x^2 + 2} dx = \int \frac{1}{u} du = \log_e u + C$$

Then substituting back the value $(x^2 + 2)$ for u we have

$$\int \frac{2x}{x^2 + 2} dx = \log_e(x^2 + 2) + C$$

This result was, of course, obvious. A more realistic example is the following.

Example 4

$$\int (4x - 3)^6 dx$$

This could be obtained by use of the binomial expansion. However,

let $u = 4x - 3$ so that $du = 4 dx$

Then $\int (4x - 3)^6 dx = \int \frac{u^6}{4} du = \frac{u^7}{28} + C$

$$= \frac{(4x - 3)^7}{28} + C$$

Other examples of integration by substitution are provided in Appendix A, sections A.6 and A.7, where trigonometric substitutions are used. Some standard integrals considered in this chapter are given in Table 6.1.

6.15   Exercises

Integrate the following expressions with respect to x.

1. $(5x^2 + 4)e^{3x}$ 2. $x^2 \log_e x$

3. $\dfrac{3}{(x + 1)(x - 1)}$ 4. $\dfrac{x}{(x - 2)(x^2 - 3)}$

5. $(4x + 3)^{10}$ 6. $\log_e x$

6.16   Revision exercises for Chapter 6 (without answers)

1. Integrate the following functions with respect to x:

(a) $y = 3x^2 - 6x + 4$

(b) $y = 2x^4 - 3x^{-3}$

TABLE 6.1

Some standard integrals
(The constants of integration are omitted.)

Function	Integral	Conditions
ax^n	$\dfrac{ax^{n+1}}{n+1}$	$n \neq -1$
$\dfrac{1}{x}$	$\log_e x$	
$\dfrac{1}{u}\dfrac{du}{dx}$	$\log_e u$	u is a function of x
$a\,e^{bx}$	$\dfrac{a\,e^{bx}}{b}$	
$\cos x$	$\sin x$	
$\sin x$	$-\cos x$	
$u\,dv$	$uv - \int v\,du$	u, v are functions of x

(c) $y = 3x/(x^2 + 5)$

(d) $y = (3x^2 - 6)/(x^3 - 6x)$

(e) $y = 3e^x - 2e^{-4x}$

(f) $y = \sin x - \cos x + \tan x$.

2. If marginal revenue is given by $MR = 50 - 2q$ where q is quantity sold, and total revenue is zero when q is zero, determine the equation of the total revenue curve. What is the elasticity of demand when $q = 25$?

3. Evaluate the area between $x = 1$ and $x = 8$ if

(a) $y = 8x - 3$

(b) $y = x^4 + 2x^{-1}$

(c) $y = 2e^{-0.4x}$

(d) $y = 2x/(2x^2 + 4)$.

4. For a product the marginal revenue is given by $MR = 170 - 3q$ where q is the level of sales. What is the extra revenue gained by increasing sales from 40 to 50 units?

5. Assuming that the marginal cost curve is linear, and that when q (output) is 5, marginal cost is 140, while q is 10, marginal cost is 130, find the total cost of increasing output from 50 to 60.

6. Sketch the following curves and hence evaluate the absolute areas between the curves, the x-axis and the given values of x:

   (a) $y = x^2 - 1$ between $x = -2$ and $x = 2$

   (b) $y = x^3$ between $x = -3$ and $x = 1$

   (c) $y = x^2 - 5x + 6$ between $x = 0$ and $x = 6$.

7. The demand curve for a product is $p = 200 - 3q - 2q^2$ and the supply curve is $p = 4 + 11q$. At the equilibrium price determine the elasticity of demand, the consumer's surplus and the producer's surplus. Sketch the curves and indicate the consumer's surplus and the producer's surplus.

8. If the marginal revenue function for a product is $MR = 300 - 4q$ and marginal cost is given by $MC = 6q + 10$, what is the maximum value of profit (ignoring fixed costs)?

9. Evaluate the integral of $y = 1/(x - 1)$ between $x = 1$ and $x = 5$ by
   (a) direct integration
   (b) using Simpson's rule with $n = 4$
   (c) using Taylor's series with four terms and compare the results.

10. Obtain the integrals of the following functions:

    (a) $y = xe^{-2x}$

    (b) $y = \log(2x - 4)$

    (c) $y = 3x/(x + 2)(x^2 - 1)$

    (d) $y = (x^2 - 1)^{15}$

    (e) $y = x \sin x$

    (f) $y = \cos 3x$.

Chapter 7
Partial Differentiation

7.1 Functions of more than one variable

So far we have been concerned with the relationship between pairs of variables, such as between y and x in calculus and between price and quantity in demand analysis. In doing this it has been necessary to make the common assumption that *ceteris paribus*, or that everything else is unchanged. However, if our methods are to be useful in the real world, we need to relax this assumption and take account of other variables which are known to be important. For example, the demand for a good depends not only on its own price but also on other prices, on consumers' incomes, on the amount of advertising and on other factors which are generally ignored such as the weather, the location of shops and the availability of credit. Thus the demand function for a particular brand of ice cream might be of the form

$$q = f(p_0, p_1, p_2, Y, A, W, \ldots)$$

where q is the quantity demanded, p_0 is the price of the brand, p_1 is the price of a competing brand, p_2 is a measure of the general price level, Y is the level of consumers' income, A is expenditure on advertising, W is an index of the weather, and so on. We need to become familiar with methods which allow us to handle such a situation.

These methods are a logical extension of the differential calculus, but to avoid confusion a new and slightly different notation is required. To introduce this notation and to illustrate its use, let us consider a three-variable model which it is possible to represent in diagrammatic form.

Let x and y be two independent variables and z the dependent variable. Then $z = f(x, y)$ and if we represent x and y on axes at right angles to each other in the horizontal plane we can imagine z as being on an axis in the vertical plane and, therefore, at right angles to both x and y. Any variation in x or y then causes a change in z and the point

257

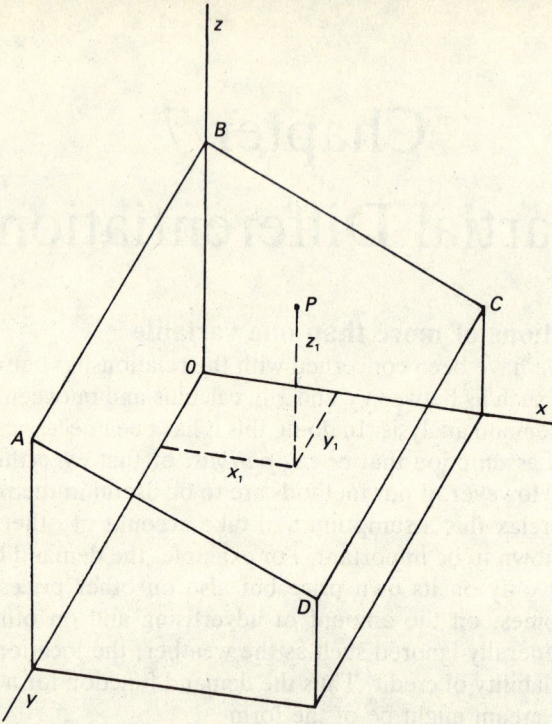

Fig. 7.1

representing z moves, in fact, in such a way as to trace out a surface above the plane xy rather like a hill or series of hills on the earth's surface.

The point P in Fig. 7.1 has a height above the xy-plane equal to z, and therefore its z-coordinate is given by $z = z_1$. To determine uniquely the location of P in the three-dimensional space it is also necessary to know the distance of the point from the yz-plane and from the xz-plane. These distances are $x = x_1$ and $y = y_1$ respectively and the point P can then be specified uniquely in relation to the three axes by the co-ordinates (x_1, y_1, z_1).

Because $z = f(x, y)$ the point P moves over the surface $ABCD$ as the values of x and y change, and we are interested in measuring the rate of change of z with respect to changes in the other variables. In fact, what is required is some value similar to the derivative, which can be used in the three-variable case to measure rates of change.

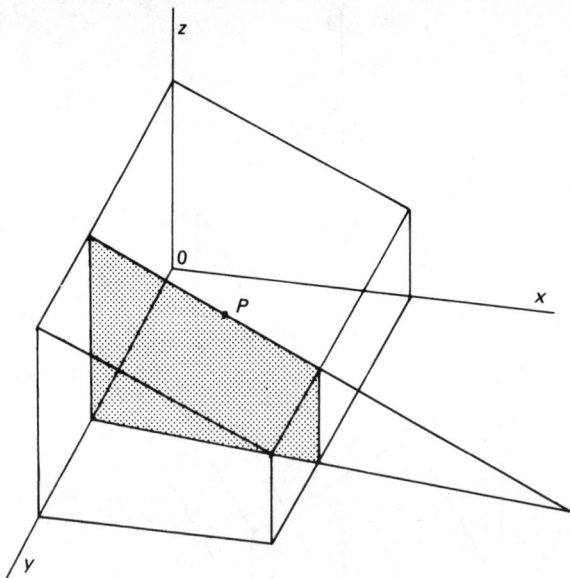

Fig. 7.2

This gives us a basis on which to work. If one of the variables, say y, is considered to be fixed, then this is equivalent to taking a section through the surface parallel to the x-axis as shown in Fig. 7.2. Any change in z on this section is then due entirely to the change in x.

Any point P on this section is, therefore, similar to the point in a two-dimensional model. A tangent can be drawn to the surface at P parallel to the x-axis and the slope of this line is a measure of the rate of change of z with respect to x when y is kept constant. It is therefore a form of derivative, but because it is the rate of change with respect to only one variable, the other variable being held constant, it is called a partial derivative, and to prevent confusion with dz/dx it is denoted by $\partial z/\partial x$ or z_x.

In a similar manner it is possible to take a section through the surface for a constant value of x and determine the rate of change of z as y changes. This ratio is written $\partial z/\partial y$ or z_y and is a measure of the slope of the section shown in Fig. 7.3.

It is not easy to obtain the value of the derivative from a diagram and, as might be expected, it is not necessary because with only minor modification, the differential calculus can be used to obtain the result

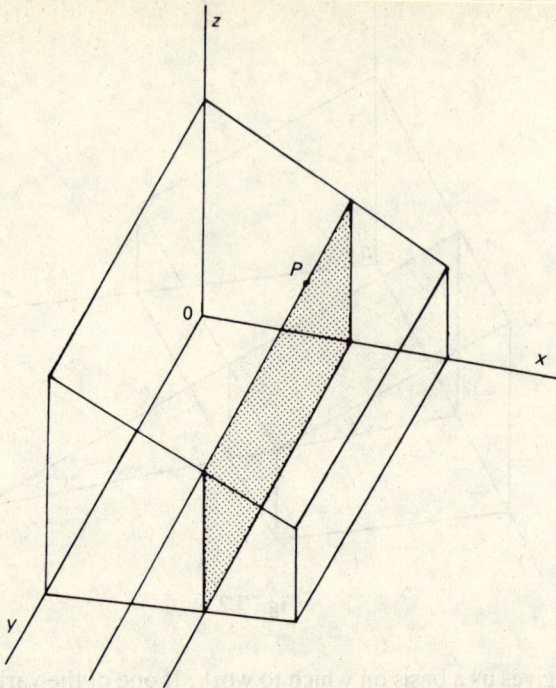

Fig. 7.3

very simply. The modification is necessary because z_x refers to the partial derivative of z with respect to the variable x and implies that all other variables in the function must be considered as constant.
For example, if

$$z = x^2 + y^2$$

$$z_x = \frac{\partial z}{\partial x} = 2x$$

This follows because y^2 is considered to be a constant and the derivative of a constant is zero. Similarly,

$$z_y = \frac{\partial z}{\partial y} = 2y$$

because x^2 is considered as a constant when the partial derivative with respect to y is required.

The same procedure applies if the expression is more complicated. For example,

$$z = x^2 + 3xy + y^2$$

To obtain the partial derivatives of z with respect to x only, we treat y as though it were a constant wherever it appears.

$$z_x = \frac{\partial z}{\partial x} = 2x + 3y$$

Similarly
$$z_y = \frac{\partial z}{\partial y} = 3x + 2y$$

The second-order partial derivatives are defined in a similar way to second-order derivatives, i.e. they are the result of differentiating the first-order derivatives. We define

$$z_{xx} = \frac{\partial^2 z}{\partial x^2} = \frac{\partial}{\partial x}(z_x)$$

$$z_{yy} = \frac{\partial^2 z}{\partial y^2} = \frac{\partial}{\partial y}(z_y)$$

$$z_{yx} = \frac{\partial^2 z}{\partial y \partial x} = \frac{\partial}{\partial y}\left(\frac{\partial z}{\partial x}\right) = \frac{\partial}{\partial y}(z_x)$$

$$z_{xy} = \frac{\partial^2 z}{\partial x \partial y} = \frac{\partial}{\partial x}\left(\frac{\partial z}{\partial y}\right) = \frac{\partial}{\partial x}(z_y)$$

Here there are four second-order derivatives, z_{xx} and z_{yy} which correspond to d^2y/dx^2 and also the two 'mixed' derivatives z_{xy} and z_{yx}. The notation used indicates the order in which the variables occur, i.e. z_{yx} is found by differentiating z with respect to x and then with respect to y. For all values of x and y for which z_{xy} and z_{yx} are continuous:

$$z_{xy} = z_{yx}$$

and the order of differentiation does not matter.

For example $z = x^2 + 3xy + 2y^3 - 5x - 4y + 20$

$$z_x = \frac{\partial z}{\partial x} = 2x + 3y - 5$$

$$z_y = \frac{\partial z}{\partial y} = 3x + 6y^2 - 4$$

$$z_{xx} = \frac{\partial}{\partial x}(z_x) = 2$$

$$z_{yy} = \frac{\partial}{\partial y}(z_y) = 12y$$

$$z_{yx} = \frac{\partial}{\partial y}(z_x) = 3$$

$$z_{xy} = \frac{\partial}{\partial x}(z_y) = 3$$

and hence $\qquad z_{xy} = z_{yx}$

The above definitions of first- and second-order partial derivatives can easily be generalised to cover the cases of more variables and higher-order derivatives.

One application of partial derivatives is in testing for functional dependence. Suppose that $w = f(x, y)$ and $z = g(x, y)$ so that the first-order partial derivatives are w_x, w_y, z_x and z_y; then, using determinants (see section 1.13), if the special determinant,

$$|\mathbf{J}| = \begin{vmatrix} w_x & w_y \\ z_x & z_y \end{vmatrix}$$

is zero, w and z are linearly or non-linearly related. Here $|\mathbf{J}|$ is called a *Jacobian determinant* or simply a *Jacobian*. For example, consider

$$w = x + 2y$$
$$z = x^2 + 4xy + 4y^2$$

then the first-order partial derivatives are

$$w_x = 1, \ w_y = 2, \ z_x = 2x + 4y, \ z_y = 4x + 8y$$

and

$$|\mathbf{J}| = \begin{vmatrix} 1 & 2 \\ 2x + 4y & 4x + 8y \end{vmatrix} = 4x + 8y - (4x + 8y) = 0$$

Since the Jacobian is zero, w and z are related and it can be seen that $z = w^2$.

7.2 Marginal analysis

This is an extension of the two-dimensional model and is best explained by the use of an example.

Let us assume a function in which the level of output is related to the quantity of inputs x and y in the following way:

$$q = f(x, y) = 50x - x^2 + 60y - 2y^2$$

then
$$\frac{\partial q}{\partial x} = 50 - 2x$$

$$\frac{\partial q}{\partial y} = 60 - 4y$$

These partial derivatives denote the marginal productivity of factor x and factor y respectively, and in this example they are a function of the amount already in use. That is, if we are at present using 10 units of each factor then

$$\frac{\partial q}{\partial x} = 50 - 2 \times 10 = 30$$

$$\frac{\partial q}{\partial y} = 60 - 4 \times 10 = 20$$

and the marginal productivity of x is 30 whilst that of y is 20.

An example which is more useful in economics is the *Cobb–Douglas production function*

$$q = AL^\alpha K^\beta$$

where
$$q = \text{output}$$
$$K = \text{capital input}$$
$$L = \text{labour input}$$

A, α and β are constants. The marginal product of labour is

$$q_L = \frac{\partial q}{\partial L} = \alpha AL^{\alpha-1}K^\beta = \frac{\alpha AL^\alpha K^\beta}{L} = \frac{\alpha q}{L}$$

and the marginal product of capital is

$$q_K = \frac{\partial q}{\partial K} = \beta AL^\alpha K^{\beta-1} = \frac{\beta AL^\alpha K^\beta}{K} = \frac{\beta q}{K}$$

It can easily be seen that these marginal products are positive if $\alpha > 0$ and $\beta > 0$, since q, K and L are all positive. Now the average product of labour is q/L and the average product of capital is q/K and so the above equations can be rearranged to give

$$\alpha = \frac{q_L}{q/L} = \frac{\text{marginal product of labour}}{\text{average product of labour}}$$

and

$$\beta = \frac{q_K}{q/K} = \frac{\text{marginal product of capital}}{\text{average product of capital}}$$

It is also easily seen that if the wage is q_L and the cost of capital is q_K, so that each factor is paid its marginal product, then the total cost is

$$q_L L + q_K K = \alpha q + \beta q = (\alpha + \beta)q$$

This will equal output, q, if $\alpha + \beta = 1$, which occurs when there are constant returns to scale (see section 7.10).

The way in which these marginal products are changing can be seen from the second-order derivatives

$$q_{LL} = \frac{\partial^2 q}{\partial L^2} = \frac{\alpha[Lq_L - q]}{L^2} = \frac{\alpha(\alpha - 1)q}{L^2}$$

$$q_{KK} = \frac{\partial^2 q}{\partial K^2} = \frac{\beta[Kq_K - q]}{K^2} = \frac{\beta(\beta - 1)q}{K^2}$$

It is usually a requirement of production functions that the marginal products of capital and labour should, at least for low values of inputs, be positive, and as the values of the inputs increase the marginal products decline. That is, $q_K > 0$, $q_L > 0$, and $q_{KK} < 0$, $q_{LL} < 0$. This occurs with the Cobb–Douglas production function if $1 > \alpha > 0$ and $1 > \beta > 0$.

7.3 Elasticity of demand

If the quantity demanded of a good (A) is determined by the price of that good and also the price of a substitute (B) then the demand function can be written as

$$q_D = f(p_A, p_B)$$

Here there are two important elasticities. The first concerns the change in the quantity of A demanded when the price of A changes and the price of B is held constant. This can be written in partial derivatives as

$$\frac{\partial q_D / q_D}{\partial p_A / p_A} \quad \text{or} \quad \frac{\partial q_D}{\partial p_A} \frac{p_A}{q_D}$$

and is referred to as the *partial elasticity of demand for good* A *with respect to the price of good* A, or the *own-price elasticity of demand for good* A. It is expected to be negative for normal goods.

For example, suppose that

$$q_D = f(p_A, p_B) = 100 - 3p_A + p_B$$

then

$$\frac{\partial q_D}{\partial p_A} = -3$$

∴ partial elasticity of demand $= -3 \dfrac{p_A}{q_D}$

If the present prices are $p_A = 10$, $p_B = 20$

$$q_D = 100 - 3(10) + 20 = 90$$

∴ partial elasticity of demand $= -3 \cdot \dfrac{10}{90} = -\frac{1}{3}$

This can be interpreted as meaning that the quantity demanded of A falls by $\frac{1}{3}\%$ if its price is increased by 1% and the price of B remains unchanged.

The second elasticity gives the reaction of the quantity of good A demanded to a change in the price of good B. Using partial derivatives this is

$$\frac{\partial q_D / q_D}{\partial p_B / p_B} \quad \text{or} \quad \frac{\partial q_D}{\partial p_B} \frac{p_B}{q_D}$$

and is called the *partial elasticity of demand for good* A *with respect to the price of good* B, or the *cross-elasticity of demand for* A *with respect to* B. If A and B are complementary goods, this cross-elasticity will be negative, while if they are substitutes it will be positive. The higher the value of the cross-elasticity the stronger is the degree of substitutability or complementarity of A and B.

In the example used previously, when $p_B = 20$,

$$\frac{\partial q_D}{\partial p_B} = 1 \text{ and } q_D = 90$$

The cross-elasticity of demand $= 1 \dfrac{(20)}{90} = 0.22$

Therefore the demand for A increases by 0.22% if its price remains unchanged and the price of B increases by 1%. Here A and B are substitutes.

This analysis extends naturally to the case of more than two independent variables. The partial derivative with respect to any one of the variables is obtained by differentiating the function with respect to this variable and treating all other variables as constants.

For example, suppose that

$$q_D = a_1 p_1 + a_2 p_2 + a_3 p_3 + a_4 p_4$$

then
$$\frac{\partial q_D}{\partial p_1} = a_1, \quad \frac{\partial q_D}{\partial p_2} = a_2$$

$$\frac{\partial q_D}{\partial p_3} = a_3, \quad \frac{\partial q_D}{\partial p_4} = a_4$$

As a further example consider the demand function

$$q = A \, p_1^\alpha \, p_2^\beta \, Y^\tau$$

where q is the quantity of a good demanded, p_1 is the price of the good, p_2 is the price of a competing good and Y is the value of consumers' income. Here it is convenient to take logarithms (to the base e), giving

$$\log q = \log A + \alpha \log p_1 + \beta \log p_2 + \tau \log Y$$

The partial derivatives are obtained as above by differentiating with respect to each variable in turn and treating the other variables as constants. The function of a function rule is used to differentiate $\log q$ (see section 5.2). The partial derivative with respect to p_1 is

$$\frac{1}{q} \frac{\partial q}{\partial p_1} = \frac{\alpha}{p_1} \quad \text{or} \quad \alpha = \frac{p_1}{q_1} \frac{\partial q}{\partial p_1}$$

and thus the own-price elasticity of demand is α. Similarly, the cross-elasticity is β and the income elasticity is τ. Since all of these are unrelated to q and the other variables, it is clear that the particular function under consideration is a *constant elasticity demand function*.

7.4   Exercises

1. Find $\partial z / \partial x$ and $\partial z / \partial y$ for the following functions:

(a) $z = xy + x^2 y + xy^2$

(b) $z = 3x^2 y + 4xy^2 + 6xy$

(c) $z = x^2 \sin y + y^2 \cos x$

(d) $z = (x + y)e^{(x+y)}$

(e) $z = \log (x^2 + y^2)$

(f) $z = \dfrac{x + y}{x - y}$

(g) $z = \dfrac{x + \sin y}{y + \sin x}$

2. For the production function

$$q = 2(K - 30)^2 + 3(L - 20)^2 + 2(K - 30)(L - 20)$$

show that the marginal products of capital (K) and labour (L) are equal when $K = 50$ and $L = 30$.

3. Let the demand for apples be given by

$$q = 240 - p_a^2 + 6p_0 - p_a p_0$$

where q is the quantity of apples demanded at price p_a, and p_0 is the price of oranges. Evaluate the partial elasticity of demand for apples when $p_a = 5$ and $p_0 = 4$ with respect to

(a) the price of apples

(b) the price of oranges.

4. For a production function $Q = f(K, L)$ the partial elasticity of output with respect to the labour input is defined as $(\partial Q / \partial L)$ (L/Q). Show that for the Cobb–Douglas function $Q = AL^\alpha K^\beta$ this partial elasticity is α.

5. In section 1.13 we saw that the condition for the linear equations

$$a_{11}x_1 + a_{12}x_2 = b_1$$

$$a_{21}x_1 + a_{22}x_2 = b_2$$

to have a unique solution (and so be independent) was that $|A| \neq 0$. That is, if $|A| = 0$ the equations are dependent. Show that this condition is the same as the requirement that the Jacobian is zero.

7.5 Differentials

In the two-variable model, $y = f(x)$, we obtained a value for the magnitude of the change in y corresponding to a small but finite change in x by using the expression

$$dy = \frac{dy}{dx} \, dx$$

This procedure can be extended when there are more variables in the model. In fact, for $z = f(x, y)$ the small change in z is equal to the small change in x multiplied by the rate of change of z with respect to x, keeping y constant, plus the small change in y multiplied by the rate of change of z with respect to y keeping x constant. This is conveniently written as

$$dz = \frac{\partial z}{\partial x} \, dx + \frac{\partial z}{\partial y} \, dy$$

This can be extended to any number of variables. For example,

suppose $z = f(p, q, r, s, \ldots)$

then $dz = \frac{\partial z}{\partial p} \, dp + \frac{\partial z}{\partial q} \, dq + \frac{\partial z}{\partial r} \, dr + \frac{\partial z}{\partial s} \, ds + \cdots$

This expression is known as the *total differential* of z.

This can be illustrated by reference to the production function relating the quantity of wheat produced to the quantity of labour (x) and land (y) which is employed.

$$q = 40x - x^2 + 60y - 2y^2$$

$$\frac{\partial q}{\partial x} = 40 - 2x$$

$$\frac{\partial q}{\partial y} = 60 - 4y$$

∴ $dq = \frac{\partial q}{\partial x} \, dx + \frac{\partial q}{\partial y} \, dy$

$$= (40 - 2x) \, dx + (60 - 4y) \, dy$$

This equation can be used to obtain an approximate value for the increased production which is possible when both labour and land are increased together. The smaller the increases in the variables dx and dy that are considered, the more accurate the answer will be. However, an approximate answer can be obtained when the supply of each factor is increased by 1 unit. If the present quantity in use is given by $x = 10$ units, $y = 10$ units, then the approximate increase will be given by

$$dq = [40 - 2(10)] + [60 - 4(10)]$$

$$= 20 + 20$$

$$= 40$$

The accurate value of the increase can be obtained by substituting $x = 11$, $y = 11$ in the production function to give $q = 737$ and substracting the value of q when $x = 10$, $y = 10$. The accurate increase is 37 units.

Differentials are also useful with *indifference curves*. If the utility function is

$$u = f(q_1, q_2)$$

where q_1 and q_2 are the quantities of two goods consumed, then a given level of utility, u^0, can be achieved with different combinations of q_1 and q_2. The expression

$$u^0 = f(q_1, q_2)$$

gives the *indifference curve* and the total differential of u^0 is

$$du^0 = \frac{\partial f}{\partial q_1} dq_1 + \frac{\partial f}{\partial q_2} dq_2$$

Since u^0 is a constant (i.e. any movement is along the indifference curve) then $du^0 = 0$ and

$$0 = f_1 \, dq_1 + f_2 \, dq_2$$

or

$$\frac{-dq_2}{dq_1} = \frac{f_1}{f_2} = \frac{\text{marginal utility of good 1}}{\text{marginal utility of good 2}} = MRS$$

This ratio is the *marginal rate of substitution* (*MRS*) or *rate of technical substitution* of q_1 for q_2 and is the negative of the slope of the indifference curve at any point. It gives the number of units of good 2 which must be given up in exchange for an extra unit of good 1 while keeping utility constant. For example, if $MRS = 2$ then we have

$$dq_2 = -2dq_1$$

and so increasing the consumption of good 1 by one unit requires a reduction of two units in the consumption of good 2.

The same concept can be applied to production functions where an *isoquant* is the locus of all technically efficient methods of producing a given level of output. If output (Q) depends on the level of capital services (K) and labour services (L) then

$$Q = f(K, L)$$

and an isoquant is

$$Q^0 = f(K, L)$$

Moving along the isoquant, so that $dQ^0 = 0$,

$$dQ^0 = Q_K \, \partial K + Q_L \, \partial L = 0$$

and solving,

$$-\frac{\partial K}{\partial L} = \frac{Q_L}{Q_K} = MRS$$

which is the marginal rate of substitution of labour for capital.

The marginal rate of substitution has the disadvantage that its size depends on the units of measurement of K and L. An alternative is to define the *elasticity of substitution* (σ) as the percentage change in the capital–labour ratio as a proportion of the percentage change in the marginal rate of substitution. Therefore,

$$\sigma = \frac{\text{percentage change in } K/L}{\text{percentage change in } MRS}$$

$$= \frac{d(K/L)/(K/L)}{d(MRS)/(MRS)}$$

To illustrate these concepts we will consider the Cobb–Douglas production function (see section 7.2)

$$Q = AL^\alpha K^\beta$$

for which the marginal products of labour and capital are

$$Q_L = \frac{\alpha Q}{L} \qquad Q_K = \frac{\beta Q}{K}$$

The marginal rate of substitution of labour for capital is

$$MRS = \frac{Q_L}{Q_K} = \frac{\alpha Q/L}{\beta Q/K} = \frac{\alpha}{\beta} \frac{K}{L}$$

Now the elasticity of substitution is

$$\sigma = \frac{d(K/L)/(K/L)}{d(MRS)/(MRS)}$$

$$= \frac{d(K/L)\ (MRS)}{d(MRS)\ K/L}$$

and rearranging the expression for *MRS*,

$$K/L = MRS\ (\beta/\alpha)$$

so that, differentiating K/L with respect to *MRS* gives

$$\frac{d(K/L)}{d(MRS)} = \frac{\beta}{\alpha}$$

and hence $$\sigma = \frac{\beta}{\alpha}\ \frac{MRS}{K/L} = \frac{\beta}{\alpha}\ \frac{(\alpha K/\beta L)}{(K/L)} = 1$$

Therefore the elasticity of substitution for a Cobb–Douglas production function is 1.

This raises the question of what kind of function has a constant elasticity of substitution (*CES*) which differs from 1. The *CES* production function is defined as

$$Q = \tau[(1 - \delta)K^{-\Phi} + \delta L^{-\Phi}]^{-\Omega/\Phi}$$

where τ, δ, Φ and Ω are parameters. Here τ is the efficiency parameter, so that the larger the value of τ the greater the level of output for given K and L, δ is related to the marginal products of K and L, Ω is the returns to scale parameter which will be discussed in section 7.10 below, and we will show that Φ is related to the elasticity of substitution. To simplify the algebra we will let $\tau = 1$ and $\Omega = 1$ so that

$$Q = [(1 - \delta)K^{-\Phi} + \delta L^{-\Phi}]^{-1/\Phi}$$

The marginal products are obtained using the 'function of a function' rule:

$$Q_L = (-1/\Phi)\delta(-\Phi)L^{-\Phi-1}[(1-\delta)K^{-\Phi} + \delta L^{-\Phi}]^{-(1/\Phi)-1}$$

$$= \frac{\delta\ Q^{1+\Phi}}{L^{1+\Phi}}$$

$$Q_K = (-1/\Phi)(1-\delta)(-\Phi)K^{-\Phi-1}[(1-\delta)K^{-\Phi} + \delta L^{-\Phi}]^{-(1/\Phi)-1}$$

$$= \frac{(1-\delta)Q^{1+\Phi}}{K^{1+\Phi}}$$

and the marginal rate of substitution is

$$MRS = \frac{Q_L}{Q_K} = \frac{\delta}{(1 - \delta)} \frac{K^{1+\Phi}}{L^{1+\Phi}}$$

and rearranging,

$$K/L = [(1 - \delta)MRS/\delta]^{1/(1+\Phi)}$$

The elasticity of substitution is

$$\sigma = \frac{d(K/L)}{d(MRS)} \frac{MRS}{K/L}$$

and since

$$\frac{d(K/L)}{d(MRS)} = \frac{1}{1 + \Phi} \frac{(1 - \delta)}{\delta} [(1 - \delta)MRS/\delta]^{-\Phi/(1+\Phi)}$$

we have

$$\sigma = \frac{1}{1 + \Phi} \frac{(1 - \delta)}{\delta} \frac{[(1 - \delta)MRS/\delta]^{-\Phi/(1+\Phi)} \, MRS}{[(1 - \delta)MRS/\delta]^{1/(1+\Phi)}}$$

and simplifying this expression gives the result

$$\sigma = \frac{1}{1 + \Phi}$$

Thus the elasticity of substitution is constant and is related to Φ.

The ideas of differentials can also be used in IS–LM models. Suppose that the IS equation is

$$Y = C(Y) + I(r) + G$$

where Y is national income, C is aggregate consumption, G is government expenditure and r is the rate of interest. We assume that the partial derivatives satisfy

$$0 < C_Y < 1 \qquad I_r < 0$$

The LM equation is $M/P = L(r, Y)$
where M is the quantity of money and P is the price level and $L_r < 0$, $L_Y > 0$. Here Y and r are endogenous and G and P are assumed to be fixed. To determine the effect on Y of an increase in M we find the total differentials of the IS and LM curves. Taking these in turn:

$$dY = C_Y dY + I_r dr + dG$$

$$\frac{dM}{P} - \frac{M}{P^2} dP = L_r dr + L_Y dY$$

Setting $dG = dP = 0$ and rearranging gives

$$(1 - C_Y)dY - I_r dr = 0$$

$$L_Y \, dY + L_r dr = dM/P$$

These can be solved by Cramer's rule (see section 1.13) or by substitution. Since we are interested in dY/dM we solve the first equation for dr and substitute into the second to give

$$L_Y \, dY + L_r(1 - C_Y)dY/I_r = dM/P$$

or

$$dY[L_Y + L_r(1 - C_Y)/I_r] = dM/P$$

and so

$$\frac{dY}{dM} = \frac{I_r/P}{L_Y I_r + L_r(1 - C_Y)}$$

This gives the effect on Y of a change in M. Now we know that $I_r < 0$, $L_Y > 0$, $L_r < 0$ and $0 < C_Y < 1$. The numerator is therefore negative (since $P > 0$). The product $L_Y I_r$ is also negative, as is $L_r(1 - C_Y)$. Therefore the ratio must be positive and so an increase in M will result in an increase in Y in this model.

7.6 Total derivatives

Many situations arise where the independent variables can all be expressed as a function of a single variable such as time. For example,

$$z = x^2 + y^2$$

with

$$x = e^t \cos t, \qquad y = e^t \sin t$$

What is required is the change in z which is brought about by the change in t as the latter affects both x and y. This can be obtained by substituting the function of t for x and y in the main equation and then differentiating this expression with respect to t. The same result can be obtained in a simpler way by making use of the results of section 7.5.

$$dz = \frac{\partial z}{\partial x} \, dx + \frac{\partial z}{\partial y} \, dy$$

Dividing through this expression by dt

$$\frac{dz}{dt} = \frac{\partial z}{\partial x}\frac{dx}{dt} + \frac{\partial z}{\partial y}\frac{dy}{dt}$$

This is similar to the function of a function or chain rule discussed in section 5.2.

For example in the case where $z = x^2 + y^2$ and $x = e^t \cos t$, $y = e^t \sin t$, we can form the derivatives

$$\frac{dx}{dt} = e^t \cos t - e^t \sin t = e^t(\cos t - \sin t)$$

$$\frac{dy}{dt} = e^t \cos t + e^t \sin t = e^t(\cos t + \sin t)$$

$$\frac{dz}{dt} = \frac{\partial z}{\partial x}\frac{dx}{dt} + \frac{\partial z}{\partial y}\frac{dy}{dt}$$

$$= 2xe^t(\cos t - \sin t) + 2ye^t(\cos t + \sin t)$$

This can be rearranged by gathering together terms containing $\cos t$ and terms containing $\sin t$ with the result

$$\frac{dz}{dt} = e^{2t}(2\cos^2 t - 2\sin t \cos t + 2\cos t \sin t + 2\sin^2 t) = 2e^{2t}$$

using result (4) of Appendix A.1.

7.7 Exercises

1. The quantity of output of a product depends on the labour (L) and capital (C) inputs in such a way that

$$q = L^3 - 3L + 4LC - C^2$$

   If $L = 10$, $C = 5$, what is the approximate increase in q for an increase of 1 in both L and C? What is the approximate increase in q if L alone increases by 1?

2. The demand for apples (a) is related to the price of oranges (o) by

$$q = 240 - p_a^2 + 6p_o - p_a p_o$$

   What is the approximate change in the demand for apples if the price of apples increased from $p_a = 5$ to $p_a = 6$ and the price of oranges changes from $p_o = 4$ to $p_o = 5$?

3. Find dx/dt, dy/dt and dz/dt for the following functions:

   (a) $z = x^2 - y^2$, $x = e^t \cos t$, $y = e^t \sin t$

   (b) $z = \log_e (x + y)$, $x = t^2 e^t$, $y = \sin t$

   (c) $z = ye^{x^2}$, $x = e^t$, $y = \log_e t$

   (d) $z = x^2 + xy + y^2$, $x = t + \dfrac{1}{t}$, $y = t - \dfrac{1}{t}$

4. The Cobb–Douglas production function can be generalised by allowing technical progress to occur as time passes. This means

that new methods of production are used resulting in a higher level of output for given levels of K and L. If these new processes are introduced gradually their effect can be represented by a variable t, known as a time trend, and if the rate of technical progress is r % per annum then output at time t is given by

$$Q_t = Ae^{rt}L_t^\alpha K_t^\beta$$

Show that (a) the marginal rate of substitution is still $(\alpha K/\beta L)$ and (b) the elasticity of substitution is still unity.

5. For the model

$$Y = C + I + G$$
$$C = a + bY$$
$$I = c - er$$
$$M/P = fY - gr$$

where a, b, c, e, f and $g > 0$, Y = income,
C = consumption, I = investment, r = interest rate,
M = quantity of money and P = price level, and the variables Y, C, I and M are endogenous:
(a) What is the effect on Y of a change in r (assuming that G is fixed)?
(b) What is the effect on M of a change in G (assuming that r and P are fixed)?

7.8 Implicit differentiation

The functions we have met have all been of the form $Y = f(x)$ or $Q = f(K, L)$ which are *explicit functions*, with the dependent variable on the left hand side. In contrast,

$$x^3 + 2x^2y = 3xy^2 + y^3$$

is an *implicit function* in which there is a complicated relationship between y and x. With implicit functions it is not easy to see how the derivative can be found. However, we know that if $z = f(x, y)$ then, allowing x to change by δx and y by δy, the total change in z is approximately,

$$\delta z = \frac{\partial z}{\partial x}\delta x + \frac{\partial z}{\partial y}\delta y$$

Dividing this throughout by δx, we have approximately

$$\frac{\partial z}{\partial x} = \frac{\partial z}{\partial x}\frac{\delta x}{\delta x} + \frac{\partial z}{\partial y}\frac{\delta y}{\delta x}$$

and as $\delta x \to 0$ this becomes

$$\frac{dz}{dx} = \frac{\partial z}{\partial x} + \frac{\partial z}{\partial y}\frac{dy}{dx}$$

If we have a relation in which $z = 0$ (identically), then dz/dx also equals 0.

This fact can be used as follows:

Suppose $z = x^3 + 2x^2y - 3xy^2 - y^3 = 0$

then $\dfrac{dz}{dx} = 0$

$$\frac{\partial z}{\partial x} = 3x^2 + 4xy - 3y^2$$

$$\frac{\partial z}{\partial y} = 2x^2 - 6xy - 3y^2$$

Substituting these values in the total derivative

$$\frac{dz}{dx} = \frac{\partial z}{\partial x} + \frac{\partial z}{\partial y}\frac{dy}{dx}$$

results in

$$0 = (3x^2 + 4xy - 3y^2) + (2x^2 - 6xy - 3y^2)\frac{dy}{dx}$$

and rearranging these terms, we have

$$\frac{dy}{dx} = -\frac{(3x^2 + 4xy - 3y^2)}{(2x^2 - 6xy - 3y^2)}$$

The above method can be generalised and used to determine the derivative of any implicit function of two variables. The function is first rearranged so that all the terms are on one side of the equality sign, i.e. it is made equal to 0. Then dz/dx is equal to zero and the following holds:

$$0 = \frac{\partial z}{\partial x} + \frac{\partial z}{\partial y}\frac{dy}{dx}$$

$$\therefore \qquad \frac{dy}{dx} = -\frac{\partial z/\partial x}{\partial z/\partial y}$$

This is known as the *implicit function rule*.
For example, suppose

$$y^2 = 3xy + x^2$$

then
$$z = y^2 - 3xy - x^2 = 0$$

$$\frac{\partial z}{\partial y} = 2y - 3x$$

$$\frac{\partial z}{\partial x} = -3y - 2x = -(3y + 2x)$$

and
$$\frac{dy}{dx} = \frac{3y + 2x}{2y - 3x}$$

An alternative approach is to differentiate each term with respect to *x*, remembering that the function of a function rule must be used on such terms as y^2 and xy. For example, suppose

$$y^2 = 3xy + x^2$$

Differentiating term by term gives

$$2y \frac{dy}{dx} = 3x \frac{dy}{dx} + 3y + 2x$$

that is
$$\frac{dy}{dx}(2y - 3x) = 3y + 2x$$

or
$$\frac{dy}{dx} = \frac{3y + 2x}{2y - 3x}$$

Similarly:
$$0 = x^3 + 2x^2y - 3xy^2 - y^3$$

Differentiating gives

$$0 = 3x^2 + 2x^2 \frac{dy}{dx} + 4xy - 6xy \frac{dy}{dx} - 3y^2 - 3y^2 \frac{dy}{dx}$$

$$= 3x^2 + 4xy - 3y^2 + \frac{dy}{dx}(2x^2 - 6xy - 3y^2)$$

∴
$$\frac{dy}{dx} = \frac{3y^2 - 3x^2 - 4xy}{2x^2 - 6xy - 3y^2}$$

7.9 Exercises

1. Determine dy/dx if

   (a) $y^3 - 4x^2y^2 + 6x^2y + 4x = 0$

   (b) $x^2y^2 + e^x - e^y = 0$

   (c) $3 \log_e(x + y) + 4x^2 - y^2 = 0$

   (d) $\sin(x + y) - 3xy = 10xy^2$

2. What is the elasticity of demand when $p = 2$ if

$$q^2 = \frac{500 - p^2}{2 + p} + pq$$

3. The relationship between output (Q), capital input (K) and labour input (L) for a particular process is

$$Q^2 = 8KL - 3Q + L^2 + 2K^2$$

Determine the marginal products of capital and labour, the marginal rate of substitution and the elasticity of substitution.

7.10 Homogeneous functions

A function is said to be *homogeneous of degree k* if

$$f(tx, ty) = t^k f(x, y)$$

for all values of t, where t and k are constants.

An example of a homogeneous function of degree one is

$$z = f(x, y) = 2x + 3y$$

Replacing x by tx and y by ty gives

$$f(tx, ty) = 2(tx) + 3(ty) = t(2x + 3y)$$
$$= tf(x, y) = tz$$

The function

$$z = f(x, y) = x^2 + 2xy + 3y^2$$

is homogeneous of degree two. This can be shown by the method above since

$$f(tx, ty) = (tx)^2 + 2(tx)(ty) + 3(ty)^2$$
$$= t^2(x^2 + 2xy + 3y^2) = t^2 z$$

It can also be seen by noticing that if we take each additive term in the expression for z and add the powers of its factors, we find that their sum is 2. By this method it is easy to see that

$$z = f(x, y) = 2 + 2x + 3y$$

is not homogeneous, since the powers of x and y in the first term are 0, and the total power in each of the other terms is 1. Also,

$$f(tx, ty) = 2 + 2tx + 3ty \neq tz$$

A simple example of a homogeneous function of degree one occurs in the production function

$$q = 3L + 5K$$

where q is the level of output resulting from inputs of L units of labour and K units of capital. If both L and K are increased by a factor t the new level of output is

$$q' = 3(tL) + 5(tK) = tq$$

and so doubling the inputs of capital and labour (that is, $t = 2$) doubles the amount of output. This is a case of *constant returns to scale*.

If the degree of homogeneity k is less than one there are *decreasing returns to scale* while if k is greater than one there are *increasing returns to scale*.

For the Cobb–Douglas production function

$$q = AL^{\alpha}K^{\beta}$$

where A, α and β are constants, if the inputs are changed by a factor t the new output is

$$q' = A (tL)^{\alpha} (tK)^{\beta} = A t^{\alpha+\beta} L^{\alpha} K^{\beta}$$
$$= t^{\alpha+\beta} (q)$$

Therefore the degree of homogeneity is $\alpha + \beta$ and if $\alpha + \beta = 1$ there are constant returns to scale. For $\alpha + \beta > 1$ there are increasing returns to scale and for $\alpha + \beta < 1$ there are decreasing returns to scale.

As we saw in section 7.2 above, if there are constant returns to scale and each factor is paid its marginal product then the total cost is the value of output since

$$q_L L + q_K K = q$$

The idea of homogeneous functions has a useful economic application in the context of *money illusion*. Consider the simple demand function

$$q = f(p, p_c, Y)$$

where q is the quantity of a good demanded, p is its price, p_c is the price of a competing product and Y is income. Money illusion occurs if behaviour changes as a result of a change in the value of currency. For example, if all prices and incomes were multiplied by 100, so that they were quoted in pence instead of pounds (or cents instead of dollars), as relative prices and real incomes are unchanged, there should be no change in behaviour, otherwise there is money illusion. Now suppose all prices and incomes change by a factor t, then the new quantity demanded is

$$q_1 = f(tp, tp_c, tY) = t^k f(p, p_c, Y)$$

where k is the degree of homogeneity of the demand function. For no money illusion, so that $q = q_1$, then $k = 0$ and the demand function must be homogeneous of degree zero. An example of such a function is

$$q = b(p/P)^\alpha \, (p_c/P)^\beta \, (Y/P)^\tau$$

where P is the general price level. Here demand depends on real prices (i.e. prices relative to the general price level) and real income. Alternatively, consider

$$q = b \, p^\alpha \, p_c^\beta \, Y^\tau$$

and let all prices and incomes change by a factor t, then

$$q_1 = t^{\alpha+\beta+\tau} \, b \, p^\alpha \, p_c^\beta \, Y^\tau = t^{\alpha+\beta+\tau} \, q$$

and the degree of homogeneity is $\alpha+\beta+\tau$ so that the condition for no money illusion is that $\alpha+\beta+\tau = 0$.

7.11 Euler's theorem

For the homogeneous function $z = f(x, y)$ *Euler's theorem* states that

$$xz_x + yz_y \equiv kz$$

where k is the degree of homogeneity and z_x and z_y are the partial derivatives of z.

For example, if

$$z = x^2 + 3y^2 + 2xy$$

then $\qquad z_x = 2x + 2y, \qquad z_y = 6y + 2x \qquad$ and $\qquad k = 2$

Therefore, $\qquad xz_x + yz_y = x(2x + 2y) + y(6y + 2x)$

$$= 2x^2 + 2xy + 6y^2 + 2xy$$

$$= 2(x^2 + 3y^2 + 2xy) = 2z$$

For a homogeneous production function, where z is the level of output and x, y are measure of inputs, z_x and z_y are the marginal products of x

and y and so Euler's theorem shows that
(quantity of x) (marginal product of x)
    + (quantity of y) (marginal product of y) = k (total output)
    Another application of Euler's theorem occurs in supply theory. If the quantity of a product supplied (q) depends on the price of the product p, and the price of a competing product r, then assuming the function is homogeneous of degree k, Euler's theorem states that

$$p \frac{\partial q}{\partial p} + r \frac{\partial q}{\partial r} = kq$$

Dividing by q gives

$$\frac{p}{q} \frac{\partial q}{\partial p} + \frac{r}{q} \frac{\partial q}{\partial r} = k$$

that is, the sum of the own price and cross elasticities of supply equals k. For example, if the supply function is

$$q = 6p^2 - 2r^2 - rp$$

$$\frac{\partial q}{\partial p} = 12p - r, \qquad \frac{\partial q}{\partial r} = -4r - p$$

Own-price elasticity is

$$\frac{p}{q} \frac{\partial q}{\partial p} = \frac{p}{q} (12p - r)$$

Cross-elasticity is

$$\frac{r}{q} \frac{\partial q}{\partial r} = \frac{r}{q} (-4r - p)$$

Sum is

$$\frac{12p^2 - rp}{q} + \frac{(-4r^2) - rp}{q} = \frac{12p^2 - 4r^2 - 2rp}{q} = \frac{2q}{q} = 2$$

which is the degree of homogeneity.
    For the Cobb–Douglas function

$$q = AL^\alpha K^\beta$$

$$q_L = \alpha AL^{\alpha-1} K^\beta = \alpha q/L$$

and

$$q_K = \beta AL^\alpha K^{\beta-1} = \beta q/K$$

Hence, by Euler's theorem,

$$Lq_L + Kq_K = L(\alpha q/L) + K(\beta q/K) = (\alpha + \beta)q$$

or
$$L\frac{q_L}{(\alpha + \beta)} + K\frac{q_K}{(\alpha + \beta)} = q$$

The total product, q, is distributed between labour and capital in the proportions

$$q_L/(\alpha + \beta) : q_K/(\alpha + \beta) \text{ or } q_L : q_K$$

and so the distribution is in proportion to the marginal products of labour and capital.

7.12 Exercises

1. Show that the following functions are homogeneous and state the degree of homogeneity:

   (a) $f(x, y) = x^3 + 3y^3 - x^2y$

   (b) $f(x, y) = \dfrac{xy + y^2}{4x^3}$

   (c) $f(x, y, z) = 2xyz + x^2y - y^2z$

2. Verify Euler's theorem for the function in 1(a), 1(b) above.
3. The demand for butter depends upon the price of butter (p) and the price of a substitute (r) according to the function

   $$q = 10r^2 - 2p^2 + 4rp$$

   Show that the sum of the own-price and cross-elasticities is equal to the degree of homogeneity.
4. Show that the production function $q = 1.1L^{0.6}K^{0.2}$ has decreasing returns to scale and that $q = 1.1L^{0.6}K^{0.5}$ has increasing returns to scale.
5. The demand for a product is $q = p_1Y/p_2^2$ where p_1 is the price of the product, p_2 is the price of a competing product and Y is income. Determine the three partial elasticities, the degree of homogeneity of the function and verify Euler's theorem.
6. A production function is said to be *homothetic* if it is homogeneous and the slope of an isoquant $(-q_L/q_K)$ is the same at the point (L^0, K^0) and (kL^0, kK^0) where $k > 0$. Show that the Cobb–Douglas function $q = AL^\alpha K^\beta$ is homothetic.

7.13 Maxima and minima

In the three-dimensional model the function z has a maximum value at the 'top of a hill' and a minimum value at the 'bottom of a valley'. These are the points which are important in economics. A third case exists when z is at a maximum, say, with respect to variation along the x-axis, but is at a minimum with respect to variation along the y-axis. This is called a saddle point since it corresponds to the shape of a saddle: viewed from the side of a horse the saddle is U-shaped, whereas from the front of a horse the saddle is arch-shaped.

To determine the position of these three points it is first necessary to obtain the partial derivative of the function with respect to each variable in turn. A maximum, minimum or saddle point can only exist at a point where each of these partial derivatives is equal to zero. This is a *necessary* condition, e.g. if $z = f(x, y)$, then for a maximum, minimum or saddle point

$$\frac{\partial z}{\partial x} = 0 \quad \text{and} \quad \frac{\partial z}{\partial y} = 0$$

The solution of these equations simultaneously determines the values of x and y, at which one of these states exists.

To decide whether a particular function has a maximum, a minimum or a saddle point, the second-order partial derivatives must then be considered in the light of the following rules:

1. The function has a maximum or a minimum point if $z_{xx}z_{yy}$ is greater than $(z_{xy})^2$. If z_{xx} and z_{yy} are both negative there is a maximum and if they are both positive there is a minimum.

2. If $z_{xx}z_{yy}$ is less than $(z_{xy})^2$ then there is a saddle point of z_{xx} and z_{yy} are both equal to zero.

   These rules can be summarised as follows:

   Maximum $z_{xx}z_{yy} > (z_{xy})^2,$ $z_{xx} < 0,$ $z_{yy} < 0$

   Minimum $z_{xx}z_{yy} > (z_{xy})^2,$ $z_{xx} > 0,$ $z_{yy} > 0$

   Saddle point $z_{xx}z_{yy} < (z_{xy})^2,$ $z_{xx} = z_{yy} = 0$

3. If $z_{xx}z_{yy}$ equals $(z_{xy})^2$ then the point may be a maximum, a minimum, a saddle point or some other type of turning point and requires further investigation.

These conditions can also be written in determinant form (see section 1.13) since

$$\begin{vmatrix} z_{xx} & z_{xy} \\ z_{xy} & z_{yy} \end{vmatrix} = z_{xx}z_{yy} - z_{xy}^2$$

which is positive for a maximum or minimum and negative for a saddle point. This type of determinant, made up of second-order partial derivatives, is known as a *Hessian determinant* or simply as a *Hessian*, and is particularly useful when more general functions are considered, as in section 7.15 below.

It can be seen from the above that there are a number of similarities with the conditions for stationary values in the two variable model. Thus, a stationary value can only exist at a point where all the first-order derivatives are equal to zero. For a saddle point the second-order derivatives must also be equal to zero.

If both z_{xx} and z_{yy} are negative (cf $d^2y/dx^2 < 0$) there is a maximum and if both z_{xx} and z_{yy} are positive (cf $d^2y/dx^2 > 0$) there is a minimum.

The procedure for finding maximum and minimum values is, therefore, very simple as the following example shows:

Suppose that $\qquad z = x^2 + 2xy + 2y^2 - 5x - 4y$

then $\qquad\qquad z_x = 2x + 2y - 5$

and $\qquad\qquad z_y = 2x + 4y - 4$

Equating these partial derivatives to zero produces two simultaneous equations in the variables x and y:

$$2x + 2y = 5$$
$$2x + 4y = 4$$

Subtracting the equations gives

$$2y = -1 \quad \text{or} \quad y = -\tfrac{1}{2} \quad \text{and hence} \quad x = 3$$

There is, therefore, a stationary point when $x = 3$, $y = -\tfrac{1}{2}$.

To determine whether this is a maximum, a minimum or a saddle point, the second-order partial derivatives must be considered:

$$z_{xx} = 2, \qquad z_{yy} = 4, \qquad z_{xy} = 2 = z_{yx}$$

$$z_{xx}z_{yy} = 8 \quad \text{and} \quad (z_{xy})^2 = 4$$

$$\therefore \qquad z_{xx}z_{yy} > (z_{xy})^2$$

The point is a maximum or a minimum and as $z_{xx} > 0$, $z_{yy} > 0$ it is a minimum.

The function has a minimum value at the point

$$x = 3, \qquad y = -\tfrac{1}{2}$$

and this value is given by

$$z = 3^2 + (2)(3)(-\tfrac{1}{2}) + 2(-\tfrac{1}{2})^2 - 5(3) - 4(-\tfrac{1}{2})$$
$$= 9 - 3 + \tfrac{1}{2} - 15 + 2 = -6\tfrac{1}{2}$$

As an economic application, consider the problem of a producer who uses L units of labour and K units of capital and has the production function

$$q = 12L + 5K - 0.2K^2 - 0.5L^2$$

and the cost of labour is 8 per unit and that of capital is 4 per unit. Assuming there is pure competition and the price of the product is 4 then the profit is

$$\pi = 4q - 8L - 4K$$
$$= 48L + 20K - 0.8K^2 - 2L^2 - 8L - 4K$$

To maximise profit we find where $\pi_K = 0 = \pi_L$ and check the second-order conditions. Now,

$$\pi_K = 20 - 1.6K - 4$$
$$\pi_L = 48 - 4L - 8$$

and setting these to zero gives $K = 10$ and $L = 10$. Here,

$$\pi_{KK} = -1.6, \ \pi_{LL} = -4 \text{ and } \pi_{KL} = 0$$

and so profits are maximised when $K = 10$ and $L = 10$, and the maximum is $\pi = 280$, $q = 100$.

7.14 Exercises

1. Find z_x, z_y, z_{xx}, z_{xy}, z_{yy} in the following cases:

   (a) $z = x^2 - xy + y^2$ (b) $z = (x + y)^3$

2. Find the maximum, minimum or saddle-point values (if any) of the following functions:

   (a) $z = y^2 + 2xy + x^3$ (b) $z = 3x + 2y - 3x^2 - 4y^2$

3. Total cost, z, depends on the output x of product A and the output y of product B.
   What is the minimum value of total cost for the following functions:

(a) $z = 50 + x^2 + y^2 - 4x - y$

(b) $z = 200 + x^3 - 3y + y^2 - 2x$

(c) $z = 250 + x^3 + y^2 - 27x - 8y$

4. For the production function

$$q = 36K + 120L - 3K^2 - 4L^2 - 10 + 2KL$$

determine the maximum value of q.

5. A monopolist produces two goods with demand functions

$$p_1 = 40 - q_1 \quad \text{and} \quad p_2 = 60 - 2q_2$$

and the joint cost function is

$$TC = q_1^2 + 3q_1 q_2 + q_2^2$$

Determine the maximum value of profits.

6. A bus company runs services during the peak period, for which the demand function is $p_1 = 105 - 25q_1$, and also in the off-peak period for which the demand function is $p_2 = 50 - 2q_2$. The cost function is $TC = 10 + 5q_1 + 2q_2$. Determine (a) the maximum value of profits if there is price discrimination, (b) the maximum value of profits if there is no price discrimination (and so $p_1 = p_2 = p$, say).

7.15 Maxima and minima for functions of more than two variables

An extension of this procedure is needed when z is a function of more than two variables because here there are more partial derivatives to deal with.

The first-order conditions for both maxima and minima require all the first partial derivatives of the function to be equal to zero. To distinguish between the maximum and minimum positions we must consider the second-order partial derivatives. This becomes a little more complicated as the number of variables increases and to help make the conditions clear they are stated in determinant notation as explained in section 1.13.

Consider the case where $z = f(x_1, x_2, x_3, \ldots, x_n)$.

The first-order conditions for a maximum or a minimum are that

$$z_1 = z_2 = z_3 = \cdots = z_n = 0$$

where $z_1 = \partial z/\partial x_1$ etc.

These equations are solved simultaneously to find positions at which the conditions are satisfied. The second-order partial derivatives z_{11}, z_{12}, z_{13} etc. are then calculated and the following determinants formed:

$$\begin{vmatrix} z_{11} \end{vmatrix} \qquad \begin{vmatrix} z_{11} & z_{12} \\ z_{12} & z_{22} \end{vmatrix} \qquad \begin{vmatrix} z_{11} & z_{12} & z_{13} \\ z_{12} & z_{22} & z_{23} \\ z_{13} & z_{23} & z_{33} \end{vmatrix}$$

The numerical values of these determinants are considered as follows

1. If all the determinants are positive there is a *minimum*.
2. If determinants of even order (i.e. 2×2, 4×4 etc.) are positive but those of odd order (i.e. 1×1, 3×3 etc.) are negative there is a *maximum*.
3. If any determinants of even order are negative there is a saddle point.

As mentioned in section 7.13, these determinants, made up of second-order partial derivatives, are referred to as Hessian determinants or Hessians.

To illustrate the use of these suppose that in a production function output, z, depends on the level of inputs x_1, x_2, and x_3 so that

$$z = f(x_1, x_2, x_3) = 100x_1 - 2x_1^2 + 50x_2 - x_2^2 + 200x_3 - 2x_3^2$$

Output is to be maximised. The first-order derivatives are

$$z_1 = 100 - 4x_1$$

$$z_2 = 50 - 2x_2$$

$$z_3 = 200 - 4x_3$$

If the function is to have a maximum or a minimum then this must occur when

$$z_1 = z_2 = z_3 = 0$$

that is $\qquad x_1 = 25, \qquad x_1 = 25, \qquad x_3 = 50$

The second-order partial derivatives are then calculated:

$$z_{11} = -4 \qquad z_{12} = 0 \qquad z_{13} = 0$$
$$z_{22} = -2 \qquad z_{23} = 0 \qquad z_{33} = -4$$

The Hessian determinants are

$$\left| -4 \right| \quad \begin{vmatrix} -4 & 0 \\ 0 & -2 \end{vmatrix} \quad \begin{vmatrix} -4 & 0 & 0 \\ 0 & -2 & 0 \\ 0 & 0 & -4 \end{vmatrix}$$

that is $\qquad -4, +8$ and -32

These values, alternately negative, positive, negative, satisfy condition 2, and the function therefore has a maximum when the variables have the values $x_1 = 25$, $x_2 = 25$, $x_3 = 50$ and the value of this maximum is given by

$$z = 100(25) - 2(25)^2 + 50(25) - (25)^2 + 200(50) - 2(50)^2 = 6875$$

As a further example, suppose the total cost (TC) function for a firm producing three goods depends on the outputs, x, y and z of these goods so that

$$TC = 1000 + 3x^2 + 2y^2 + 2z^2 - 2xy - 40z - 20x$$

and assume that total cost is to be minimised. The first-order derivatives are

$$TC_x = 6x - 2y - 20$$
$$TC_y = 4y - 2x$$
$$TC_z = 4z - 40$$

and setting these to zero (the necessary conditions for a minimum) gives $x = 4$, $y = 2$ and $z = 10$.

The second-order derivatives are

$$TC_{xx} = 6, \ TC_{xy} = -2, \ TC_{xz} = 0, \ TC_{yy} = 4, \ TC_{yz} = 0, \ TC_{zz} = 4$$

and letting x, y and z correspond to x_1, x_2 and x_3 in the previous formulae, the Hessians are

$$|z_{11}| = |TC_{xx}| = |6| = 6$$

$$\begin{vmatrix} z_{11} & z_{12} \\ z_{12} & z_{22} \end{vmatrix} = \begin{vmatrix} TC_{xx} & TC_{xy} \\ TC_{xy} & TC_{yy} \end{vmatrix} = \begin{vmatrix} 6 & -2 \\ -2 & 4 \end{vmatrix} = 24 - 4 = 20$$

$$\begin{vmatrix} z_{11} & z_{12} & z_{13} \\ z_{12} & z_{22} & z_{23} \\ z_{13} & z_{23} & z_{33} \end{vmatrix} = \begin{vmatrix} TC_{xx} & TC_{xy} & TC_{xz} \\ TC_{xy} & TC_{yy} & TC_{yz} \\ TC_{xz} & TC_{yz} & TC_{zz} \end{vmatrix} = \begin{vmatrix} 6 & -2 & 0 \\ -2 & 4 & 0 \\ 0 & 0 & 4 \end{vmatrix} = 4(24 - 4) = 80$$

Since each of these is positive we have a minimum when $x = 4$, $y = 2$ and $z = 10$ and $TC = 760$.

7.16 Constrained maxima and minima

The situation frequently arises when a function has a definite maximum which is not attainable because of some restriction on the value which the variables can assume. This can happen in the case of a production function where it may be impossible to reach the level of output which would yield the maximum revenue due to an insufficient supply of the factors of production. In these cases it is important to determine the maximum which can actually be achieved. This can be done using the method of Lagrangian multipliers.

Let us consider the function

$$z = 100x_1 - 2x_1^2 + 60x_2 - x_2^2$$

To find the maximum value of this without restriction the first-order partial derivatives are obtained and equated to zero.

$$z_1 = 100 - 4x_1 = 0$$

$$z_2 = 60 - 2x_2 = 0$$

$$\therefore \qquad x_1 = 25, \qquad x_2 = 30$$

The second-order partial derivatives are

$$z_{11} = -4 \qquad z_{22} = -2 \qquad z_{12} = 0$$

$$\therefore \qquad z_{11}z_{22} > z_{12}^2$$

and $$z_{11} < 0 \qquad z_{22} < 0$$

The function, therefore, has a maximum value when $x_1 = 25$, $x_2 = 30$ and $z = 100(25) - 2(25)^2 + 60(30) - (30)^2 = 2150$.

If there is a limited amount of capital available to buy resources then a constraint is applied to the system. Let this be represented by

$$x_1 + x_2 = 40$$

One method of maximising z subject to this constraint is to substitute directly for x_1, using $x_1 = 40 - x_2$, into z so that

$$z = 100(40 - x_2) - 2(40 - x_2)^2 + 60x_2 - x_2^2$$

$$= 4000 - 100x_2 - 3200 + 160x_2 - 2x_2^2 + 60x_2 - x_2^2$$

$$= 800 + 120x_2 - 3x_2^2$$

This function can be maximised as in section 5.7 by

$$\frac{dz}{dx_2} = 120 - 6x_2 \quad \text{and} \quad \frac{dz}{dx_2} = 0 \quad \text{when} \quad x_2 = 20$$

$$\frac{d^2z}{dx_2^2} = -6$$

therefore $x_2 = 20$ gives a maximum and

$$z = 800 + 120(20) - 3(20)^2 = 2000$$

is the maximum value of z.

In this particular case the method of substituting for x_1 from the constraint into the function for z has worked out well. However, in general such a substitution can be awkward to carry out and also does not produce as much information as an alternative method, using *Lagrangian multipliers*, which is now considered.

This involves forming a new expression

$$z' = z + \lambda u$$

where $u = (40 - x_1 - x_2)$ and λ is a variable known as the *Lagrangian multiplier*. It follows, that by optimising z' we are effectively optimising z and at the same time satisfying the constraint condition. The expression for z' has three unknowns, x_1, x_2 and λ, and the procedure is to find the stationary values by setting the first-order partial derivatives to zero. Here

$$z' = 100x_1 - 2x_1^2 + 60x_2 - x_2^2 + \lambda(40 - x_1 - x_2)$$

$$z_1' = 100 - 4x_1 - \lambda = 0$$

$$z_2' = 60 - 2x_2 - \lambda = 0$$

$$z_\lambda' = 40 - x_1 - x_2 = 0$$

The points at which the function has a maximum or minimum value are obtained by solving this set of simulataneous equations to give $x_1 = 20$, $x_2 = 20$, $\lambda = 20$. To check that this is a maximum value it is necessary to know the values of the second-order partial derivatives of the original function and the first-order partial derivatives of the constraint equation. Again we denote partial derivatives with respect to x_1 by the subscript 1, and partial derivatives with respect to x_2 by the subscript 2. There is a maximum value if

$$2u_1u_2z_{12} > u_1^2z_{22} + u_2^2z_{11}$$

and the minimum value if

$$2u_1u_2z_{12} < u_1^2z_{22} + u_2^2z_{11}$$

provided u is linear in both x_1 and x_2.

These conditions can be written in the form of a special Hessian determinant which has u_1 and u_2 on the borders:

$$\begin{vmatrix} z_{11} & z_{12} & u_1 \\ z_{12} & z_{22} & u_2 \\ u_1 & u_2 & 0 \end{vmatrix} \begin{matrix} > 0 \text{ for a maximum} \\ < 0 \text{ for a minimum} \\ = 0 \text{ for inconclusive} \end{matrix}$$

In the example

$$z_{11} = -4 \qquad z_{12} = 0 \qquad z_{22} = -2$$
$$u_1 = 1 \qquad u_2 = 1$$

Hence

$$2u_1u_2z_{12} = 0$$

and

$$u_1^2z_{22} + u_2^2z_{11} = -6$$

and so we have a maximum value when $x_1 = 20$ and $x_2 = 20$. This maximum value is

$$z = 100(20) - 2(20)^2 + 60(20) - (20)^2 = 2000$$

The maximum output which could be obtained without restriction on x_1 and x_2 was 2,150. Thus, in general, the consequence of placing a constraint upon any of the variables is to reduce the maximum value attainable. An exception would be the case where the two maxima were equal but under these circumstances the constraint is not really effective.

The economic interpretation to be placed on the value of λ (20 here) is of interest. It is the marginal product of resources, i.e. it is a measure of the increase in z which would be produced by an increase in the level of the constraint by one unit. For example, if $x_1 + x_2$ can be increased to 41 the function to be maximised is changed to

$$z'' = 100x_1 - 2x_1^2 + 60x_2 - x_2^2 + \lambda(41 - x_1 - x_2)$$

and in this case it would be found that the maximum value of z'' is greater than z' by approximately 20 units (see Section 7.18, Exercise 1).

As a further example suppose that an individual has a utility function,

$$u(x_1, x_2) = \log(x_1 x_2)$$

which depends on the quantities consumed of the two goods x_1 and x_2. In the absence of any constraint, utility is maximised when x_1 and x_2 are infinite. If the total income is M and the prices are p_1 and p_2 then, ignoring the possibility of debt, expenditure is constrained to be equal to income. That is, the budget constraint is

$$M = p_1 x_1 + p_2 x_2$$

The problem is to determine the values of x_1 and x_2 which maximise utility, subject to the budget constraint. That is, maximise

$$z' = \log(x_1 x_2) + \lambda(M - p_1 x_1 - p_2 x_2)$$

Notice that we have written the constraint as $M - p_1 x_1 - p_2 x_2$ when it could have been written as $p_1 x_1 + p_2 x_2 - M$. This has no effect on the optimisation procedure but changes the sign on λ. Keeping to our previous notation, the partial derivatives are

$$z_1' = (1/x_1) + \lambda(-p_1)$$

$$z_2' = (1/x_2) + \lambda(-p_2)$$

$$z_\lambda' = M - p_1 x_1 - p_2 x_2$$

and setting to zero and solving these gives $x_1 = M/2p_1$, $x_2 = M/2p_2$, $\lambda = 2/M$

Letting $u = M - p_1 x_1 - p_2 x_2$, $u_1 = -p_1$, $u_2 = -p_2$ and the second-order derivatives are, with $z = \log(x_1 x_2)$, $z_{11} = -1/x_1^2$, $z_{22} = -1/x_2^2$, $z_{12} = 0$. The bordered Hessian determinant is

$$\begin{vmatrix} z_{11} & z_{12} & u_1 \\ z_{12} & z_{22} & u_2 \\ u_1 & u_2 & 0 \end{vmatrix} = \begin{vmatrix} -1/x_1^2 & 0 & -p_1 \\ 0 & -1/x_2^2 & -p_2 \\ -p_1 & -p_2 & 0 \end{vmatrix} = \frac{8p_1^2 p_2^2}{M^2}$$

which is positive and so we have a maximum, given by

$$u(x_1, x_2) = \log(M^2/4p_1 p_2)$$

We also note that the budget constraint is satisfied at the maximum.

7.17   A linear expenditure system

The linear expenditure system was proposed by R. Stone and arises from maximising a special form of utility function subject to a budget constraint. Consider a simple model with two goods and let x and y be the quantities of each purchased by a consumer. The utility derived from the goods depends on x and y and we will take the particular form of the utility function to be

$$U(x, y) = (x - a)^\alpha (y - b)^\beta$$

where it is assumed that $x > a > 0$ and $y > b > 0$. Notice that this utility function is similar to the Cobb–Douglas production function discussed above. Here, a and b are the minimum quantities that will be consumed and actual consumption will be greater than the minimum. We assume that all the income (M) is spent. Below we will show that α and β are the marginal budget shares which indicate by how much expenditure on the product will increase if income increases by one unit. Since changes in expenditure equal changes in income, $\alpha + \beta = 1$. Also, $\alpha > 0$ and $\beta > 0$.

Taking logarithms to the base e,

$$\log U = \alpha \log(x - a) + \beta \log(y - b)$$

and the budget constraint is

$$M = p_1 x + p_2 y$$

To maximise utility subject to the budget constraint we maximise

$$F = \alpha \log(x - a) + \beta \log(y - b) + \lambda(M - p_1 x - p_2 y)$$

where λ is a Lagrangian multiplier. The partial derivatives are

$$\frac{\partial F}{\partial x} = \frac{\alpha}{x - a} - \lambda p_1$$

$$\frac{\partial F}{\partial y} = \frac{\beta}{y - b} - \lambda p_2$$

$$\frac{\partial F}{\partial \lambda} = M - p_1 x - p_2 y$$

The first-order condition for a maximum is that these are all zero. Solving for λ gives

$$\frac{\alpha}{p_1(x - a)} = \frac{\beta}{p_2(y - b)}$$

or

$$\alpha p_2(y - b) = \beta p_1(x - a)$$

and so

$$p_1 x = \{\alpha p_2(y - b) + a\beta p_1\}/\beta$$

Substituting in

$$0 = M - p_1 x - p_2 y$$

gives $$0 = \beta M - \alpha p_2(y - b) - a\beta p_1 - \beta p_2 y$$

and so $$p_2 y = \beta M - a\beta p_1 + b\alpha p_2$$

using $\alpha + \beta = 1$. Finally, replacing α by $1 - \beta$ gives

$$p_2 y = bp_2 + \beta\{M - ap_1 - bp_2\}$$

Now $p_2 y$ is the total expenditure on the second good and since b is the minimum quantity of y that is consumed, bp_2 is called the *subsistence expenditure* on the second good. The final term is the *supernumerary expenditure* on the second good, being β, the marginal budget share, times the income remaining after the subsistence expenditure. Notice that the expenditure on the second good is a linear function of p_1, p_2 and M, which explains why it is known as a linear expenditure system. The corresponding result for the first good is

$$p_1 x = ap_1 + \alpha\{M - ap_1 - bp_2\}$$

and α is the marginal budget share.

This linear expenditure function can be transformed into a demand function by dividing by p_1 to give

$$x = a + \alpha\{M - ap_1 - bp_2\}/p_1$$

The own-price elasticity of demand is from

$$\frac{\partial x}{\partial p_1} = \alpha\{(-ap_1) - [M - ap_1 - bp_2]\}/p_1^2$$

$$= -\alpha[M - bp_2]/p_1^2$$

and so, $$E = \frac{p_1 \partial x}{x \partial p_1} = \frac{-\alpha[M - bp_2]}{p_1 x}$$

This is negative since $\alpha > 0$, $M > bp_2$ and $p_1 x > 0$. The (absolute) elasticity is less than 1 if

$$\alpha[M - bp_2] < p_1 x$$

and since, from above,

$$p_1 x = ap_1 + \alpha\{M - ap_1 - bp_2\}$$

the condition becomes

$$0 < ap_1 + \alpha(-ap_1) = ap_1(1 - \alpha),$$

which is satisfied if $\alpha < 1$. Therefore the elasticity is less than 1 if $0 < \alpha < 1$, and similarly, for the second good, $0 < \beta < 1$.

7.18 Exercises

1. Find the maximum value of the function
$$z = 100x_1 - 2x_1^2 + 60x_2 - x_2^2$$
subject to the constraint $x_1 + x_2 = 41$.

2. Find the maximum or minimum value of
$$z = 3x^2 + y^2 - 12x - 14y + 150$$

Find the restricted maximum or minimum value of z when there is a constraint requiring $2x + y = 9$.

3. Find the maximum or minimum value of the following functions subject to the given constraints:

   (a) $z = x^2 + y^2$ with $5x + 4y = 40$

   (b) $z = 18 - \dfrac{x^2}{2} - y^2$ with $2x + 3y = 12$

   (c) $z = 3x^2y^2$ with $x + 3y = 18$

4. Let the production function for a firm be given by
$$x = 20l + 40c - 2l^2 - 3c^2$$

and the cost of labour (l) and of capital (c) be 4 and 5 units respectively.

   Find the maximum value of x if the total cost is equal to 28 units.

5. The utility function of a consumer is $U = 6xy$ where x and y are the quantities consumed of two goods. The price of x is 5 per unit and that of y is 10 per unit. If total expenditure is limited to 100, what is the maximum value of U?

7.19 Revision exercises for Chapter 7 (without answers)

1. Determine z_x, z_y, z_{xx}, z_{yy}, and z_{xy} for

   (a) $z = 2xy - x^2 - y^2$

   (b) $z = ye^{-x} + xe^{-y}$

   (c) $z = \log(2x + 3y)$

   (d) $z = x/(x^2 - y^2)$

   (e) $z = x \sin y$.

2. The demand for a product is given by

$$q^d = 500 - 3p + 10p^c - p^2 + 0.2Y$$

where p is the price of the product, p^c is the price of a competing product and Y is income. Determine the own-price, competing price and income partial elasticities if $p = 10$, $p^c = 12$ and $Y = 100$. Would revenue increase if p was reduced to 8?

3. For the production function $Q = 3K^2 + 4L^2 - 15KL$ determine the marginal rate of substitution of capital (K) for labour (L) and the elasticity of substitution.

4. If $z = xy - x^2 - y^2$ and $x = e^t$, $y = \log t$ then determine dz/dt.

5. Determine dy/dx if (a) $y^4 + x^4 - 6xy = 3x^2y^3$ (b) $xe^y = ye^x$.

6. Check whether the following are homogenous functions and, if they are, verify Euler's theorem.

   (a) $z = x^2y - y^2x$

   (b) $z = xe^{-y} + ye^{-x}$

   (c) $z = (x^2 - y^2)/6xy$

7. Determine the maximum, minimum or saddle-point values (if any) of the following functions:

   (a) $z = x^2 + y^2 - 6xy$

   (b) $z = 200 - x^3 - 2y^3$

   (c) $z = x^2 - 4x - xy + 2y^2$

8. The production function for a product is

$$q = 6K + 8L - 0.4K^2 - 0.2L^2$$

and labour (L) costs 3 per unit while capital (K) costs 2 per unit. The product sells for 15 per unit. What level of output will maximise profits?

9. Find the maximum value of $z = x^2y^2$ subject to the constraint that $x + y = 16$. If the constraint changes to $x + y = 17$ what is the change in the maximum value of z?

10. The utility of a consumer is given by $U = x^2 + y^2$. If the price of x is 3, the price of y is 5 and the income is 100, what is the maximum value of U? What is the effect on the maximum of increasing income to 101?

Chapter 8
Linear Programming

8.1 Introduction

We saw in Chapter 7 how it is possible to find the maximum of a given function when there are constraints on the values which some or all of the variables can assume. To do this we made use of the differential calculus and the method of Lagrangian multipliers. This is, however, only applicable when the constraint is in the form of an equality, that is, one which must be exactly met.

There are many situations where the restriction is not so precise and in particular where the constraint is expressed in the form of an inequality. This might be the case, for example, where a company is concerned with maximising the profitability of its operations, subject to the limitations imposed upon it by the availability of production capacity, whether in the form of machine time, skilled labour or financial resources. This type of optimisation problem cannot generally be handled with the help of the differential calculus, but there are mathematical tools available which can provide a solution in a simple way. To illustrate this we take the case where all the interrelationships between the variables and all the constraints can be expressed in linear form and linear programming can be used. The following example explains the method, and illustrates the principles by means of a graphical approach. This might be used in simple situations, but these are unlikely to occur in practice and so a mathematical method is explained which has more general applicability and can be computerised easily.

8.2 A product-mix problem

A small company has two machines U and V which are both required to produce two products A and B. The problem is one of deciding how much of each product should be manufactured in order to provide the maximum contribution to the profitability of the company. To obtain a solution it is necessary to have information about the

TABLE 8.1

	No. of hours required per unit of product		Total no. of hours available
	A	B	
Machine U	1	2	30
Machine V	2	1	30
Contribution per unit of product	2	3	

capacity of each machine, the amount of time which is required on each machine by each product, and the contribution which each product makes to profitability. Such a situation might lead to the information provided in Table 8.1.

The contribution per unit of product is equal to the selling price less the variable cost of production, and in order to use the linear-programming approach for solving this product-mix problem it is necessary to assume that this remains constant over the levels of output which are under consideration.

If we let x_1 and x_2 be the number of units of products A and B respectively which the company should manufacture, the relevant information can be neatly summarised in algebraic form as follows:

Maximise
$$2x_1 + 3x_2 = z \tag{1}$$

subject to the constraints

$$x_1 + 2x_2 \leqslant 30 \tag{2}$$

$$2x_1 + x_2 \leqslant 30 \tag{3}$$

$$x_1 \geqslant 0 \tag{4}$$

$$x_2 \geqslant 0 \tag{5}$$

The first relationship is an equation which represents the total contribution which the company can expect from the two products, i.e. it is the sum of the contribution per unit multiplied by the number of units which will be manufactured of each product. This is called the *objective function* and the purpose of the exercise is to obtain the values of x_1 and x_2 for which this is a maximum.

The second relationship is an inequality which states that the number of hours used making x_1 units of product A on machine U plus the number of hours used making x_2 units of product B on machine U must not exceed the maximum number of machine hours available. The third relationship is again an inequality and describes the constraint on machine V in a similar manner. Inequalities (4) and (5) merely state that the output of either product cannot be negative. These are obvious practical constraints on the system.

Any values of the variable x_1 and x_2 which satisfy these inequalities are said to constitute a *feasible solution* to the problem. What is required is a method for determining from all the possible feasible solutions that one which yields the maximum value of z as specified in the objective function. This gives the values of the variables which produce the maximum profit and is known as the *optimal solution*.

8.3 A graphical approach

The situation being considered can be analysed using a graphical method because there are just two variables. The two axes represent the variables x_1 and x_2. The lines corresponding to the equations $x_1 + 2x_2 = 30, 2x_1 + x_2 = 30$ are then drawn on the graph as shown in Fig. 8.1.

These equations correspond to the upper limit of machine capacity and therefore any combinations of A and B which it is possible to manufacture must lie on the side of the lines closest to the origin. To satisfy both constraints they must, in fact, lie within the area which is doubly cross-hatched in the diagram. This is known as the *feasible region*. We now superimpose on this graph the profit line $2x_1 + 3x_2 = z$ for profits of 24 and 60 units, that is $2x_1 + 3x_2 = 24$ and 60 respectively.

These lines are parallel to each other because the slope of the line $z = 2x_1 + 3x_2$ is independent of the value of z. They are shown as broken lines in Fig. 8.2.

The further away from the origin this line is taken the higher the value of z, but the upper limit which is allowable is determined by the constraints on the system. It is immediately apparent that no point on the line $z = 60$ lies within the feasible region, and therefore this value of z is unattainable. In fact, the highest value z which satisfies the constraints is reached when the profit line passes through one of the extreme points of the feasible region. This will always be so if the constraints are linear except in the special case where the profit function is

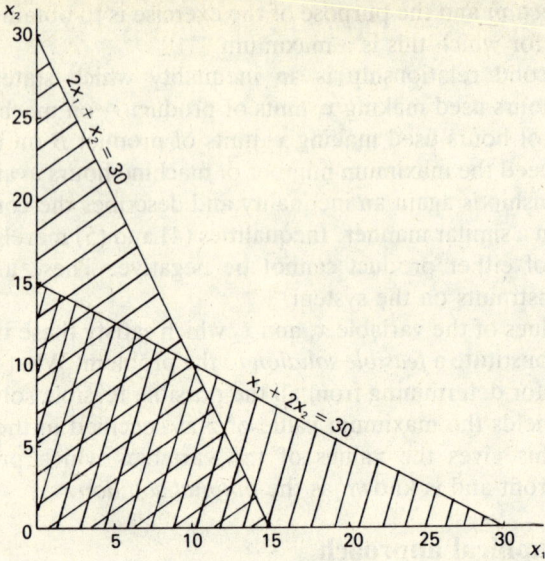

Fig. 8.1

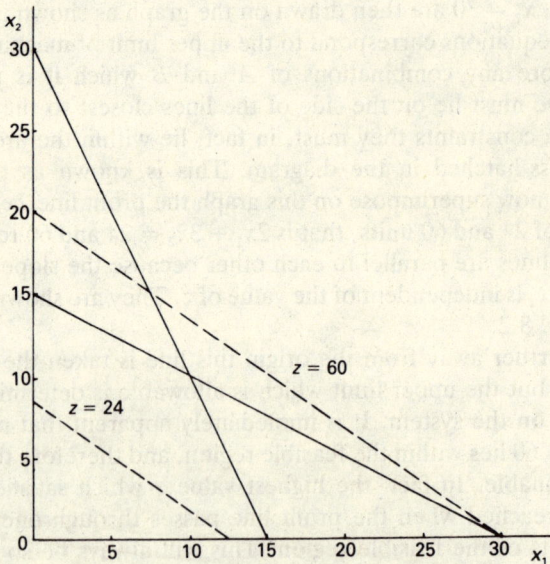

Fig. 8.2

parallel to one of the constraints, and in this case, the problem has a multitude of solutions corresponding to the points on the constraint line.

To obtain the solution graphically we move the line with slope $-\frac{2}{3}$ outwards until it reaches the limit of the feasible region, and this point determines the values of x_1 and x_2 corresponding to the outputs of A and B. In this case it is at the point where the two constraint equations intersect, and hence the optimum combination of products to manufacture can be easily determined by the usual method of solution:

$$2x_1 + x_2 = 30 \tag{6}$$

$$x_1 + 2x_2 = 30 \tag{7}$$

Making use of determinants, we have

$$x_1 = \frac{\begin{vmatrix} 30 & 1 \\ 30 & 2 \end{vmatrix}}{\begin{vmatrix} 2 & 1 \\ 1 & 2 \end{vmatrix}} = \frac{60 - 30}{4 - 1} = 10$$

$$x_2 = \frac{\begin{vmatrix} 2 & 30 \\ 1 & 30 \end{vmatrix}}{\begin{vmatrix} 2 & 1 \\ 1 & 2 \end{vmatrix}} = \frac{60 - 30}{4 - 1} = 10$$

The result is that 10 units should be manufactured of both products A and B. It follows that the contribution will equal

$$10 \times 2 + 10 \times 3 = 50$$

It is easy to check that the constraints on machine capacity are not violated by substituting in the original equations.

$$x_1 + 2x_2 = 30$$

$$2x_1 + x_2 = 30$$

These equations are exactly satisfied and therefore both machines are fully utilised, i.e. there is no spare or slack capacity.

8.4 The simplex method

The simplex method employs an *iterative procedure* which progresses in a series of steps ultimately leading to the optimal solutions. The rules of operation are described below, using the same example for illustrative purposes.

Step 1. Convert the inequalities (2) and (3) into equations by the insertion of new variables x_3 and x_4.

$$x_1 + 2x_2 + x_3 = 30 \tag{8}$$

$$2x_1 + x_2 + x_4 = 30 \tag{9}$$

The only qualification which must be added here is that the new variables must only take on positive values:

$$x_3 \geqslant 0 \qquad x_4 \geqslant 0$$

This is necessary because they represent the difference between the time which is utilised on each machine and the total time available on the machines. They are therefore known as *slack variables* and cannot, by definition, be negative.

Step 2. Obtain a first feasible solution to the system. The insertion of the slack variables proves useful here because an immediately obvious solution is given by

$$x_1 = 0, \qquad x_2 = 0, \qquad x_3 = 30, \qquad x_4 = 30$$

A unique solution to two equations in four unknowns can only be obtained if two of the unknowns are given defined values. For convenience they are assigned the value zero in this exercise.

This solution corresponds to zero output and slack or spare capacity on both machines and also produces zero profit. It corresponds to the origin in the graphical method and hence it is on the edge of the feasible region. The reason it is chosen is that it is an easily found solution. The non-zero variables x_3 and x_4 are taken as a starting point for the iterative procedure and are known as the *basic elements* and form a *basis* in a tableau which is drawn up in Table 8.2.

The elements in the tableau are obtained from the initial data for the system. They correspond to the coefficients of equations (8) and (9) and of the objective function equated to zero.

TABLE 8.2

	x_1	x_2	x_3	x_4	p
x_3	1	2	1	0	30
x_4	2	1	0	1	30
z	-2	-3	0	0	0

Equation (8) $\qquad x_1 + 2x_2 + x_3 + 0x_4 = 30$

Equation (9) $\qquad\qquad 2x_1 + x_2 + 0x_3 + x_4 = 30$

Objective function $\qquad z - 2x_1 - 3x_2 + 0x_3 + 0x_4 = 0$

In the tableau the basic elements x_3 and x_4 are written to the left of the equation in which they appear.

The first feasible solution is obtained by letting $x_1 = 0$ and $x_2 = 0$ with the result that

$$x_3 = 30 \qquad \text{(from (8))}$$

$$x_4 = 30 \qquad \text{(from (9))}$$

and the profit, $\qquad\qquad z_1 = 0 \qquad \text{(from the objective function)}$

It is obvious that this solution can be improved upon and that any improvements can only come from the manufacture of either A or B. This could be shown in the tableau by either x_1 or x_2 appearing in the basis with the result that either x_3 or x_4 must become zero and leave the basis.

The method consists of

(a) determining the variable to enter the basis, known as the *entering* variable, and,

(b) determining the variable to leave the basis, known as the *departing* variable, and

(c) calculating the profit which results from the interchange.

One such change is made at each stage in the iterative procedure and if the rules are followed carefully the method leads progressively to the optimal solution and indicates clearly when this optimum is reached.

Step 3. Select the entering variable by considering the elements in the lowest row of the tableau. These elements correspond to the marginal profit which would be obtained by the manufacture of any products other than those at present in the basis. In the first tableau the elements are -2 and -3, which correspond to the contribution per unit from products A and B respectively. It is possible to introduce only one new variable into the basis at a time and it seems logical to select that variable which offers the largest marginal profit, in this case x_2, or product B. Thus we have

Rule 1. The entering variable is chosen as that variable which has the largest negative element in the bottom row of the tableau.

Step 4. Select the departing variable. Before doing this, it is important to remember that the variable leaving the basis becomes zero and the two *non-basic* zero variables determine the values of the basic variables. For example,

if	$x_3 = 0$ and $x_1 = 0$	
then	$2x_2 = 30$	(from (8))
and	$x_2 + x_4 = 30$	(from (9))
That is	$x_2 = 15 \qquad x_4 = 15$	
But if	$x_4 = 0$ and $x_1 = 0$	
then	$2x_2 + x_3 = 30$	(from (8))
	$x_2 = 30$	(from (9))
That is	$x_2 = 30 \qquad x_3 = -30$	

This second result is not feasible because the values of the variables in a linear programming problem must all be positve and $x_3 = -30$ violates this constraint.

It follows therefore that the departing variable can only be x_3.

In the case of large problems, it would be a laborious task to check each variable independently to see whether the resulting interchange violates the non-negativity constraints. It is not, in fact, necessary because the departing variable can be found by considering the ratio p_i/a_{ij} for each basic variable. p_i is the element in the last column of the row in which the basic variable is written and a_{ij} is the element in the column of the entering variable and the row of the basic variable.

The calculation for the present example is as follows:

	x_1	x_2	x_3	x_4	p	$\dfrac{p_i}{a_{ij}}$
x_3	1	2	1	0	30	$\dfrac{30}{2}$
x_4	2	1	0	1	30	$\dfrac{30}{1}$
z	-2	-3	0	0	0	

↑
entering variable

In order to ensure that all the variables will have positive values at each state of the procedure, it is necessary to specify

Rule 2. The departing variable is that one for which the ratio p_i/a_{ij} has the *smallest positive* value.

The element at the intersection of the column containing the entering variable and the row containing the departing variable is known as the *pivot* and plays an important part in the following step.

In this example the entering variable is x_2, the departing variable is x_3 and the pivot element is 2.

	x_1	x_2	x_3	x_4	p	
x_3	1	2	1	0	30	← departing variable
x_4	2	1	0	1	30	
z	-2	-3	0	0	0	

↑
entering variable

We now require to form the new set of equations from which we can determine the value of x_2 when x_1 and x_3 have the value zero. This will tell us how many units of product A to manufacture and how many units of spare capacity there will be. This information, along with the profit which will be produced by this course of action, can be obtained in the following way.

Step 5. Rewrite the tableau with the entering variable replacing the departing variable in the basis and form the new matrix of coefficients as follows:

(a) Divide all the elements in the pivot row by the pivot element.
(b) Add or subtract such multiples of the element in the pivot row from corresponding elements of the other rows to reduce all other elements in the pivot column to zero.

In this example all the elements in the pivot row are divided by two to form a new row. Multiples of this new row are added or subtracted from the other rows as follows:

(1) The elements of the new row are subtracted from the corre-

sponding elements of row two of the old tableau to form row
two of the new tableau.

(2) Three times the elements of the new row are added to the
corresponding elements of row three of the old tableau to form
row three of the new tableau.

	x_1	x_2	x_3	x_4	p
x_2	$\frac{1}{2}$	1	$\frac{1}{2}$	0	15
x_4	$(2 - \frac{1}{2})$	$(1 - 1)$	$(0 - \frac{1}{2})$	$(1 - 0)$	$(30 - 15)$
z	$(-2 + \frac{3}{2})$	$(-3 + 3)$	$(0 + \frac{3}{2})$	$(0 + 0)$	$(0 + 45)$

This can be rewritten as

	x_1	x_2	x_3	x_4	p
x_2	$\frac{1}{2}$	1	$\frac{1}{2}$	0	15
x_4	$\frac{3}{2}$	0	$-\frac{1}{2}$	1	15
z	$-\frac{1}{2}$	0	$\frac{3}{2}$	0	45

This tableau corresponds to the solution

$$x_1 = 0, \quad x_2 = 15, \quad x_3 = 0, \quad x_4 = 15, \quad z = 45$$

that is, the company could manufacture 15 units of B and as each unit of
B contributes 3 units to profits, the total contribution would amount to
45 units.

It can also be noted that this product mix uses all the available
capacity of machine U (slack variable $x_3 = 0$) but leaves spare capacity
on machine V (slack variable $x_4 = 15$). Check: 15 units of product B
require $15 \times 2 = 30$ hours on machine U but only $15 \times 1 = 15$ hours on
machine V.

The second tableau gives a solution which is an improvement on
that given by the first tableau. But is it the optimal solution? The answer
is no, and this is indicated by the presence in the bottom row of the
tableau of a negative element. These elements correspond to the marginal profit which could be obtained by the introduction into the basis of
a non-basic element. Therefore, in this example, profit is being foregone
by not having x_1 in the basis, i.e. by not manufacturing product A. This

profit is equal to $\frac{1}{2}$ unit for each unit of x_1 introduced. It should be noted here that this marginal profit is no longer equal to the direct contribution per unit from A because the introduction of this product necessarily means that we must reduce the quantity of product B which is produced in order not to violate the capacity constraint on machine U. (There was no spare capacity on this machine with the first product mix.) The negative figures in the bottom row of the tableau therefore correspond to the marginal profit which is obtained after the product mix has been changed by introducing product A and adjusting the level of manufacture of the other product (or products) to satisfy the constraints on the system.

The variable with the *largest negative element* in the bottom row of the tableau is selected for the entering variable (in this case x_1 which has the only negative element). The departing variable is found by considering the ratio p_i/a_{ij} and the interchange carried out as before.

Steps 3 and 4 repeated

	x_1	x_2	x_3	x_4	p	$\dfrac{p_i}{a_{ij}}$
x_2	$\frac{1}{2}$	1	$\frac{1}{2}$	0	15	$15/\frac{1}{2}$
x_4	$\boxed{\frac{3}{2}}$	0	$-\frac{1}{2}$	1	15	$15/\frac{3}{2}$ ← departing variable
z	$-\frac{1}{2}$	0	$\frac{3}{2}$	0	45	

$\uparrow$
entering variable

Step 5 repeated

	x_1	x_2	x_3	x_4	p
x_2	$(\frac{1}{2} - \frac{1}{2})$	$(1 - 0)$	$(\frac{1}{2} + \frac{1}{6})$	$(0 - \frac{1}{3})$	$(15 - 5)$
x_1	1	0	$-\frac{1}{3}$	$\frac{2}{3}$	10
z	$(-\frac{1}{2} + \frac{1}{2})$	$(0 + 0)$	$(\frac{3}{2} - \frac{1}{6})$	$(0 + \frac{1}{3})$	$(45 + 5)$

which is rewritten as

	x_1	x_2	x_3	x_4	p
x_2	0	1	$\frac{2}{3}$	$-\frac{1}{3}$	10
x_1	1	0	$-\frac{1}{3}$	$\frac{2}{3}$	10
z	0	0	$\frac{4}{3}$	$\frac{1}{3}$	50

This tableau corresponds to the solution

$$x_1 = 10, \quad x_2 = 10, \quad x_3 = 0, \quad x_4 = 0, \quad z = 50$$

that is, the company would manufacture 10 units of A, 10 units of B, and the total contribution from this combination of output would be 50. The last iteration has again improved the profit of the company and at the same time has arrived at a combination which utilises all the available capacity of both the scarce resources (slack variables x_3 and x_4 are both equal to zero). It can also be seen that none of the elements in the bottom row of the tableau are negative and this indicates that no increase in profit can be obtained by a change in the product mix. The final tableau therefore provides the optimal solution to the problem.

8.5   A summary of the simplex method

To explain and carry out the iterative procedure at one and the same time may give the impression that it is lengthy and difficult. To show that this is not so, the necessary stages are repeated here with all the detail that would be required to carry out the calculations in a practical case. The problem itself has been altered slightly to one in which the company has more capacity available on machine U than on machine V. The data for the problem are as shown in Table 8.3.

Mathematically this can be stated as follows.

Maximise $\qquad\qquad 2x_1 + 3x_2 = z$

subject to the constraints

$$x_1 + 2x_2 \leqslant 40$$
$$2x_1 + x_2 \leqslant 30$$
$$x_1 \geqslant 0$$
$$x_2 \geqslant 0$$

The slack variables, x_3 and x_4 are inserted in the inequalities with the qualification that $x_3 \geqslant 0$, $x_4 \geqslant 0$

TABLE 8.3

	A	B	Total no. of hours available
	No. of hours required per unit of product		
Machine U	1	2	40
Machine V	2	1	30
Contribution per unit of product	2	3	

$$x_1 + 2x_2 + x_3 = 40$$
$$2x_1 + x_2 + x_4 = 30$$

The procedure is as shown in Tableaux 1–3. The optimal solution is to make $6\frac{2}{3}$ units of A and $16\frac{2}{3}$ units of B; the total contribution from this combination would be $63\frac{1}{3}$. This solution assumes that it is possible to manufacture fractional parts of a product and in many cases this is realistic because it is the average output per week which is being discussed, and this is likely to include fractional units. When such a situation is not permissible it is possible to continue the procedure further and use the method of integer programming which will produce a solution in which the variables are restricted to whole units.

Tableau 1

	x_1	x_2	x_3	x_4	p	$\dfrac{p_i}{a_{ij}}$	
x_3	1	[2]	1	0	40	40/2	← departing variable
x_4	2	1	0	1	30	30/1	
z	−2	−3	0	0	0		
		↑					
		entering variable					

Tableau 2

	x_1	x_2	x_3	x_4	p	$\dfrac{p_i}{a_{ij}}$	
x_2	$\frac{1}{2}$	1	$\frac{1}{2}$	0	20	$20/\frac{1}{2}$	
x_4	$\boxed{\frac{3}{2}}$	0	$-\frac{1}{2}$	1	10	$10/\frac{3}{2}$	← departing variable
z	$-\frac{1}{2}$	0	$\frac{3}{2}$	0	60		
	↑						

entering variable

Tableau 3

	x_1	x_2	x_3	x_4	p
x_2	0	1	$\frac{2}{3}$	$-\frac{1}{3}$	$\frac{50}{3}$
x_1	1	0	$-\frac{1}{3}$	$\frac{2}{3}$	$\frac{20}{3}$
z	0	0	$\frac{4}{3}$	$\frac{1}{3}$	$\frac{190}{3}$

8.6   Incremental values

In the problem under discussion we have assumed that the capacity of the machines U and V is fixed and cannot be increased. However, the capacity might be increased by, say, buying new machines or introducing overtime working on the present machines. This raises the question of whether it is worthwhile increasing the capacity of machine U or machine V, or of both of them. There are two ways of deciding on this. The first is by reworking the problem with new constraints while the second uses information from the final simplex tableau.

Taking the first of these, since it helps in understanding what is happening: in the original problem, if the total number of hours available on machine U, is increased from 30 to 31, (2) of section 8.2 becomes

$$x_1 + 2x_2 = 31$$

and together with (3),

$$2x_1 + x_2 = 30$$

the graphical method gives the new solution as $x_1 = 29/3$ and $x_2 = 32/3$. The contribution to profit is $z = 51.33$ compared with the previous value of 50. Increasing capacity on machine U by 1 hour will therefore increase profits by 1.33. Similarly, if the total hours available on machine V are increased from 30 to 31, with machine U being limited to 30 hours, the new solution is $x_1 = 32/3$ and $x_2 = 29/3$ giving $z = 50.33$, an increase of 0.33 on the previous value. The value of one extra hour of time on machine U is thus 1.33 while that of one extra hour on machine V is 0.33. These are called the *incremental worth*, *opportunity cost*, *dual price* or *shadow price* of machine time, since they indicate how much it is worth paying for the extra time. That is, if one hour of time on machine U can be bought for less than 1.33 it is worth buying since using it will increase profits by 1.33. Here it is clear that one hour of time on machine U is more valuable than one hour of time on machine V.

Turning now to the second way of checking whether it is worthwhile increasing capacity on the machines, we can use some of the information which is obtained as a by-product of the simplex method. This appears in the final tableau from the simplex method, as shown as in tableau 4.

Tableau 4

	x_1	x_2	x_3	x_4	p
x_2	0	1	$\frac{2}{3}$	$-\frac{1}{3}$	10
x_1	1	0	$-\frac{1}{3}$	$\frac{2}{3}$	10
z	0	0	$\frac{4}{3}$	$\frac{1}{3}$	50

The bottom row contains zero elements in the columns corresponding to the variables in the basis, that is, x_1 and x_2, but positive elements in the columns corresponding to the non-basic variables, that is, x_3 and x_4. These positive elements correspond to the incremental value of one hour of machine time on U and V respectively. This means that one extra hour of available machine time on U would increase the total contribution by $\frac{4}{3}$ to $51\frac{1}{3}$ and one extra machine hour on V would increase it by $\frac{1}{3}$ to $50\frac{1}{3}$.

In many cases, of course, extra capacity cannot be added in single units, e.g. hours, but must be added in fairly large steps. This would occur where the purchase of an extra machine was contemplated with a resulting increase of capacity of, say, 30 hours. Occasionally the structure of the problem allows the increased contribution to be determined directly from the incremental value. In the second example, an increase in capacity of 10 hours in machine U resulted in an increased contribution of $13\frac{1}{3}$ units, and this is exactly equal to $10 \times \frac{4}{3}$, i.e. the extra contribution was directly proportional to the number of additional units of capacity. This is not generally the case, and therefore it is better to re-run the linear programme if the effect of changes in any of the constraints is required.

An interesting point which arises from the solution of the linear programme is, as economists might expect, that at the optimum the marginal revenue equals the marginal cost for all those products which the company should produce.

In the example, the incremental value of machine capacity of U and V is respectively $\frac{4}{3}$ and $\frac{1}{3}$, and this must therefore be the value to the company of time on these machines at the margin. By multiplying these values by the hours which a unit of A and B uses on the machines we will obtain a value for the marginal cost of the two products.

Product A: marginal cost $= 1 \times \frac{4}{3} + 2 \times \frac{1}{3} = \frac{6}{3} = 2$

The contribution per unit of A, which is in fact its marginal revenue, is also equal to 2.

Product B: marginal cost $= 2 \times \frac{4}{3} + 1 \times \frac{1}{3} = \frac{9}{3} = 3$

This is equal to the contribution per unit or marginal revenue from B. It can be quite easily shown that the marginal revenue equals the marginal cost for all products that appear at a positive level in the optimal solution. Those products which do not appear in the final solution have a marginal cost greater than their marginal revenue, which is a result that would be expected on theoretical grounds.

8.7   The general problem

The linear programming problem discussed above may be written more generally as

Max $z = c_1 x_1 + c_2 x_2$

subject to $a_{11} x_1 + a_{12} x_2 \leqslant b_1$

$$a_{21}x_1 + a_{22}x_2 \leqslant b_2$$

with $\qquad\qquad x_1 \geqslant 0, x_2 \geqslant 0$

This can be extended to the case of n variables:

Max $\qquad\qquad z = c_1x_1 + c_2x_2 + \cdots + c_nx_n$

subject to $\qquad a_{11}x_1 + a_{12}x_2 + \cdots a_{1n}x_n \leqslant b_1$

$$a_{21}x_1 + a_{22}x_2 + \cdots a_{2n}x_n \leqslant b_2$$

$$\cdots\cdots$$

$$a_{m1}x_1 + a_{m2}x_2 + \cdots a_{mn}x_n \leqslant b_m$$

with $x_1 \geqslant 0, x_2 \geqslant 0, \cdots x_n \geqslant 0$. Notice that the number of constraints is m and this need not be the same as n, the number of variables.

Examples of applications of this are

(a) maximising profits (z) from selling x_i units of output each giving a profit c_i, subject to restrictions on the capacity (b_i) of the machines used in production;
(b) maximising the return on a portfolio where x_i is the quantity held of security of type i, subject to restrictions on the spread of the securities;
(c) maximising utility (z) from consuming x_i units of good i subject to an income constraint and limits to the time that can be used consuming the goods;
(d) maximising revenue (z) from selling x_i units of good i subject to restrictions on the expenditure on advertising each product in different media (b_i);
(e) maximising the consumption of vitamins (x_i) subject to the total cost and the way the vitamins are combined in different foods.

These more general problems can be tedious to solve, particularly if n is large and nowadays computer packages (such as LINDO/PC) are available to obtain the solution.

A rather different complication occurs when the problem is one of minimisation subject to constraints, rather than maximisation. This occurs when the objective is to minimise costs, subject to achieving minimum values of the variables. For example, minimise the cost of a particular diet with the requirement that a basic level of each type of vitamin is consumed or, given a set of production targets, how different machines can be used to produce the output at a minimum cost. The solution procedure using the simplex method is basically the same and so is not discussed here. However, it turns out that every linear-programming maximisation problem (generally referred to as the *primal*) has an equivalent minimisation problem (known as the *dual*). Returning to our two-variable maximisation problem we have:

Max $$z = c_1 x_1 + c_2 x_2$$

subject to $$a_{11} x_1 + a_{12} x_2 \leqslant b_1$$

$$a_{21} x_1 + a_{22} x_2 \leqslant b_2$$

with $$x_1 \geqslant 0, x_2 \geqslant 0$$

The equivalent minimisation problem can be written

Min $$z' = b_1 u_1 + b_2 u_2$$

subject to $$a_{11} u_1 + a_{21} u_2 \geqslant c_1$$

$$a_{12} u_1 + a_{22} u_2 \geqslant c_2$$

with $u_1 \geqslant 0$, $u_2 \geqslant 0$. The variables u_1 and u_2 are the *dual variables*. To help in the interpretation of these we recall that in the maximisation example z is the total profit, c_1 and c_2 are the profit per unit of x_1 and x_2, and b_1 and b_2 are the total numbers of hours available on each machine. The problem is to choose the values of x_1 and x_2 which maximise total profit while satisfying the constraints; that is, in the numerical example,

Max $$z = 2x_1 + 3x_2$$

subject to $$x_1 + 2x_2 \leqslant 30$$

$$2x_1 + x_2 \leqslant 30$$

with $$x_1 \geqslant 0, x_2 \geqslant 0$$

Putting the numbers into the dual we have

Min $$z' = 30u_1 + 30u_2$$

subject to $$u_1 + 2u_2 \geqslant 2$$

$$2u_1 + u_2 \geqslant 3$$

with $$u_1 \geqslant 0, u_2 \geqslant 0$$

Now u_1 is associated with the first constraint and is in fact the opportunity cost or shadow price of one hour of time on machine U for the decision maker. We saw in the previous section that tableau 4 includes these shadow prices and if the first constraint is changed marginally, say from 30 to 31 hours, the maximum value of total profit will increase by u_1 units. Similarly, u_2 is the shadow price of one hour of time on machine V. If the cost of the unit of resource is less than the shadow price it is worthwhile buying more of it. However, remember that the shadow price applies only at the margin and not necessarily if there is a large change in the value of a constraint.

The importance of the dual is that the optimal value of the objective function of the primal always equals the optimal value of the objective function of the dual. Therefore, if the solution to the primal problem is known, the solution to the dual can be found. More generally, the primal problem with n variables and m constraints can be written:

Max
$$z = c_1 x_1 + c_2 x_2 + \cdots + c_n x_n$$

subject to
$$a_{11} x_1 + a_{12} x_2 + \cdots a_{1n} x_n \leqslant b_1$$
$$a_{21} x_1 + a_{22} x_2 + \cdots a_{2n} x_n \leqslant b_2$$
$$\cdots$$
$$a_{m1} x_1 + a_{m2} x_2 + \cdots a_{mn} x_n \leqslant b_m$$

with
$$x_1 \geqslant 0, x_2 \geqslant 0, \cdots, x_n \geqslant 0$$

The corresponding dual problem has m variables and n constraints and is

Min
$$z' = b_1 u_1 + b_2 u_2 + \cdots + b_m u_m$$

subject to
$$a_{11} u_1 + a_{21} u_2 + \cdots + a_{m1} u_m \geqslant c_1$$
$$a_{12} u_1 + a_{22} u_2 + \cdots + a_{m2} u_m \geqslant c_2$$
$$\cdots$$
$$a_{1n} u_1 + a_{2n} u_2 + \cdots + a_{mn} u_m \geqslant c_n$$

with
$$u_1 \geqslant 0, u_2 \geqslant 0, \cdots, u_m \geqslant 0$$

If $n < m$ it is usually easier to solve the primal problem.

8.8 Exercises

1. Use the simplex method to establish the product mix which will provide the maximum contribution in the example shown in Table 8.4. Calculate the additional contribution which will be obtained if one additional hour of capacity became available on either machine U or machine V and show that this is equal to the

TABLE 8.4

	No. of hours required per unit of product		Total no. of hours available
	A	B	
Machine U	2	3	25
Machine V	4	1	35
Contribution per unit of product	9	7	

incremental value of machine capacity derived from the final tableau of the simplex procedure.

2. Using the basic data provided in Exercise 1, determine the number of units of A and B which should be manufactured if either of the following conditions applies.
   (a) the total number of available hours of machine V is increased to (i) 40 and (ii) 45 in successive stages;
   (b) the contribution per unit of A is reduced to 4;
   (c) the number of hours required by product B on machine U is reduced to 2.

3. A farmer can grow two types of crop, A and B. The profit per unit of A is 3 and that for B is 2. Each unit of A requires one unit of labour, two units of land and three units of machine time, while for B the requirements are four of labour, one of land and one of machine time. There are 60 units of labour, 40 units of land and 50 units of machine time available. (a) What should the farmer produce to maximise profits? (b) Is it worthwhile renting 10 extra units of land at a total cost of 5?

4. Suppose that for a healthy diet an adult has a daily requirement of at least 10 units of energy and 15 units of protein and the following foods are available:

	Bread	Cheese	Meat
Price per unit	2	4	5
Energy per unit	1	2	1
Protein per unit	1	3	6

(a) Set up the linear-programming problem for minimising the total cost of satisfying the daily requirements.

(b) Form the dual of the minimisation problem and obtain the optimal solution.

8.9 Revision exercises for Chapter 8 (without answers)

1. Use the graphical method to determine the maximum

   value of $\qquad z = 10x_1 + 5x_2$

   subject to $\qquad 5x_1 + 5x_2 \leqslant 21$

   $\qquad\qquad\qquad 4x_1 + 8x_2 \leqslant 30$

   and $\qquad\qquad x_1 \geqslant 0, x_2 \geqslant 0$

2. Use the simplex method to answer Question 1. What is the shadow price of the first constraint?

3. Use the graphical method to determine the maximum

   value of $\qquad z = 2x_1 + x_2$

   subject to $\qquad x_1 + x_2 \leqslant 30$

   $\qquad\qquad\qquad 2x_1 + x_2 \leqslant 40$

   and $x_1 \geqslant 0$, $x_2 \geqslant 0$, and comment on the solution.

4. A publisher produces hardback and paperback versions of a book. To produce one hardback copy requires 5 units of labour and 3 of machine time, and to produce one paperback copy requires 3 units of labour and 2 of machine time. The profit from one hardback is 5 and that from one paperback is 2. There is a limit of 40 units of labour and 30 of machine time. What combination of books should be produced in order to maximise profits? Is it worthwhile buying extra labour at a cost of 2 per unit?

5. Write down and solve the dual of the following problem:

   min $\qquad\qquad z = 2x_1 + 3x_2 + x_3$

   subject to $\qquad 2x_1 + x_2 + x_3 \geqslant 4$

   $\qquad\qquad\qquad x_1 + 2x_2 + 4x_3 \geqslant 6$

   with $\qquad\qquad x_1 \geqslant 0, x_2 \geqslant 0$

6. A workshop can make chairs, tables and desks. The profit is 5 from each chair, 15 from each table and 20 from each desk. A chair needs 3 units of labour, 2 of machine time and 1 of polishing. A

table needs 4 units of labour, 4 of machine time and 3 of polishing. A desk needs 8 units of labour, 6 of machine time and 6 of polishing. Labour is limited to 35 units, machine time to 30 units and polishing to 30 units. What should be produced? What are the opportunity costs of labour, machine time and polishing?

7. Determine the maximum value of

$$z = 4x_1 + 4x_2 + 7x_3 + x_4$$

subject to

$$x_1 + 3x_2 + 3x_3 + x_4 \leqslant 50$$
$$3x_1 + x_2 + 2x_3 + 2x_4 \leqslant 60$$
$$x_1 + x_2 + 4x_3 + 3x_4 \leqslant 60$$
$$2x_1 + 2x_2 + x_3 + x_4 \leqslant 50$$

with

$$x_i \geqslant 0$$

Chapter 9
Differential Equations

9.1 Introduction

In Chapter 6 we saw that equations of the form

$$\frac{dy}{dx} = f(x)$$

could be solved by integration. Such an equation is known as a *differential equation* since it expresses the relationship between the derivative of y with respect to x and the value of x itself. It contains only the first-order derivative and is, therefore, known as a *first-order differential equation* and because this term is only raised to the power one it is a *first-degree* equation.

$(dy/dx)^2 = 6x + 2$ is a differential equation of *first-order and second-degree* because it contains only a first-order derivative and this is raised to the power two.

$d^2y/dx^2 = 6$ is a differential equation of *second-order and first-degree* because it contains a second-order derivative which is raised to the power one.

$$\frac{d^2y}{dx^2} + \left(\frac{dy}{dx}\right)^2 + y = 0 \qquad \text{and} \qquad \frac{d^2y}{dx^2} - y = 0$$

are both differential equations of second-order and first-degree because the highest-order derivative that is present in each equation is the second and this is raised to the power one in both cases.

In general, a differential equation contains one or more derivatives, all or some of which may be raised to a power other than zero.

Differential equations are classified into groups by

1. *The order*. This is the same as that of the highest-order derivative which is present in the equation.

319

2. *The degree*. This is determined by the power to which the highest-order derivative is raised.

The general differential equation

$$\left(\frac{d^n y}{dx^n}\right)^{\varkappa} + \left(\frac{d^{n-1} y}{dx^{n-1}}\right)^{\theta} + \cdots + y + c = 0$$

is one of the *n*th order and the $\varkappa$th degree, even if θ is greater than $\varkappa$.

The equations which occur mostly frequently are those of first-and second-order, both usually being of the first degree. The problem is to find a solution which satisfies the following conditions.

1. It must be free of terms containing derivatives.
2. It must satisfy the differential equation.

This can be illustrated by reference to the example

$$\frac{d^2 y}{dx^2} = 6$$

Integration of this second-order equation leads to the first-order equation

$$\frac{dy}{dx} = 6x + A$$

which on further integration leads to

$$y = 3x^2 + Ax + B$$

This is the solution of the second-order equation since it does not contain derivatives and it satisfies the original equation (as can be checked by differentiation). In obtaining the solution the result

$$\int \frac{dy}{dx} \, dx = y$$

was used and since

$$\int 1 dy = y$$

we are effectively treating dy/dx as the ratio of two differentials. This separation of dy and dx is particularly useful in the solution of certain differential equations, as will be seen below.

We now consider some of the standard techniques of solution of first-and second-order differential equations.

9.2   First-order differential equations

The general form of first-order and first-degree differential equations is

$$\frac{dy}{dx} = F(x, y)$$

We will consider various special cases before looking at a general method of solution of this equation.

SEPARABLE VARIABLES
An equation with separable variables is of the form

$$\frac{dy}{dx} = \frac{f_1(x)}{f_2(y)}$$

or, using differentials

$$f_1(x)dx = f_2(y)dy$$

where $f_1(x)$ contains only terms in x, and $f_2(y)$ contains only terms in y. Integrating gives

$$\int f_1(x)dx = \int f_2(y)dy$$

For example, if

$$\frac{dy}{dx} = \frac{x}{y}$$

then $y\, dy = x\, dx$.
The solution can be found by integrating both sides of this equation

$$\int y\, dy = \int x\, dx$$

$$\therefore \qquad \tfrac{1}{2}y^2 = \tfrac{1}{2}x^2 + C$$

or $\qquad y^2 - x^2 = 2C = \text{constant} = A$

A check can be made that this solution does satisfy the differential equation by differentiating, in the following manner. (For discussion see section 7.8.)

Let $\qquad z = y^2 - x^2 - A$

then $\qquad \dfrac{\partial z}{\partial x} = -2x \qquad \dfrac{\partial z}{\partial y} = 2y$

and
$$\frac{dy}{dx} = -\frac{\partial z/\partial x}{\partial z/\partial y} = -\frac{(-2x)}{2y} = \frac{x}{y}$$

the solution does therefore, satisfy the equation. The arbitrary constant, A, arises because the process is simply one of integration. To determine the value of this constant requires some further information about the values of x and y. This is frequently in the form of *initial conditions*, where the starting values are given (such as revenue is zero when sales are zero), or *boundary conditions*, where the maximum or minimum values are stated (such as x is positive or y has a maximum of 100). In the example we assume that the initial condition is

$$y = 2 \quad \text{when} \quad x = 1$$

then
$$2^2 - 1^2 = A \quad \text{or} \quad A = 3$$

and the unique solution to the differential equation

$$\frac{dy}{dx} = \frac{x}{y} \quad \text{given} \quad y = 2 \quad \text{when} \quad x = 1$$

is
$$y^2 - x^2 = 3$$

Example 1

What is the general form of the demand equation which has a constant elasticity of -1?

Let q be the quantity demanded at price p. Then elasticity, e, is defined by

$$e = \frac{p}{q}\frac{dq}{dp}$$

For $e = -1$, we require

$$\frac{p}{q}\frac{dq}{dp} = -1 \quad \text{or} \quad \frac{dq}{q} = -\frac{dp}{p}$$

Integrating, we have

$$\log_e (q) = -\log_e (p) + \log_e (A)$$

where A is a constant.

$$\therefore \qquad \log_e (q) = \log_e (p^{-1}) + \log_e (A) = \log_e (Ap^{-1})$$

and
$$q = Ap^{-1} \quad \text{or} \quad pq = A$$

Differential Equations with Homogeneous Coefficients

The above method breaks down if the variables are not separable but if the differential equation

$$f_1(x, y) \, dx + f_2(x, y) \, dy = 0$$

has homogeneous coefficients it can be converted to the separable-variables type by a suitable change of variable. (For a discussion of homogeneity see section 7.10.)

For example, if

$$\frac{dy}{dx} = \frac{x + y}{x}$$

then

$$(x + y) \, dx - x \, dy = 0$$

In this case the variables are not separable but the coefficients of dx and dy are homogeneous of degree one. This can be converted into the separable-variables type by the substitutions

$$y = vx \qquad \text{and} \qquad dy = v \, dx + x \, dv$$

After carrying out the substitution we have

$$(x + vx) \, dx - x(v \, dx + x \, dv) = 0$$

which can be rearranged into

$$x(1 - v) \, dx + x(v \, dx - x \, dv) = 0$$
$$x[(1 - v) \, dx + (v \, dx - x \, dv)] = 0$$

If this is to be valid for all values of x then

$$(1 - v) \, dx + (v \, dx - x \, dv) = 0$$

that is

$$dx - v \, dx + v \, dx - x \, dv = 0$$

or

$$dx - x \, dv = 0$$

This is now in the separable-variables form and can be treated as such

$$dx = x \, dv$$

$$\frac{dx}{x} = dv$$

$$\therefore \qquad \int \frac{dx}{x} = \int dv$$

and hence $\log_e x = v + c$

To obtain the solution in terms of the variables x and y it is necessary to substitute for v in this equation.

$$v = \frac{y}{x}$$

$\therefore$ $\log_e x = \dfrac{y}{x} + c$ or $y = x(\log_e x - c)$

This is, therefore, the solution of the linear differential equation

$$\frac{dy}{dx} = \frac{x + y}{x}$$

The result can be checked by differentiation.

Example 2

The total cost, y, of producing x units of output of a product is related to the marginal cost of production by the equation

$$\frac{dy}{dx} = \frac{y - 2x}{2y - x}$$

What is the total cost function?
The equation can be rewritten as

$$(2y - x)\, dy + (2x - y)\, dx = 0$$

The coefficients of this equation are homogeneous of degree 1, and so we substitute $y = vx$ and $dy = v\, dx + x\, dv$

$$(2vx - x)(v\, dx + x\, dv) + (2x - vx)\, dx \;= 0$$

$$(2v^2 x - xv + 2x - vx)\, dx + (2vx^2 - x^2)\, dv = 0$$

Dividing through by x gives

$$(2v^2 - 2v + 2)\, dx + x(2v - 1)\, dv = 0$$

that is $\dfrac{dx}{x} + \dfrac{2v - 1}{2v^2 - 2v + 2}\, dv = 0$

Integrating gives

$$\log_e x + \tfrac{1}{2}\log_e (2v^2 - 2v + 2) = \log_e A$$

or
$$x(2v^2 - 2v + 2)^{1/2} = A$$

Substituting $v = y/x$ and squaring gives

$$x^2\left(\frac{2y^2}{x^2} - \frac{2y}{x} + 2\right) = A^2$$

or
$$2y^2 - 2xy + 2x^2 = A^2$$

This is the relationship between total cost, y, and output, x. The value of A can be found if some extra information on x and y is available. Suppose that the initial condition is that when $x = 0$, $y = 100$ so that the fixed cost is 100. Substituting in the solution gives $A^2 = 20,000$.

DIFFERENTIAL EQUATIONS WITH NON-HOMOGENEOUS COEFFICIENTS
  These are of the form

$$(a_1x + b_1y + c_1)\, dy + (a_2x + b_2y + c_2)\, dx = 0$$

The coefficients are not homogeneous but they can be made so by the substitutions

$$a_1x + b_1y + c_1 = Y \qquad (1)$$
$$a_2x + b_2y + c_2 = X \qquad (2)$$

This results in

$$X\, dx + Y\, dy = 0$$

or
$$\frac{dy}{dx} = -\frac{X}{Y} \qquad (3)$$

It is now necesary to differentiate (1) and (2)

$$\frac{dY}{dx} = a_1 + b_1\frac{dy}{dx} = a_1 + b_1\left(-\frac{X}{Y}\right) \qquad (4)$$

$$\frac{dX}{dx} = a_2 + b_2\frac{dy}{dx} = a_2 + b_2\left(-\frac{X}{Y}\right) \qquad (5)$$

Dividing (4) by (5), we obtain

$$\frac{dY/dx}{dX/dx} = \frac{dY}{dX} = \frac{a_1 + b_1(-X/Y)}{a_2 + b_2(-X/Y)}$$

$$\therefore \qquad \frac{dY}{dX} = \frac{a_1Y - b_1X}{a_2Y - b_2X}$$

or $\qquad (a_1Y - b_1X)\, dX - (a_2Y - b_2X)\, dY = 0$

This equation is homogeneous in X and Y and it is possible to obtain a solution as above. The exercise is left to the reader to complete as the procedure is no different from that previously demonstrated (section 9.4, Exercise 1).

Example 3

$$(x + 2y + 1)\, dx + (2x + y + 2)\, dy = 0$$

Let $\qquad\qquad\qquad x + 2y + 1 = X \qquad\qquad\qquad (1a)$

$$2x + y + 2 = Y \qquad\qquad\qquad (2a)$$

and $\qquad\qquad\qquad \dfrac{dy}{dx} = -\dfrac{X}{Y} \qquad\qquad\qquad (3a)$

Differentiating (1a) and (2a), we have

$$\frac{dX}{dx} = 1 + \frac{2dy}{dx} = 1 - \frac{2X}{Y} = \frac{Y - 2X}{Y} \qquad (4a)$$

$$\frac{dY}{dx} = 2 + \frac{dy}{dx} = 2 - \frac{X}{Y} = \frac{2Y - X}{Y} \qquad (5a)$$

Dividing (5a) by (4a) gives

$$\frac{dY/dx}{dX/dx} = \frac{dY}{dX} = \frac{(2Y - X)/Y}{(Y - 2X)/Y} = \frac{2Y - X}{Y - 2X}$$

$\therefore \qquad\qquad (2Y - X)\, dX - (Y - 2X)\, dY = 0$

This is homogeneous in X and Y, so let

$$Y = VX \qquad \text{and} \qquad dY = V\, dX + X\, dV$$

then $\quad (2VX - X)\, dX - (VX - 2X)(V\, dX + X\, dV) = 0$

$$X[(2V - 1)\, dX - (V - 2)(V\, dX + X\, dV)] = 0$$

$\therefore \qquad\qquad (2V - 1)\, dX - (V - 2)(V\, dX + X\, dV) = 0$

$$2V\, dX - dX - V^2\, dX - VX\, dV + 2V\, dX + 2X\, dV = 0$$

$$(2V - 1 - V^2 + 2V)\, dX - (VX - 2X)\, dV = 0$$

$$(-V^2 + 4V - 1)\, dX - X(V - 2)\, dV = 0$$

This is in the separable-variables form and can be written

$$\frac{dX}{X} = \left(\frac{V - 2}{-V^2 + 4V - 1}\right) dV$$

The integral of the left hand side is simply $\log_e X$. The integral of the right hand side is obtained as follows:

$$\int \frac{V - 2}{-V^2 + 4V - 1} dV = -\int \frac{\frac{1}{2}(2V - 4)}{V^2 - 4V + 1} dV$$

$$= -\frac{1}{2} \int \frac{2V - 4}{V^2 - 4V + 1} dV$$

$$= -\frac{1}{2} \log_e (V^2 - 4V + 1) + C$$

Combining these two results

$$\log_e X = -\frac{1}{2} \log_e (V^2 - 4V + 1) + C$$

It is more convenient in examples such as this to let the constant assume a logarithmic form, i.e. let $C = \log_e A$. From the properties of logarithms (see section 4.14) it is possible to write this as

$$\log_e X = \log_e A - \log_e \sqrt{(V^2 - 4V + 1)}$$

$$= \log_e \frac{A}{\sqrt{(V^2 - 4V + 1)}}$$

$$\therefore \qquad X = \frac{A}{\sqrt{(V^2 - 4V + 1)}}$$

or $\qquad X^2(V^2 - 4V + 1) = A^2 = \text{constant} = B$

The result must be expressed in terms of the variables x and y and this is done in two stages

1. Replace V by Y/X

then $X^2\left(\dfrac{Y^2}{X^2} - \dfrac{4Y}{X} + 1\right) = B$

that is $Y^2 - 4YX + X^2 = B$

2. Replace X by $(x + 2y + 1)$ and Y by $(2x + y + 2)$

then $(2x + y + 2)^2 - 4(x + 2y + 1)(2x + y + 2)$
$+ (x + 2y + 1)^2 = B$

that is,

$$4x^2 + y^2 + 4 + 4xy + 4y + 8x - 8x^2 - 4xy - 8x - 16xy - 8y^2 - 16y$$

$$- 8x - 4y - 8 + x^2 + 4y^2 + 1 + 4xy + 4y + 2x = B$$

or $\qquad - 3x^2 - 3y^2 - 3 - 12xy - 12y - 6x = B$

$$- 3(x^2 + y^2 + 4xy + 4y + 2x + 1) = B$$

$\therefore \qquad x^2 + y^2 + 4xy + 4y + 2x = -\dfrac{B}{3} - 1 = \text{constant}$

$\therefore$ the solution to the differential equation

$$(x + 2y + 1)\, dx + (2x + y + 2)\, dy = 0$$

is given by

$$x^2 + y^2 + 4xy + 4y + 2x = C$$

EXACT DIFFERENTIAL EQUATIONS

An *exact differential equation* involving x and y is one which may immediately (i.e. without multiplying through by any factor) be expressed in terms of derivatives of functions of x and y, where these functions do not themselves involve derivatives.

$$3x^2y + x^3 \frac{dy}{dx} = 0$$

is an exact differential equation because it is formed by differentiating the equation $x^3y = A$.

It should be possible to find the solution to an exact differential equation quickly if it is recognised as such. This is not always obvious at first sight but the following rule always holds.

$$P\, dx + Q\, dy = 0$$

is an exact differential equation if

$$\frac{\partial P}{\partial y} = \frac{\partial Q}{\partial x}$$

For example, consider the equation

$$3x^2y + x^3 \frac{dy}{dx} = 0$$

$$(3x^2y)\, dx + (x^3)\, dy = 0$$

$$\therefore \qquad \frac{\partial(3x^2y)}{\partial y} = 3x^2 \qquad \text{and} \qquad \frac{\partial(x^3)}{\partial x} = 3x^2$$

These two partial derivatives are equal and therefore the equation is exact. The solution in this case is fairly obvious when the values of *P* and *Q* are considered.

$$P = 3x^2y \qquad Q = x^3$$

and the solution is $x^3y = C$.

When the solution is not so obvious it can be found by using the following relationships:

If
$$z = f(x, y) = 0$$

Then
$$dz = \frac{\partial z}{\partial x}dx + \frac{\partial z}{\partial y}dy = 0$$

(See section 7.5)

To reverse this procedure and find *z* it is necessary to compare the coefficients in the equation with the partial derivatives

$$\frac{\partial z}{\partial x} \qquad \text{and} \qquad \frac{\partial z}{\partial y}$$

For example, in the exact equation

$$(2x + 3y)\, dx + (3x - 2y)\, dy = 0$$

$$\frac{\partial z}{\partial x} = 2x + 3y \qquad\qquad\qquad (1)$$

and
$$\frac{\partial z}{\partial y} = 3x - 2y \qquad\qquad\qquad (2)$$

It is now possible to suggest the function from which these two partial derivatives were formed.

From (1),   *z* must contain the terms $x^2 + 3xy + C$

and from (2),   *z* must contain the terms $3xy - y^2 + C$

It must be concluded that the function will include all these terms without double counting any that are common to both. This leads to the solution

$$z = x^2 + 3xy - y^2 + C = 0$$

and it is easily checked that this is the solution to the differential equation.

This differential equation has homogeneous coefficients and could have been solved as such with the same result but with a good deal more effort. It is, therefore, extremely useful to be able to recognise an exact differential equation when it occurs and to apply the above method to obtain a solution.

A GENERAL METHOD FOR FIRST-ORDER LINEAR EQUATIONS

It is generally possible to make a non-exact differential equation into an exact one by multiplying by an *integrating factor*. For example,

let

$$\frac{dy}{dx} = x + y$$

then

$$\frac{dy}{dx} - y = x$$

The left hand side of the equation is not exact and it is difficult to integrate as it stands. But if all the terms are multiplied by e^{-x} the equation becomes

$$e^{-x}\frac{dy}{dx} - e^{-x}y = e^{-x}x$$

The left hand side is now exact and is the derivative of $e^{-x}y$ and the right hand side of the equation can be integrated by parts (see section 6.14).

$$\int xe^{-x}\,dx = x(-e^{-x}) - \int (-e^{-x})\,dx$$
$$= -xe^{-x} - e^{-x} + C = e^{-x}(-x - 1) + C$$

This is equal to the integral of the left hand side, i.e. the exact part, of the equation

∴

$$e^{-x}y = e^{-x}(-x - 1) + C$$

Dividing throughout by e^{-x}, we have

$$y = -x - 1 + \frac{C}{e^{-x}} = -x - 1 + Ce^x$$

therefore the solution to the differential equation

$$\frac{dy}{dx} = x + y \quad \text{is} \quad y = Ce^x - x - 1$$

This can be checked by differentiation:

$$\frac{dy}{dx} = Ce^x - 1 \quad \text{and} \quad Ce^x = x + y + 1$$

$$\therefore \quad \frac{dy}{dx} = x + y + 1 - 1 = x + y$$

In general, the integrating factor is equal to e^θ where $\theta = \int P \, dx$ for all differential equations of the form

$$\frac{dy}{dx} + Py = Q$$

where P and Q are functions of x only, including the case where they are constants. This method is useful where both P and the product of the integrating factor and Q are reasonably easy to integrate.

Example 4

$$\frac{dy}{dx} + 4y = 2x$$

In this case $\qquad P = 4 \qquad \text{and} \qquad \int P \, dx = 4x$

Therefore the integrating factor is e^{4x}, and multiplying by this gives

$$e^{4x} \frac{dy}{dx} + 4e^{4x}y = 2xe^{4x}$$

The left hand side is now the derivative of ye^{4x}, and integrating the right hand side gives

$$\int 2xe^{4x} \, dx = \frac{2xe^{4x}}{4} - \int \frac{e^{4x}}{4} 2 \, dx = \frac{xe^{4x}}{2} - \frac{e^{4x}}{8} + A$$

$$\therefore \qquad ye^{4x} = \frac{xe^{4x}}{2} - \frac{e^{4x}}{8} + A \qquad \text{or} \qquad y = \frac{x}{2} - \frac{1}{8} + Ae^{-4x}$$

9.3   Applications of first-order differential equations

We now examine a number of applications of first-order differential equations in economics and business.

THE DOMAR DEBT MODEL

Suppose that the national debt, D, varies as time (t) passes and its rate of change is related to the level of national income such that

$$\frac{dD}{dt} = ay \qquad (a > 0)$$

where y varies with t. Also, suppose that y is increasing at a constant rate b, so that

$$\frac{dy}{dt} = b \qquad (b > 0)$$

Solving for y, $y = bt + A$
where A is a constant of integration. Therefore,

$$\frac{dD}{dt} = abt + aA$$

and integrating, $D = 0.5abt^2 + aAt + B$
where B is a constant of integration. The ratio of debt to national income is given by

$$\frac{D}{y} = \frac{0.5abt^2 + aAt + B}{bt + A} = 0.5at + \frac{0.5aAt + B}{bt + A}$$

As $t \to \infty$, $0.5at \to \infty$ and so $D/y \to \infty$. That is, with this model, the ratio of debt to national income increases without limit as time passes.

One unrealistic aspect of this model is the assumption that y increases at a constant rate. Suppose instead that y increases at a constant percentage rate,

$$\frac{dy}{dt} = by \qquad (b > 0)$$

Now the solution for y is from

$$\frac{dy}{y} = b \, dt \qquad \text{or} \qquad \log y = bt + \log C$$

and $y = C \, e^{bt}$ where C is a constant.

The relationship for the change in debt is now

$$\frac{dD}{dt} = aC\,e^{bt}$$

and integrating gives

$$D = \frac{aC}{b}\,e^{bt} + E$$

where E is a constant. The ratio of debt to national income is now

$$\frac{D}{y} = \frac{(a/b)C\,e^{bt} + E}{C\,e^{bt}} = (a/b) + (E/C)e^{-bt}$$

As $t \to \infty$, $D/y \to (a/b)$, which is a positive constant. That is, the debt becomes proportional to the level of national income.

COST FUNCTIONS

The costs of many activities vary as time passes. For example, the running costs of a car generally increase as the car gets older and at the same time the value of a car declines. Suppose that the relationship between running costs (C) and resale value (V) is given by

$$\frac{dC}{dt} = \frac{a}{V} \quad (a > 0)$$

where a is a constant. Also, suppose that the value changes according to

$$\frac{dV}{dt} = -bV \quad (b > 0)$$

where b is a constant. There are two obvious initial conditions here: when $t = 0$ the running costs, C, are zero and the resale value, V, is the purchase price of the car, V_0, say. Thus, when $t = 0$, $C = 0$ and $V = V_0$. Solving the second equation:

$$\frac{dV}{V} = -b\,dt \quad \text{or} \quad \log V = -bt + \log A$$

where A is a constant. Rearranging gives

$$V = Ae^{-bt}$$

The initial condition $t = 0$, $V = V_0$ gives $V_0 = A$ since $e^0 = 1$ so that the

the solution for the value is

$$V = V_0 e^{-bt}$$

Substituting into the cost equation,

$$\frac{dC}{dt} = \frac{a}{V_0 e^{-bt}} = \frac{a e^{bt}}{V_0}$$

Integrating,

$$C = \frac{a e^{bt}}{V_0 b} + B$$

where B is a constant. Using the initial condition $t = 0$ and $C = 0$,

$$0 = \frac{a}{V_0 b} + B \quad \text{or} \quad B = \frac{-a}{V_0 b}$$

The solution is

$$C = \frac{a\,(e^{bt} - 1)}{V_0 b}$$

MARKET EQUILIBRIUM

In some models of demand and supply a dynamic term is included to allow for adjustments of behaviour. For example the market for a product might be described by

$$\text{demand: } q = 150 - 6p + 5\frac{dp}{dt}$$

$$\text{supply: } q = 4p - 50 + 20\frac{dp}{dt}$$

Here both demand and supply are higher if prices are rising. In equilibrium:

$$150 - 6p + 10\frac{dp}{dt} = 4p - 50 + 20\frac{dp}{dt}$$

or

$$-10\frac{dp}{dt} = 10p - 200$$

Rearranging:

$$\frac{-10\,dp}{10p - 200} = dt$$

Integrating: $\qquad -\log (10p - 200) = t - \log A$

or $\qquad\qquad\qquad 10p - 200 = Ae^{-t}$

where A is a constant. The value of A can be determined from an initial condition. Suppose that when $t = 0$, $p = 21$ then $A = 10$ and the time path of prices is given by

$$10p = 200 + 10e^{-t} \qquad \text{or} \qquad p = 20 + e^{-t}$$

WALRASIAN ADJUSTMENT

In the study of markets, the equilibrium price is that for which the quantity demanded (q^d) and the quantity supplied (q^s) are equal. The analysis is static and time has no role to play. However, the basic laws of demand and supply state that if demand exceeds supply for any good there is a tendency for the price to rise, while if supply exceeds demand the price will fall. These effects can be modelled by the Walrasian adjustment mechanism

$$\frac{dp}{dt} = a(q^d - q^s) \qquad (a > 0)$$

where a is a constant, in which the change in price is positive if $q^d > q^s$. Here it is assumed that prices do not adjust instantaneously and that the rate of adjustment is larger the greater the difference between q^s and q^d. Now suppose that the demand and supply curves are

$$q^d = 60 - 3p$$

$$q^s = -15 + 2p$$

and that $a = 0.4$, then substituting into the differential equation gives

$$\frac{dp}{dt} = 0.4(60 - 3p + 15 - 2p) = 0.4(75 - 5p) = 30 - 2p$$

While this is a non-homogeneous equation the solution can still be found as previously. The equation can be written

$$\frac{dp}{30 - 2p} = dt$$

and integrating,

$$-0.5 \log (30 - 2p) = t + A$$

This can be rearranged as

$$30 - 2p = Be^{-2t} \quad \text{or} \quad p = 15 - 0.5Be^{-2t}$$

Notice that as $t \to \infty$, $p \to 15$ which is the equilibrium price. In this case the price converges to the equilibrium price. However, it is easily seen that if, say, $a = -0.4$ then the solution is

$$p = 15 + 0.5Be^{2t}$$

and as $t \to \infty$, $p \to \infty$ (if $B > 0$) so that equilibrium is not achieved.

CONTINUOUSLY COMPOUNDED INTEREST

The continuously compounded interest model of section 4.10 can be obtained by using differential equations. Let S be the amount in a savings account and suppose that initially $S = 100$. If interest is compounded continuously at 5% per annum then the change in S is 0.05 of S or

$$\frac{dS}{dt} = 0.05\, S$$

This can be solved by writing it as

$$\frac{dS}{S} = 0.05\, dt$$

and integrating to give

$$\log S = 0.05t + \log A$$

or

$$S = Ae^{0.05t}$$

The initial condition is when $t = 0$, $S = 100$ and so $A = 100$ and the solution is that the value in the account at time t is

$$S_t = 100e^{0.05t}$$

9.4   Exercises

1. Find the solution to the equation

$$(a_1 Y - b_1 X)\, d\,X - (a_2 Y - b_2 X)\, d\,Y = 0.$$

2. What is the general form of the demand equation which has an elasticity of $-n$?

3. The total cost of production, y, and the level of output x are related to the marginal cost of production by the equation

$$\frac{dy}{dx} = \frac{24x^2 - y^2}{xy}$$

What is the total cost function if $x = 2$ when $y = 4$?

4. Determine the consumption function for which the marginal propensity to consume is

$$\frac{dc}{dx} = a - bx$$

where c is consumption and x is income.

5. With a Walrasian adjustment process, if the quantities supplied and demanded differ then the rate of change of prices is given by

$$\frac{dp}{dt} = kz$$

where t is time, z is excess demand and p is the deviation of price from its equilibrium level. Solve the differential equation for p when

(a) $z = ap$ (b) $z = ap + bp^2$

6. The Harrod–Domar growth model can be formulated as follows:

$$S = aY \tag{1}$$

$$I = b\frac{dY}{dt} \tag{2}$$

$$I = S \tag{3}$$

Where S is savings, Y is income and I is investment. Substitute from (1) and (2) into (3) and solve the differential equation for Y.

7. Solve the following equations

(a) $x\dfrac{dy}{dx} = x - y$ (b) $\dfrac{dy}{dx} = \dfrac{y + 1}{2y + x}$

(c) $\dfrac{dy}{dx} = x(1 + x) - 2y$ (d) $\dfrac{dy}{dx} + 4y = x^3$

8. If supply and demand are given by

$$q^s = -a + bp \text{ and } q^d = e - fp + \frac{dp}{dt} \quad (a, b, e, f > 0)$$

assuming that the market clears at every point in time determine the time path of p.

9.5 Second-order homogeneous differential equations

The general form of linear second-order differential equation with constant coefficients is

$$a\frac{d^2y}{dx^2} + b\frac{dy}{dx} + cy = f(x)$$

where a, b and c are constants. When $f(x)$ is equal to zero the equation is said to be *homogeneous* and one solution of the homogeneous equation

$$a\frac{d^2y}{dx^2} + b\frac{dy}{dx} + cy = 0$$

is of the form $y = e^{mx}$ where m is a constant. To determine m let $y = e^{mx}$.

Then

$$\frac{dy}{dx} = me^{mx} \qquad \frac{d^2y}{dx^2} = m^2e^{mx}$$

Substituting these values in the homogeneous equation, we obtain

$$am^2e^{mx} + bme^{mx} + ce^{mx} = 0$$

$$e^{mx}(am^2 + bm + c) = 0$$

$$\therefore \qquad (am^2 + bm + c) = 0$$

This quadratic in m is called the *characteristic* or *auxiliary equation* and the roots can be found by the usual method (see section 3.3). Therefore

$$m = \frac{-b \pm \sqrt{(b^2 - 4ac)}}{2a}$$

or

$$m_1 = \frac{-b + \sqrt{(b^2 - 4ac)}}{2a} \qquad m_2 = \frac{-b - \sqrt{(b^2 - 4ac)}}{2a}$$

This suggests two possible alternatives for the solution to the differential equation:

$$y = e^{m_1x} \qquad \text{and} \qquad y = e^{m_2x}$$

But the solution to a second-order differential equation must contain two constants of integration. This is achieved by combining the two solutions in the form

$$y = k_1 e^{m_1 x} + k_2 e^{m_2 x}$$

where k_1 and k_2 are the two arbitrary constants. This value of y satisfies the differential equation as can be shown by substitution.

Example 1

$$\frac{d^2y}{dx^2} - \frac{5\,dy}{dx} + 6y = 0$$

Then if
$$y = e^{mx}$$

$$m^2 e^{mx} - 5m e^{mx} + 6e^{mx} = 0$$

$$e^{mx}(m^2 - 5m + 6) = 0 \quad \text{or} \quad m^2 - 5m + 6 = 0$$

$\therefore m_1 = 2$ and $m_2 = 3$ and the solution is

$$y = k_1 e^{2x} + k_2 e^{3x}$$

Check:
$$\frac{dy}{dx} = 2k_1 e^{2x} + 3k_2 e^{3x}$$

$$\frac{d^2y}{dx^2} = 4k_1 e^{2x} + 9k_2 e^{3x}$$

Hence

$$\frac{d^2y}{dx^2} - 5\frac{dy}{dx} + 6y = 4k_1 e^{2x} + 9k_2 e^{3x} - 10k_1 e^{2x} - 15k_2 e^{3x} + 6k_1 e^{2x} + 6k_2 e^{3x}$$

and this is equal to 0.

The term e^{mx} will always be ignored after the function has been differentiated and it is not necessary to include it once we are familiar with its purpose. If it is ignored the above procedure can be conveniently summarised in the following set of rules.

1. From the differential equation form a new equation in which d^2y/dx^2 is replaced by m^2, dy/dx is replaced by m and y is replaced by unity:

$$am^2 + bm + c = 0$$

2. Find the roots of this quadratic equation, m_1 and m_2.
3. Form the solution to the differential equation

$$y = k_1 e^{m_1 x} + k_2 e^{m_2 x}$$

The values of the constants k_1 and k_2 can be found if two sets of data are available about the system.

Returning to the particular differential equation discussed above, let us assume that the initial conditions are when $x = 0$, $y = 1$ and when $x = 0$, $dy/dx = 4$. This information can be used to find the values of k_1 and k_2.

For

$$y = k_1 e^{2x} + k_2 e^{3x}$$

$$\frac{dy}{dx} = 2k_1 e^{2x} + 3k_2 e^{3x}$$

Substituting the initial conditions into these equations

$$1 = k_1 e^0 + k_2 e^0 = k_1 + k_2$$

$$4 = 2k_1 e^0 + 3k_2 e^0 = 2k_1 + 3k_2$$

The solution of this pair of simultaneous equations is

$$k_1 = -1 \qquad k_2 = 2$$

The solution to the differential equation

$$\frac{d^2 y}{dx^2} - \frac{5\,dy}{dx} + 6y = 0$$

with $\qquad x = 0 \qquad y = 1; \qquad x = 0, \qquad dy/dx = 4$

is given by

$$y = -1e^{2x} + 2e^{3x} = 2e^{3x} - e^{2x}$$

The above method must be modified when the roots of the auxiliary equation are equal, i.e. $m_1 = m_2 = m$. The solution is given by

$$y = (k_1 + k_2 x)e^{mx}$$

If this modification is not applied the result contains only one constant of integration and this could not form the complete solution to a second-order differential equation:

$$y = k_1 e^{mx} + k_2 e^{mx} = (k_1 + k_2)e^{mx} = ke^{mx}$$

The basic result does, however, hold when the roots of the auxiliary equation are complex. For example,

$$m_1 = a + bi \qquad m_2 = a - bi$$

In this case the solution is given by

$$y = k_1 e^{(a+bi)x} + k_2 e^{(a-bi)x}$$

This can be expressed in a more convenient form if the following expansions are used (see section 5.16):

$$e^{ix} = \cos x + i \sin x \qquad e^{-ix} = \cos x - i \sin x$$

Replacing x by bx gives

$$e^{bix} = \cos bx + i \sin bx \qquad e^{-bix} = \cos bx - i \sin bx$$

By algebraic manipulation

$$y = k_1 e^{(a+bi)x} + k_2 e^{(a-bi)x}$$

becomes $y = k_1 e^{ax} e^{bix} + k_2 e^{ax} e^{-bix} = e^{ax}[k_1 e^{bix} + k_2 e^{-bix}]$

$$= e^{ax} [k_1(\cos bx + i \sin bx) + k_2(\cos bx - i \sin bx)]$$

and by collecting together the terms in $\cos bx$ and $\sin bx$

$$y = e^{ax}[(k_1 + k_2) \cos bx + i(k_1 - k_2) \sin bx]$$

$$= e^{ax}[k_3 \cos bx + k_4 \sin bx]$$

where $\qquad k_3 = k_1 + k_2 \qquad$ and $\qquad k_4 = i(k_1 - k_2)$

An even more useful form can be obtained by considering a right-angled triangle with sides equal to k_3, k_4 and $\sqrt{(k_3^2 + k_4^2)}$ as shown in Fig. 9.1. Let the angle indicated be ϵ.

Then $\qquad \cos \epsilon = \dfrac{k_3}{\sqrt{(k_3^2 + k_4^2)}} \qquad$ and $\qquad \sin \epsilon = \dfrac{k_4}{\sqrt{(k_3^2 + k_4^2)}}$

These results can be used if the solution to the differential equation is written in a slightly different form.

$$y = e^{ax}[k_3 \cos bx + k_4 \sin bx]$$

$$= e^{ax} \sqrt{(k_3^2 + k_4^2)} \left[\cos bx \frac{k_3}{\sqrt{(k_3^2 + k_4^2)}} + \sin bx \frac{k_4}{\sqrt{(k_3^2 + k_4^2)}} \right]$$

$$= e^{ax} \sqrt{(k_3^2 + k_4^2)}[\cos bx \cos \epsilon + \sin bx \sin \epsilon]$$

$$= e^{ax} A[\cos bx \cos \epsilon + \sin bx \sin \epsilon]$$

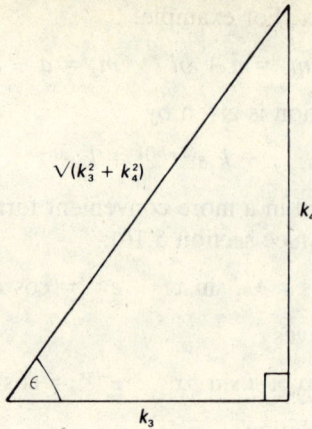

Fig. 9.1

where $A = \sqrt{(k_3^2 + k_4^2)}$ and is a constant.

Using the trigonometrical identity

$$\cos (bx - \epsilon) = \cos bx \cos \epsilon + \sin bx \sin \epsilon$$

(see Appendix A, section A.2) the solution can now be written in the form

$$y = Ae^{ax} \cos (bx - \epsilon)$$

where A and ϵ are the two necessary constants with values

$$A = \sqrt{(k_3^2 + k_4^2)} \qquad \text{and} \qquad \epsilon = \tan^{-1} (k_4/k_3)$$

The values of both can be obtained from the initial data about the system. Thus the solution to a second-order homogeneous differential equation whose auxiliary equation has complex roots involves trigonometrical functions.

Example 2

The relationship between national income, y, and time, x, is given by

(a) $\dfrac{d^2y}{dx^2} - 3 \dfrac{dy}{dx} + 2y = 0$

(b) $\dfrac{d^2y}{dx^2} - 10 \dfrac{dy}{dx} + 25y = 0$

(c) $\dfrac{d^2y}{dx^2} + 6\dfrac{dy}{dx} + 10y = 0$

In each case, obtain y as a function of x if $x = 0$ when $y = 0$ and $dy/dx = 2$ when $x = 0$.

(a) $\dfrac{d^2y}{dx^2} - 3\dfrac{dy}{dx} + 2y = 0.$

Let $y = e^{mx}$. Then

$$\frac{dy}{dx} = me^{mx} \qquad \text{and} \qquad \frac{d^2y}{dx^2} = m^2 e^{mx}$$

Substituting in the equation gives

$$e^{mx}(m^2 - 3m + 2) = 0$$

and so $\qquad m = \dfrac{+3 \pm \sqrt{(9-8)}}{2} = 2$ or $1.$

$\therefore \qquad\qquad y = Ae^{2x} + Be^x$

The initial conditions are $x = 0$, $y = 0$ and $x = 0$, $dy/dx = 2$. Substituting gives two equations in A and B:

$$0 = A + B$$

and since $dy/dx = 2Ae^{2x} + Be^x$, $2 = 2A + B$. Therefore, solving for A and B gives $A = -B$ and so $A = 2$, $B = -2$.

$\therefore \qquad\qquad y = 2e^{2x} - 2e^x$

(b) $\dfrac{d^2y}{dx^2} - 10\dfrac{dy}{dx} + 25y = 0$

Let $y = e^{mx}$. Then

$$\frac{dy}{dx} = me^{mx} \qquad \text{and} \qquad \frac{d^2y}{dx^2} = m^2 e^{mx}$$

Substituting in the equation gives

$$e^{mx}(m^2 - 10m + 25) = 0$$

and so $\qquad m = \dfrac{10 \pm \sqrt{(100-100)}}{2} = 5$

$\therefore \qquad\qquad y = (A + Bx)e^{5x}$

$$\frac{dy}{dx} = (A + Bx)5e^{5x} + Be^{5x} = (5A + B + 5Bx)e^{5x}$$

When $\qquad\qquad x = 0, \qquad y = 0 \qquad$ and $\qquad A = 0$

When $\qquad x = 0, \qquad dy/dx = 2, \qquad$ so $\qquad 2 = 5A + B$

$\therefore \qquad\qquad A = 0$ and $B = 2$, and the solution is $y = 2xe^{5x}$

(c) $\dfrac{d^2y}{dx^2} + 6\dfrac{dy}{dx} + 10y = 0$

Let $y = e^{mx}$. Then

$$\frac{dy}{dx} = me^{mx} \qquad \text{and} \qquad \frac{d^2y}{dx^2} = m^2e^{mx}$$

Substituting into the equation gives

$$e^{mx}(m^2 + 6m + 10) = 0$$

and so $\qquad\qquad m = \dfrac{-6 \pm \sqrt{(36 - 40)}}{2} = -3 \pm i$

We know that if $m = a \pm bi$ then

$$y = Ae^{ax} \cos(bx - \epsilon)$$

and so since $m = -3 \pm i$, $a = -3$ and $b = 1$.

$$y = Ae^{-3x} \cos(x - \epsilon)$$

The initial conditions are $x = 0$, $y = 0$ which gives

$$0 = A\cos(-\epsilon) \qquad\qquad\qquad (1)$$

and $\qquad\qquad\qquad x = 0, \qquad \dfrac{dy}{dx} = 2$

Now $\qquad \dfrac{dy}{dx} = -3Ae^{-3x}\cos(x - \epsilon) - Ae^{-3x}\sin(x - \epsilon)$

$\therefore \qquad\qquad 2 = -3A\cos(-\epsilon) - A\sin(-\epsilon) \qquad\qquad (2)$

From (1) either $A = 0$ or $\cos(-\epsilon) = 0$, that is $\epsilon = \pi/2$

From (2): $\qquad 2 = -A[3(0) + \sin(-\pi/2)] = -A(-1)$

$\therefore \qquad A = +2, \qquad \epsilon = \pi/2 \qquad$ and $\qquad y = 2e^{-3x}\cos(x - \pi/2)$

9.6 Exercises

Solve the following differential equations:

1. $\dfrac{d^2y}{dx^2} - \dfrac{dy}{dx} - 12y = 0$

2. $\dfrac{d^2y}{dx^2} - 3\dfrac{dy}{dx} + 2y = 0$

3. $\dfrac{d^2y}{dx^2} - 6\dfrac{dy}{dx} + 9y = 0$

4. $\dfrac{d^2y}{dx^2} - 2\dfrac{dy}{dx} + y = 0$

5. $\dfrac{d^2y}{dx^2} - 16y = 0$

6. $\dfrac{d^2y}{dx^2} + 16y = 0$

7. $\dfrac{d^2y}{dx^2} + 2\dfrac{dy}{dx} + 10y = 0$

9.7 Second-order non-homogeneous differential equations

The *non-homogeneous* second-order differential equation with constant coefficients has the general form

$$a\frac{d^2y}{dx^2} + b\frac{dy}{dx} + cy = f(x)$$

where $f(x)$ is either a constant or a function of x only, and a, b, c are constants.

It can be shown that the solution to this type of equation is the sum of two parts which are obtained independently. These are called the *complementary function* (*CF*) and the *particular integral* (*PI*). The complementary function is simply the solution of the homogeneous part of the equation formed by letting $f(x) = 0$ and this can be obtained by the method of section 9.5. The particular integral is any solution which satisfies the differential equation and its form depends upon the form of $f(x)$.

(a) $f(x)$ is a constant. For example,

$$\frac{d^2y}{dx^2} - 5\frac{dy}{dx} + 6y = 18$$

If we let $y = K = $ a constant then

$$\frac{dy}{dx} = 0, \qquad \frac{d^2y}{dx^2} = 0$$

Substituting these values in the differential equation

$$0 - 0 + 6K = 18 \qquad \text{or} \qquad K = 3$$

Thus the solution $y = 3$ is a particular integral because it satisfies the differential equation.

The complementary function was obtained in section 9.5 when the solution to

$$\frac{d^2y}{dx^2} - 5\frac{dy}{dx} + 6y = 0$$

was shown to be $y = k_1e^{2x} + k_2e^{3x}$. Therefore the complete solution to the differential equation

$$\frac{d^2y}{dx^2} - 5\frac{dy}{dx} + 6y = 18$$

is the sum of the complementary function and the particular integral and is given by

$$y = k_1e^{2x} + k_2e^{3x} + 3$$

This can be verified by differentiation and substitution in the usual way. Any initial conditions are applied to the complete solution to give the values of the arbitrary constants.

(b) $f(x)$ is linear in x. For example,

$$\frac{d^2y}{dx^2} - 5\frac{dy}{dx} + 6y = 18x$$

Let us try a solution of the form $y = K_1x + K_2$. Then

$$\frac{dy}{dx} = K_1, \qquad \frac{d^2y}{dx^2} = 0$$

Substituting these values in the differential equation

$$0 - 5K_1 + 6(K_1x + K_2) = 18x$$

or

$$6K_1x - 5K_1 + 6K_2 = 18x$$

For this equation to be satisfied it must be true for all values of x

$$\therefore \qquad 6K_1 = 18$$

TABLE 9.1

$f(x)$	Particular integral
$ax^2 + bx + c$	$K_1x^2 + K_2x + K_3$
e^x	K_1e^x
$\sin x$	$K_1 \sin x + K_2 \cos x$

and

$$-5K_1 + 6K_2 = 0$$

∴

$$K_1 = 3 \quad \text{and} \quad K_2 = \tfrac{5}{2}$$

The particular integral is then

$$y = 3x + \tfrac{5}{2}$$

and the complete solution to the differential equation

$$\frac{d^2y}{dx^2} - 5\frac{dy}{dx} + 6y = 18x$$

is

$$y = k_1e^{2x} + k_2e^{3x} + 3x + \tfrac{5}{2}$$

From the above results it would seem reasonable to expect the particular integral to assume the form that $f(x)$ has in the differential equation and this is indeed so in simple cases. Table 9.1 gives a few examples. The reason that the last of these functions contains both sine and cosine terms is that the differentiation of one leads to the other and therefore a combination of two terms is used.

For example

$$\frac{d^2y}{dx^2} - 5\frac{dy}{dx} + 6y = 18 \sin x$$

Let the particular integral be represented by

$$y = K_1 \sin x + K_2 \cos x$$

then

$$\frac{dy}{dx} = K_1 \cos x - K_2 \sin x$$

$$\frac{d^2y}{dx^2} = -K_1 \sin x - K_2 \cos x$$

Substituting these values in the differential equation, we obtain

$$(-K_1 \sin x - K_2 \cos x) - 5(K_1 \cos x - K_2 \sin x)$$

$$+ 6(K_1 \sin x + K_2 \cos x) = 18 \sin x$$

Equating the coefficients of the sine and cosine terms

$$\sin x(-K_1 + 5K_2 + 6K_1) = 18 \sin x$$

$$\cos x(-K_2 - 5K_1 + 6K_2) = 0$$

that is

$$5K_1 + 5K_2 = 18$$

$$-5K_1 + 5K_2 = 0$$

From which $K_2 = \frac{9}{5}$ and $K_1 = \frac{9}{5}$

∴ The solution to the differential equation

$$\frac{d^2y}{dx^2} - 5\frac{dy}{dx} + 6y = 18 \sin x$$

is

$$y = k_1 e^{2x} + k_2 e^{3x} + \tfrac{9}{5}(\sin x + \cos x)$$

In some cases the particular integral may not be easy to find using the above procedure.

For example, difficulty is experienced in the case where the term in y is not present in the left-hand side of the equation and $f(x)$ is a constant. For example,

$$\frac{d^2y}{dx^2} - 3\frac{dy}{dx} = 16$$

The auxiliary equation is

$$m^2 - 3m = 0$$

that is

$$m(m - 3) = 0$$

∴

$$m_1 = 0 \quad \text{and} \quad m_2 = 3$$

and

$$CF = k_1 e^{0x} + k_2 e^{3x} = k_1 + k_2 e^{3x}$$

For the particular integral let $y = K$. Then

$$\frac{dy}{dx} = 0, \quad \frac{d^2y}{dx^2} = 0$$

and substitution of these values leads to $0 - 0 = 16$, which is a contradiction.

In such a case we can try $y = Kx$. Then

$$\frac{dy}{dx} = K, \quad \frac{d^2y}{dx^2} = 0$$

and on substitution $\quad\quad 0 - 3K = 16$

and $\quad\quad\quad\quad\quad\quad\quad K = -\frac{16}{3}$

and $\quad\quad\quad\quad\quad\quad\quad PI = -\frac{16}{3}x$

The solution to the differential equation

$$\frac{d^2y}{dx^2} - 3\frac{dy}{dx} = 16$$

is $\quad\quad\quad\quad\quad y = k_1 + k_2e^{3x} - \frac{16}{3}x$

In general, the particular integral can be found by this trial and error method. It is also possible, and sometimes better, to find it by the method of operators. This is not, however, discussed here because very little advantage is gained by its use in simple cases. The reader should consult a specialised text for a discussion of the method and its application to more difficult problems.

9.8 Exercises

1. Solve the following differential equations with the given initial conditions:

(a) $\dfrac{d^2y}{dx^2} - 10\dfrac{dy}{dx} + 16y = 2x$ with $x = 0$, $y = -5/64$ and

$x = 0$, $\dfrac{dy}{dx} = 1$

(b) $\dfrac{d^2y}{dx^2} - 16y = e^x$ with $x = 0$, $y = 14/15$ and $x = 0$,

$\dfrac{dy}{dx} = 29/15$

(c) $\dfrac{d^2y}{dx^2} - 2\dfrac{dy}{dx} + y = e^{2x}$ with $x = 0$, $y = 2$ and $x = 1$,

$y = e^2$

(d) $\dfrac{d^2y}{dx^2} + 2\dfrac{dy}{dx} + 5y = x$ with $x = 0$, $y = -2/25$ and $x = 0$,

$\dfrac{dy}{dx} = 6/5$

(e) $\dfrac{d^2y}{dx^2} - \dfrac{dy}{dx} = 2\cos x$ with $x = 0$, $y = 1$ and $x = 0$,

$dy/dx = 2$

2. The relationship between output (x), and total cost (y) for a firm is given by

$$\frac{d^2y}{dx^2} + \frac{dy}{dx} - 6y = 18\,x^2$$

where dy/dx is the marginal cost, and d^2y/dx^2 is the rate of change of marginal cost. Given that $y = 0$ and $dy/dx = 1$ when $x = 0$, obtain the total cost as a function of output only.

9.9   Graphical presentation

The solution of a differential equation expresses one of the variables as a function of the other. In many economic problems the independent variable is time and it is of particular interest to consider the variations which occur in the dependent variable over a period of time. A convenient way of presenting this information is by a graph with time (t) on the horizontal axis. It is impossible to discuss all the types of differential equations, but a few examples are given here to indicate the method to be used in obtaining a graph of the solution in any particular case.

LINEAR FIRST-ORDER DIFFERENTIAL EQUATIONS
The solution to an equation of this form is

1. a polynomial in t, no general form exists. The graph can be obtained by using the methods discussed in section 5.18.
2. an exponential function $y = ke^{at}$. In this case various possibilities exist:

    (a) k is positive and a is positive (Fig 9.2) e^{at} continuously increases with time and at an ever-increasing rate. Therefore the function becomes very large and is said to be explosive. The value of k determines the starting position, i.e. the value of the function at time $t = 0$.
    (b) k is positive and a is negative (Fig 9.3). This function continuously decreases with time and is said to be damped.

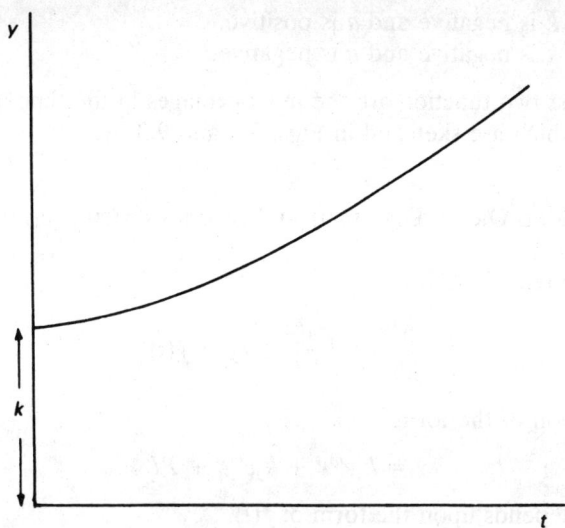

Fig. 9.2

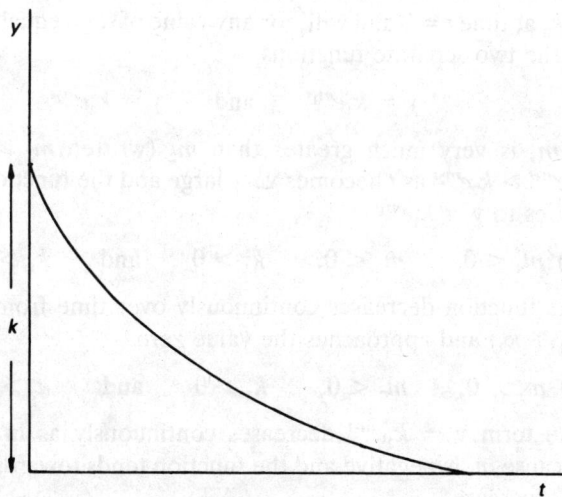

Fig. 9.3

(c) k is negative and a is positive.
(d) k is negative and a is negative.

The last two function are the mirror images in the time axis of the functions which are sketched in Figs. 9.2 and 9.3.

LINEAR SECOND-ORDER DIFFERENTIAL EQUATIONS WITH CONSTANT COEFFICIENTS

The general equation

$$a\frac{d^2y}{dt^2} + b\frac{dy}{dt} + cy = f(t)$$

has a solution of the form

$$y = k_1 e^{m_1 t} + k_2 e^{m_2 t} + PI$$

where PI depends upon the form of $f(t)$.

1. The complementary function is the sum of two exponential terms and the graph will depend upon the values of m_1 and m_2, the roots of the auxiliary equation.

(a) $m_1 > 0 \qquad m_2 > 0, \qquad k_1 > 0 \qquad$ and $\qquad k_2 > 0$

The function will increase continuously from the value $y = k_1 + k_2$ at time $t = 0$ and will, for any value of t, be equal to the sum of the two separate functions

$$y = k_1 e^{m_1 t} \qquad \text{and} \qquad y = k_2 e^{m_2 t}$$

If m_1 is very much greater than m_2 (written $m_1 \gg m_2$) then $k_1 e^{m_1 t} \gg k_2 e^{m_2 t}$ as t becomes very large and the function approximates to $y = k_1 e^{m_1 t}$.

(b) $m_1 < 0, \qquad m_2 < 0, \qquad k_1 > 0 \qquad$ and $\qquad k_2 > 0$

The function decreases continuously over time from the value $(k_1 + k_2)$ and approaches the value zero.

(c) $m_1 > 0, \qquad m_2 < 0, \qquad k_1 > 0 \qquad$ and $\qquad k_2 > 0$

The term $y = k_2 e^{m_2 t}$ decreases continuously as in (b) above because m_2 is negative and the function tends towards $y = k_1 e^{m_1 t}$ as t increases.

The same reasoning can be applied to other possible com-

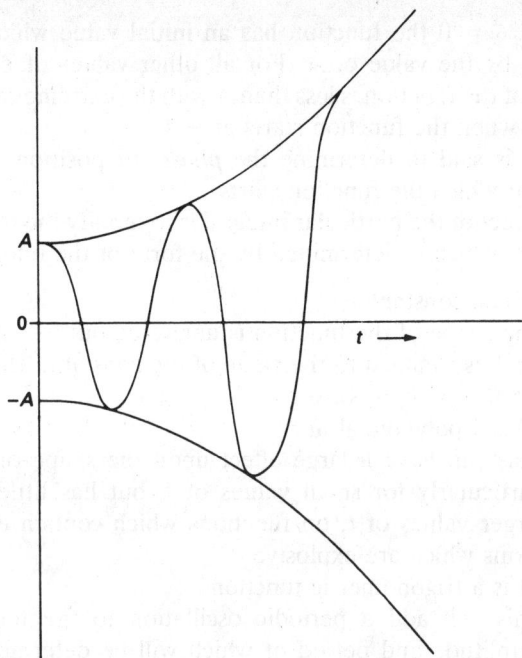

Fig. 9.4

binations of the constants in the complementary function and the shape of the function determined.

(d) Complex roots

The solution is of the form

$$y = Ae^{at} \cos (kt - \epsilon)$$

and this can be thought of as two parts. Ae^{at} has a form similar to one of those discussed above, and $\cos (kt - \epsilon)$ is a function which can take on all values between -1 and $+1$ but has no values outside these limits. If $A > 0$, $a > 0$, the solution oscillates about the time axis with extreme values Ae^{at} and $-Ae^{at}$ (Fig. 9.4).

The amplitude of the oscillations increases at each successive cycle and the function is explosive. The constant ϵ helps to determine the starting point in the cycle because when $t = 0$

$$y = A \cos (-\epsilon) = A \cos \epsilon$$

If $\epsilon = 0$ the function has an initial value which is determined by the value of A. For all other values of ϵ the initial value of the function is less than A with the extreme case of $\cos \epsilon = -1$ when the function starts at $-A$.

ϵ is said to determine the *phase*, or position, within the cycle at which the function starts.

2. The effect of the particular integral is to modify the function in a manner which is determined by the form of the integral.

(a) *PI* is a constant
The shape of the function is unaltered but its value at time $t = 0$ is changed to the value of the constant. That is, when $t = 0$, $y = k_1 + k_2 + c$.

(b) *PI* is a polynomial in t
This can have a large effect upon the shape of the curve particularly for small values of t, but has little effect for larger values of t, on functions which contain exponential terms which are explosive.

(c) *PI* is a trigonometric function
This will add a periodic oscillation to the function, the amplitude and period of which will be determined by the trigonometric function.

9.10 Exercises

Sketch the graphs of the solutions to Exercises 9.8, Questions 1 and 2.

9.11 An inflation–unemployment model

A popular way of analysing the interactions between inflation, unemployment and monetary policy is based on the expectations-augmented Phillips curve. Here we will examine the model in continuous time using differential equations. The equivalent model using discrete time will be discussed in section 10.11.

The original Phillips curve can be written

$$w = a - bU \qquad (a, b > 0)$$

where w is the rate of change of money wages and U is the level of unemployment. This implies that the higher is unemployment then the lower will be wage changes. The equation is usually modified in two ways. First, w is replaced by $w - p^e$, where p^e is expected inflation, so that it is expected real wages which are related to unemployment. Secondly, the rate of inflation, p, is related to w by

$$p = w - c \quad (c > 0)$$

where c is the rate of productivity increase (assumed to be constant). Together these result in the expectations-augmented Phillips curve being written as

$$p = a - bU - c + p^e \quad (a, b, c > 0) \tag{1}$$

To explain inflation expectations a simple adaptive expectations equation can be used:

$$\frac{dp^e}{dt} = k(p - p^e) \quad (1 \geqslant k > 0) \tag{2}$$

This states that the change in the expectation is a proportion, k, of the error in expectations in the current period. Finally, we assume that the change in unemployment is related to the rate of growth of real money, or

$$\frac{dU}{dt} = -e(m - p) \quad (e > 0) \tag{3}$$

where m is the nominal rate of growth of money and p is the rate of inflation. We will also assume that m is a constant.

The three equations (1)–(3) explain p, p^e and U and can be combined to give differential equations in any of these variables. Initially we will reduce the model to a differential equation in p.

First we differentiate (1) with respect to time, t, to give

$$\frac{dp}{dt} = -b \frac{dU}{dt} + \frac{dp^e}{dt} \tag{4}$$

since a and c are constants. Next we substitute from (2) and (3) to remove the derivatives of U and p^e:

$$\frac{dp}{dt} = be(m - p) + k(p - p^e) = bem + (k - be)p - kp^e$$

Since this includes p^e we differentiate it to give

$$\frac{d^2p}{dt^2} = (k - be) \frac{dp}{dt} - k \frac{dp^e}{dt} \tag{5}$$

Now we can use (4) to eliminate the dp^e/dt term

$$\frac{d^2p}{dt^2} = (k - be)\frac{dp}{dt} - k\left(\frac{dp}{dt} + b\frac{dU}{dt}\right)$$

$$= -be\frac{dp}{dt} - bk\frac{dU}{dt}$$

Finally, using (3),

$$\frac{d^2p}{dt^2} = -be\frac{dp}{dt} + bek(m - p)$$

which can be written:

$$\frac{d^2p}{dt^2} + be\frac{dp}{dt} + bekp = bekm \tag{6}$$

This is a second-order non-homogeneous differential equation. The particular integral is found by trying $p = Z$, say, a constant, so that

$$\frac{dp}{dt} = 0 \text{ and } \frac{d^2p}{dt^2} = 0 \text{ and } bekZ = bekm \text{ giving } Z = m$$

Therefore in this model the equilibrium rate of inflation equals the rate of growth of nominal money.

The complementary function has the auxiliary equation, from putting $p = e^{ft}$,

$$f^2 + be f + bek = 0$$

and this has roots which depend on b, e and k. They will be real if $(be)^2 \geq 4bek$ or, since $be > 0$, $be \geq 4k$. Taking an example, suppose that $k = 1$ so that in (2) expectations adjust immediately to the actual rate of inflation, the rate of monetary expansion, $m = 4$ and also $be = 5$. The auxiliary equation is

$$f^2 + 5f + 5 = 0$$

and the roots are $f_1 = -1.382$ and $f_2 = -3.618$. The negative signs indicate that the solution will converge towards the equilibrium value. The full solution is

$$p = Ae^{-1.382t} + Be^{-3.618t} + 4$$

If we assume the initial conditions are $t = 0$, $p = 5$, and $t = 1$, $p = 6$, then

$$t = 0, p = A + B + 4 = 5$$
$$t = 1, p = 0.2511A + 0.0268B + 4 = 6$$

giving $A = 8.8$, $B = -7.8$ with the result that

$$p = 8.8e^{-1.382t} - 7.8e^{-3.618t} + 4$$

As t increases, p rapidly approaches the equilibrium value of 4. For example, when $t = 5$, $p = 4.009$.

To see the effect of changing k, which reflects the speed of adjustment of expectations to actual inflation, suppose that $k = 0.4$ and $be = 5$. The auxiliary equation is now

$$f^2 + 5f + 2 = 0$$

which gives $f_1 = -0.4384$ and $f_2 = -4.5615$. Again the negative signs indicate that the solution will converge on the equilibrium value. The full solution is

$$p = Ae^{-0.4384t} + Be^{-4.5615t} + 4$$

and with the same initial conditions as above,

$$t = 0, p = A + B + 4 = 5$$
$$t = 1, p = 0.6451A + 0.0104B + 4 = 6$$

resulting in $A = 3.1347$ and $B = -2.1347$. The solution is

$$p = 3.1347e^{-0.4384t} - 2.1347e^{-4.5615t} + 4$$

In this case for $t = 5$, $p = 4.35$, for $t = 10$, $p = 4.04$ and for $t = 15$, $p = 4.004$ so that the model takes longer to reach equilibrium.

Finally, suppose that $be = 3$ and $k = 1$ making the auxiliary equation

$$f^2 + 3f + 3 = 0$$

with the roots $-1.5 \pm 0.866i$. From section 9.5 the solution will be of the form

$$p = Ae^{at} \cos(bt - \epsilon) + 4$$

or
$$p = Ae^{-1.5t} \cos(0.866t - \epsilon)$$

The initial conditions give

$$t = 0, p = A \cos(-\epsilon) + 4 \qquad\qquad = 5$$
$$t = 1, p = A(0.2231) \cos(0.866 - \epsilon) + 4 = 6$$

Since $\cos(-\epsilon) = \cos(\epsilon)$ the first gives $A = 1/\cos(\epsilon)$ and substituting into the second and simplifying,

$$2 = \frac{0.2231 \cos(0.866 - \epsilon)}{\cos(\epsilon)}$$

Now from (9) in Appendix A, section A.2,

$$\cos(x - y) = \cos x \cos y + \sin x \sin y$$

and so using this,

$$2 = \frac{0.2231 [\cos(0.866) \cos \epsilon + \sin(0.866) \sin \epsilon]}{\cos \epsilon}$$

where the angles are measured in radians. Since 0.866 radians is 49.6°,

$$2 = 0.2231 [0.6481 + 0.7615 \tan \epsilon]$$

or $\tan \epsilon = 10.92$ and so $\epsilon = 1.48$ radians. Therefore, $A = 1/\cos \epsilon = 11.0288$. The full solution is

$$p = 11.0288 \, e^{-1.5t} \cos(0.866t - 1.48) + 4$$

The cosine term results in a cyclical pattern in the behaviour of p but convergence towards the equilibrium value of 4 is rapid (for $t = 5$, $p = 3.99$).

The expectations-augmented Phillips curve model can also be analysed as a differential equation in either unemployment or in expected prices. The equations of the model are

$$p = a - bU - c + p^e \quad (a, b, c > 0) \tag{1}$$

$$\frac{dp^e}{dt} = k(p - p^e) \quad (1 \geqslant k > 0) \tag{2}$$

$$\frac{dU}{dt} = -e(m - p) \quad (e > 0) \tag{3}$$

To obtain the differential equation in U, first notice that putting (1) into (2) gives

$$\frac{dp^e}{dt} = k(a - bU - c) \tag{7}$$

and putting (1) into (3),

$$\frac{dU}{dt} = -em + ea - ebU - ec + ep^e$$

Differentiating results in

$$\frac{d^2U}{dt^2} = -eb\frac{dU}{dt} + e\frac{dp^e}{dt}$$

and using (7),

$$\frac{d^2U}{dt^2} = -eb\frac{dU}{dt} + ek(a - bU - c)$$

This can be written as

$$\frac{d^2U}{dt^2} + be\frac{dU}{dt} + bekU = (a - c)ek \tag{8}$$

which has the same coefficients as (6) above, the differential equation in p, except for the constant term. The behaviour of the solutions for U is therefore the same as those for p, apart from the equilibrium solution which depends on the constant term. This is to be expected since there is no reason for the equilibrium level of unemployment (which is likely to be affected by labour market institutions and demographic factors) and the equilibrium level of inflation (which will depend on productivity and the rate of increase in nominal money) to be identical. We also note that the rate of monetary expansion, m, does not occur in (8) and so changing m would not affect unemployment. The equilibrium solution of (8), found by putting $U = Z$, say, so that $dU/dt = 0$ and $d^2U/dt^2 = 0$, is $U = (a - c)/b$ which is also independent of m. This is known as the *natural rate of unemployment* and is unrelated to the rate of inflation. Thus, in the long run, there is no trade-off between unemployment and inflation in this model.

Finally, to get the differential equation in p^e, differentiate (1),

$$\frac{dp}{dt} = -b\frac{dU}{dt} + \frac{dp^e}{dt}$$

and using (3),

$$\frac{dp}{dt} = be(m - p) + \frac{dp^e}{dt}$$

Differentiating (2),

$$\frac{d^2p^e}{dt^2} = \frac{kdp}{dt} - \frac{kdp^e}{dt}$$

and substituting for dp/dt,

$$\frac{d^2p^e}{dt^2} = kbe(m - p)$$

and then for p from (2),

$$\frac{d^2p^e}{dt^2} = kbem - kbep^e - be\frac{dp^e}{dt}$$

This can be written

$$\frac{d^2p^e}{dt^2} + be\frac{dp^e}{dt} + bekp^e = bekm$$

and again the coefficients on the left hand side are the same as in (6) and (8). The equilibrium value of p^e, from putting $p^e = Z$, a constant, is $p^e = m$ and so expected inflation equals the rate of monetary expansion.

9.12 Exercises

1. In section 9.11 the original Phillips curve was modified to give real wages, $(w - p^e)$, as a function of unemployment. If there is *money illusion* and workers do not get full compensation for inflation, the expectations-augmented Phillips curve can be written

$$p = a - bU - c + fp^e \qquad (a, b, c, f > 0) \qquad (1)$$

where in the special case $f = 1$ we have the previous formulation. Using the other two equations of the model,

$$\frac{dp^e}{dt} = k(p - p^e) \qquad (k > 0) \qquad (2)$$

$$\frac{dU}{dt} = -e(m - p) \qquad (e > 0) \qquad (3)$$

obtain the differential equation in p and compare it with (6) of section 9.11.

2. The market for a product is given by

$$q^s = ap - b + c \frac{dp}{dt} + f \frac{d^2p}{dt^2} \quad (a, b, c, f > 0)$$

$$q^d = g - hp \qquad\qquad\qquad (g, h > 0)$$

where the quantity supplied is affected by both the first- and second-order derivatives of price. Assuming

$$q^s = q^d,$$

(a) derive the condition for the time path of prices to be cyclical, and
(b) if $a = 10$, $b = 200$, $c = 8$, $f = 1$, $g = 1600$ and $h = 5$, and the initial conditions are $t = 0$, $p = 130$ and $t = 0$, $dp/dt = -2$, determine the solution for p.

9.13 Revision exercises for Chapter 9 (without answers)

1. Solve the following differential equations:

(a) $\dfrac{dy}{dx} = 6x^2$ (b) $\dfrac{dy}{dx} = 4xy$

(c) $\dfrac{dy}{dx} = \dfrac{x + 2y}{2x + y}$ (d) $\dfrac{dy}{dx} = \dfrac{x^2 - y^2}{xy}$

(e) $(2x - y + 3)\,dx = (y - x - 1)\,dy$

(f) $(x^2 + y^2)\,dx = (x - y)\,dy$

(g) $y\,dx + x\,dy = 0$

(h) $(x + 2)\,dy = (4y + x^2)\,dx$

(i) $x^2\,dy = (3x^2 - 4 - 3y)\,dx$

2. If the marginal cost for an output of q is given by

$$MC = 25 + 3q^2$$

determine the total cost function if $TC = 1500$ when $q = 10$.

3. The elasticity of demand for a product varies with the price (p) and quantity demanded (q) and is $-2p/q$. Determine the demand function if, when $q = 30$, $p = 10$.

4. An economist finds that the marginal propensity to consume is given by $MPC = 0.5 + 0.2 \log Y$ where Y is income. What is the equation of the consumption function if $C = 50$ when $Y = 100$?

5. Solve the following differential equations with the given initial conditions:

(a) $\dfrac{d^2y}{dx^2} + \dfrac{dy}{dx} + 2y = 2x + 3 \quad x = 0,\ y = 0,\ \dfrac{dy}{dx} = 0$

(b) $\dfrac{d^2y}{dx^2} - 2y = 3 \qquad\qquad x = 0,\ y = 0,\ \dfrac{dy}{dx} = 0$

(c) $\dfrac{d^2y}{dx^2} + 2y = 2x \qquad\qquad x = 0,\ y = 0,\ \dfrac{dy}{dx} = 0$

(d) $\dfrac{d^2y}{dx^2} - \dfrac{4dy}{dx} + 5y = 1 \quad x = 0,\ y = 0,\ \dfrac{dy}{dx} = 0$

6. For a particular economy the path of national income is given by

$$\frac{d^2Y}{dt^2} - \frac{2dY}{dt} + 2Y = 2$$

The initial conditions $t = 0$, $Y = 100$ and $dY/dt = 5$. Solve the equation and plot the graph of Y against t. What happens to Y as $t \to \infty$?

7. If the inflation rate in an economy is constant, what is happening to prices?

8. In a particular country inflation is increasing according to

$$\frac{d^2P}{dt^2} = 2t$$

where P = prices and t = time. If when $t = 0$, $P = 100$ and $dP/dt = 0$, plot the graph of prices against time.

9. For a production process the marginal cost, MC, satisfies

$$\frac{dMC}{dq} = 2 + q$$

where q is output. Determine the total cost function if when $q = 0$, total cost $= 1000$ and $MC = 5$. Sketch the graph of total cost against q.

10. The market for a product is given by

$$q^s = 3p - 16 + 0.5\frac{d^2p}{dt^2}$$

$$q^d = 500 - p - \frac{dp}{dt}$$

At the equilibrium, what is the time path of prices?

Chapter 10

Difference Equations

10.1 Introduction

In Chapter 9 we saw that differential equations express the relationship between two variables (e.g. x and y) and also the rate of change of one variable with respect to the other, (i.e. dy/dx). We know that this rate of change (or derivative) relates to variables which change continuously. In this chapter we will be considering variables which change in discrete steps and in particular variables which are measured at or over certain periods of time. Examples of such variables are the level of stocks at the end of a month, the national income earned during a calendar year, the daily closing price of shares on the stock exchange, and the value of a bank deposit at the end of a year.

For example, if £C is invested in a bank at an interest rate of i per cent per year compounded annually, then the value of this investment after one year, C_1, is given by

$$C_1 = (1 + i)C$$

and the value after two years, C_2, is

$$C_2 = (1 + i)C_1 = (1 + i)^2 C$$

In general, the value of the investment at the end of year t is

$$C_t = (1 + i)^t C = a^t C \qquad \text{where} \qquad a = 1 + i$$

Now
$$C_t = (1 + i)C_{t-1} = aC_{t-1}$$

where a relates the value of the investment after t years to its value after $(t - 1)$ years. This expression is known as a *difference equation*.

It is a *first-order* difference equation because the time periods involved are only one period apart. The value in any time period can be determined if the value in the preceding time period is known.

Thus
$$C_t = aC_{t-1}$$
$$C_{t+1} = aC_t = a^2C_{t-1}$$
$$C_{t+2} = aC_{t+1} = a^3C_{t-1}$$

In particular, if $C_0 = C$, then $C_t = a^tC$.

This equation is said to be the *solution* to the difference equation $C_t = aC_{t-1}$ since it enables us to determine the value of the investment at any time period directly from its value initially.

For example, if $C = £100$ and the rate of interest is 5 per cent per annum then $a = 1 + 0.05 = 1.05$ and $C_t = (1.05)^t \times 100$ so that after 20 years, $C_{20} = 100(1.05)^{20} = 265.33$.

In general, the necessary requirements for a solution to a difference equation are as follows.

1. It expresses the value of the variable in time period t in terms of the given data, i.e. the value of the variable in the base period. It can therefore be used to determine the value of the variable in any time period directly without calculating all the intermediate values as would be necessary if the original difference equation were used.
2. It satisfies the original difference equation.

For the example used above, the second condition can be checked as follows:

Let $C_t = a^tC_0$ be the solution to the difference equation $C_t = aC_{t-1}$; then $C_{t+1} = a^{t+1}C_0$ is obtained by substituting $t + 1$ for t whenever it occurs in the solution. Now

$$C_{t+1} = aC_t = a(a^tC_0) = a^{t+1}C_0$$

and so $C_t = a^tC_0$ satisfies the original difference equation.

10.2 Compound interest and the addition of capital at yearly intervals

It is possible to construct a difference equation to represent the situation in which the capital sum has interest added to it annually along with a further injection of capital. For example, an initial capital sum C with interest i per cent compounded annually has a further sum of A added at the end of each year.

Then with $a = 1 + i$ as above,

$$C_0 = C$$

$$C_1 = C_0(1 + i) + A = aC_0 + A$$

$$C_2 = (aC_0 + A)(1 + i) + A = a^2C_0 + aA + A$$

$$C_3 = (a^2C_0 + aA + A)(1 + i) + A = a^3C_0 + a^2A + aA + A$$

and in general,

$$C_t = a^tC_0 + A(a^{t-1} + a^{t-2} + \cdots + 1)$$

The first term on the right hand side of the equation is identical to that in the previous example where A was not added at the end of each year. The second term is a geometric progression the sum of which is quite easily obtained (see section 4.4) and the result is

$$C_t = a^tC_0 + A\left(\frac{a^t - 1}{a - 1}\right)$$

This is a solution because

(a) it enables us to determine the value of the capital sum at any time period from a knowledge of C, a and A, the three given quantities;

(b) it satisfies the difference equation

$$C_{t+1} = C_t(1 + i) + A$$

$$= aC_t + A$$

For example, let £100 be invested at an interest rate of 10% per year compounded annually and let the capital added each year be £20.

Then

$$C_0 = 100$$

$$C_1 = 100(1 + 0.10) + 20$$

$$C_2 = 100(1 + 0.10)^2 + 20(1 + 0.10) + 20$$

$$\vdots$$

The value after any time period can be determined by considering the solution

$$C_t = a^tC_0 + A\left(\frac{a^t - 1}{a - 1}\right)$$

where

$$a = 1 + i = 1 + 0.10 = 1.1$$

then
$$C_t = 1.1^t(100) + 20\left(\frac{1.1^t - 1}{1.1 - 1}\right)$$

$$= 100(1.1)^t + 20\left(\frac{(1.1)^t - 1}{0.1}\right)$$

$$= 100(1.1)^t + 200(1.1^t - 1)$$

After 5 years, $t = 5$ and
$$C_5 = 100(1.1)^5 + 200[(1.1)^5 - 1]$$
$$= 100(1.61) + 200(1.61 - 1) = 161 + 200(0.61)$$
$$= 161 + 122 = 283$$

The solution in this case is made up of two terms, the second of which arises from the presence of the additional term in the difference equation.

There are a number of other applications of this simple difference equation model, including any process such as an insurance policy, a bank loan, a house mortgage or a sinking fund where regular payments are made and the rate of interest can be assumed to be approximately constant. The solution can always be obtained from first principles, as above, but this is not necessary since, by classifying difference equations in the same way as that used in Chapter 9 for differential equations, some general solution methods can be found.

10.3 First-order linear difference equations

HOMOGENEOUS
These can be written in general
$$Y_{t+1} = \lambda Y_t$$
where λ is a constant and the solution is given by
$$Y_t = k(\lambda)^t$$

k is a constant whose value is determined by the initial conditions. For example, if $Y_t = C$ when $t = 0$, then $C = k\lambda^0 = k$ and the solution is $Y_t = C(\lambda)^t$. As with differential equations, the solution to every first-order difference equation contains one arbitrary constant whose value can be determined only from other information about the system.

Notice that, for $k > 0$, as t increases, Y will tend to zero if $-1 < \lambda < 1$, while if $\lambda > 1$, Y will tend to infinity.

NON-HOMOGENEOUS

These can be written in general as

$$Y_{t+1} = \lambda Y_t + f(t)$$

or

$$Y_{t+1} - \lambda Y_t = f(t)$$

The solution to such an equation is obtained in two parts. The first is the solution of the homogeneous form $Y_{t+1} = \lambda Y_t$ and this is obtained as above. It is known as the *transient solution*. The second part is a *particular solution* to the difference equation and its form depends upon the form of $f(t)$. It is known as the *equilibrium solution*. The complete solution is the sum of these two parts.

When $f(t) = K$, where K is a constant, the equilibrium solution is, from section 10.2:

$$Y_t = K\frac{1 - \lambda^t}{1 - \lambda}$$

The complete solution to the difference equation

$$Y_{t+1} - \lambda Y_t = K$$

is

$$Y_t = Y_0(\lambda)^t + K\frac{1 - \lambda^t}{1 - \lambda} = A\lambda^t + \frac{K}{1 - \lambda}$$

where A is a constant.

The behaviour of Y_t as t changes is determined by the values of A, K and λ. The most interesting of these is λ. Assuming that $A > 0$ and $K > 0$, then if $-1 < \lambda < 1$, as t increases, $\lambda^t \to 0$ and Y_t converges on the value $K/(1 - \lambda)$. The difference equation is said to be *stable*. If $\lambda > 1$ then Y_t will 'explode' and tend to infinity, while if $\lambda < -1$ Y_t will become larger and larger but will oscillate between positive and negative values.

First-order difference equations in which $f(t)$ is not constant will be considered along with second-order equations in which the same condition holds.

10.4 The cobweb model

In section 1.7, we discussed the situation in which the quantity demanded and supplied of a commodity are equated under conditions of perfect competition. Thus,

$$q_D = f(p) \qquad \text{demand function}$$

$$q_s = f(p) \qquad \text{supply function}$$

and $\qquad q_D = q_s$

But in many cases the response to changes in demand is not instantaneous. Thus a price increase does not immediately result in an increased supply. This is particularly so of such things as agricultural products, the supply of which cannot be altered to any great extent in a short time. The rate of change in the quantity which can be supplied is determined by the time cycle of growth or production of the good. Once a crop is planted its size is governed by weather conditions etc., and an increase in price during the growth and harvesting period has little effect on this year's supply. Similarly industries which are working at full capacity cannot increase the supply without putting in hand a capital-investment programme which is likely to take some time to implement. This problem can, of course, be overcome to some extent by allowing goods to be transferred into and out of inventory, i.e. by having a buffer stock in which fluctuations can take place instantaneously. If, initially, this possibility is ignored it seems reasonable to assume that a change in the price which can be obtained for a good has a delayed effect upon the quantity of the good supplied.

The simplest case we consider is one in which the delay, or lag is for a single time period and applies to the supply function only. This can be written as follows:

$$q_t = f_D(p_t) \qquad \text{demand function}$$

$$q_t = f_s(p_{t-1}) \qquad \text{supply function}$$

This model states that the price in any time period determines the quantity supplied in the subsequent time period and this latter quantity when considered in relation to the demand function determines the price in that period. This price then determines the quantity supplied in the next period and so on.

If the price in two successive periods is different, then the quantity supplied in these two periods is different, and it is not difficult to visualise a situation where the price and the supply continue to change from period to period. If this occurs it is of interest to see whether there is an equilibrium position for the system, and to ask how the price and

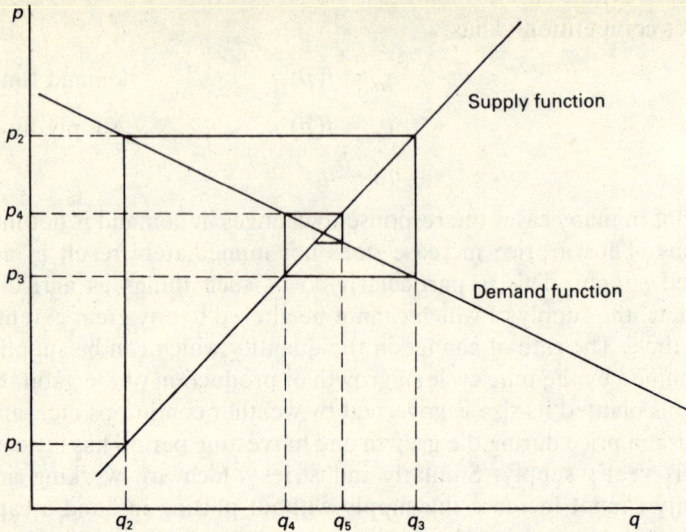

Fig. 10.1

quantity supplied change over a period of time. The situation can be represented graphically and the time path of the system determined.

An example where the demand and supply functions are both linear is illustrated in Fig. 10.1. Given that the system starts with a price p_1 in the initial time period, the supply function enables us to determine the quantity q_2 which will be supplied at this price in the next time period. But if q_2 is supplied in the period 2 it can demand a price p_2 which is considerably higher than that in the preceding period. This figure can be found on the graph by moving vertically from the supply function to the demand function.

This high price results in an increase in the quantity supplied in the following time period. In this case it is equal to q_3. This figure can be found on the graph by moving horizontally from the demand function to the supply function.

From this point onwards, the procedure is to repeat the above steps, i.e. moving alternately from the supply to the demand function in vertical steps and from the demand to the supply function in horizontal steps. The result is that the changes in price and quantity between successive time periods become smaller and smaller and the system approaches an equilibrium value which is located at the intersection of the demand and supply functions.

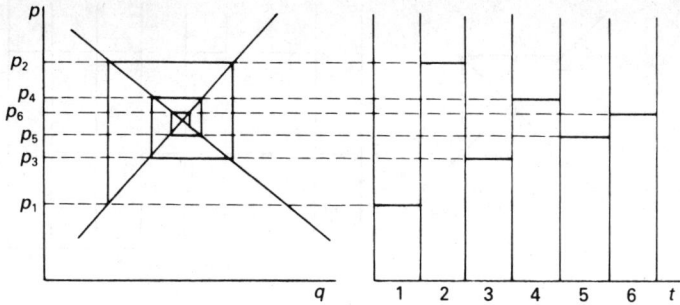

Fig. 10.2

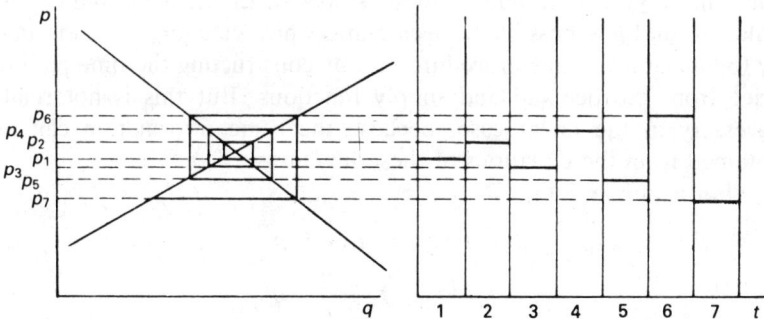

Fig. 10.3

It is interesting to show the changes in price which occur by transferring the information to a graph where time is represented on the horizontal axis (Fig. 10.2). The price fluctuates between time periods, the size of the fluctuations decreasing with time. The system tends to an equilibrium value and is, therefore, said to be *stable* or *converging*.

Fig. 10.3 illustrates a situation in which the size of the fluctuations increases with time. This system is said to be *unstable*, *diverging* or *explosive*, and if it remained unchanged (which is highly unlikely) it would oscillate violently and the price of the good would fluctuate wildly from year to year.

A third situation is possible in which the price changes between successive time periods but alternates in such a way that it assumes either one or other of only two values, i.e. the system *oscillates* (Fig. 10.4).

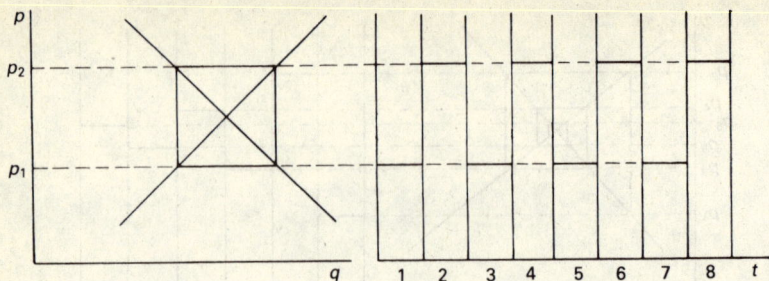

Fig. 10.4

Any linear system conforms to one of these three types. The particular type is determined by the slopes of the demand and supply functions and it is possible to decide into which category a system falls by following the above procedure, i.e. by constructing the time path of price from the demand and supply functions. But this is not really necessary in the linear case, because the same information can be obtained from the equations of these functions.

For example, let

$$q_t = f_D(p_t) = a_1 + b_1 p_t$$

and
$$q_t = f_s(p_{t-1}) = a_2 + b_2 p_{t-1}$$

Then if we assume that the quantity supplied equals the quantity demanded in any time period,

$$f_s(p_{t-1}) = f_D(p_t)$$

that is
$$a_2 + b_2 p_{t-1} = a_1 + b_1 p_t$$

This is a first-order difference equation

$$b_1 p_t = b_2 p_{t-1} + a_2 - a_1$$

or
$$p_t = \left[\frac{b_2}{b_1}\right] p_{t-1} + \frac{a_2 - a_1}{b_1}$$

The solution to the homogeneous form

$$p_t = \frac{b_2}{b_1} p_{t-1} \quad \text{is given by} \quad p_t = k\left(\frac{b_2}{b_1}\right)^t$$

(see section 10.3) and it can be shown that the particular solution is

$$p_t = \frac{(a_2 - a_1)/b_1}{1 - (b_2/b_1)} = \frac{a_2 - a_1}{b_1 - b_2}$$

The complete solution is therefore,

$$p_t = k \left[\frac{b_2}{b_1} \right]^t + \frac{a_2 - a_1}{b_1 - b_2}$$

where k is an arbitrary constant, the value of which is determined from the initial conditions.

The particular solution is a constant and has no influence on the changes which occur in the time path of price. These are determined by the solution to the homogeneous form. The manner in which the price changes between successive time periods is determined by the term b_2/b_1, where b_1 is the slope of the demand curve and is a function of the elasticity of demand which can be assumed, in general, to be negative. Similarly b_2 is a function of the elasticity of supply which is positive. The term b_2/b_1 must, therefore, be negative, $(b_2/b_1)^2$ is positive and $(b_2/b_1)^3$ is negative and so on. Its value is alternately positive and negative in each successive time period. It follows from the fact that the particular solution is constant that the function will oscillate about the value $p_t = (a_2 - a_1)/(b_1 - b_2)$ and the size of the oscillation depends upon the ratio b_2/b_1. Three distinct cases arise.

1. $|b_2/b_1| < 1$, or $|b_2| < |b_1|$, that is the modulus or absolute value of b_2 is less than that of b_1. In this case $(b_2/b_1)^t$ decreases in value as t increases and the size of the oscillations decreases over time. This corresponds to the stable system and the equilibrium price is $(a_2 - a_1)/(b_1 - b_2)$.

2. $|b_2/b_1| = 1$, that is $|b_2| = |b_1|$. In this case $(b_2/b_1)^t$ has an absolute value of unity but is alternately positive and negative as t changes. This corresponds to the situation in which the price takes on only two values

$$\frac{a_2 - a_1}{b_1 - b_2} + k \quad \text{and} \quad \frac{a_2 - a_1}{b_1 - b_2} - k$$

3. $|b_2/b_1| > 1$, that is $|b_2| > |b_1|$.

In this case $(b_2/b_1)^t$ increases in absolute value as t increases, changing sign between successive time periods. This corresponds to the unstable system because the size of the oscillations becomes larger and larger.

It is apparent, from this analysis, that the stability or otherwise of any demand–supply system depends upon the relative slopes of the two functions. In most practical situations these functions are not linear and therefore the slopes will be functions of price. In such cases it is possible to visualise a system which is unstable over one price range but stable for changes which are outside this price range. This can be demonstrated by the graphical approach and the result confirmed algebraically.

10.5　The Harrod–Domar growth model

This assumes that

1. The amount of resources saved by the community during any time period is a function only of the income received by the community during that period.

   That is　　　　　　　　　$S_t = aY_t$

   where　S_t represents savings in time period t
   　　　　Y_t represents income in time period t
   　　　　a is a positive constant which is less than 1

2. The amount of investment which occurs in any time period will be a function of the increase of income in that period over the income of the previous time period.

   That is　　　　　　　　　$I_t = g(Y_t - Y_{t-1})$

   where g is a constant which is less than 1.

3. The amount which will be available for investment in any time period will be equal to the amount saved in the same time period.

   That is　　　　　　　　　$I_t = S_t$

Using these three assumptions it is possible to construct a model of the system as follows.

$$S_t = aY_t \tag{1}$$

$$I_t = g(Y_t - Y_{t-1}) \tag{2}$$

$$I_t = S_t \tag{3}$$

If (1) and (2) are substituted into (3) a first-order linear difference equation is formed.

$$g(Y_t - Y_{t-1}) = aY_t$$

$$Y_t(g - a) = gY_{t-1}$$

$$Y_t = \left(\frac{g}{g-a}\right)Y_{t-1}$$

The solution to this is

$$Y_t = k\left(\frac{g}{g-a}\right)^t = k\lambda^t$$

where k is equal to the value of income in time period $t = 0$

Since

$$Y_0 = k\left(\frac{g}{g-a}\right)^0 = k$$

then

$$Y_t = Y_0\left(\frac{g}{g-a}\right)^t$$

From assumptions 1 and 2, a and g must both be positive and therefore the sign of λ depends on whether g is greater than or less than a. If $g = a$ then λ is infinitely large and the model breaks down.

10.6 A consumption model

A further example of a simple economic model is provided by the equations

$$C_t = aY_{t-1} + B \tag{1}$$

$$Y_t \equiv C_t + I_t \tag{2}$$

where C_t = consumption expenditure

Y_t = national income

I_t = investment expenditure

If I_t is constant in each year – for example, $I_t = F$ – then substituting from (2) into (1) gives

$$C_t = aC_{t-1} + (aF + B)$$

which is a first-order non-homogeneous difference equation. The solution of the homogeneous part is

$$C_t = ka^t$$

and the equilibrium solution is

$$C_t = \frac{(aF + B)}{1 - a}$$

so that the complete solution is

$$C_t = ka^t + \frac{(aF + B)}{1 - a}$$

10.7   Exercises

1. Solve the following difference equations and sketch the time path of Y_t for $t = 0$ to $t = 5$.

   (a) $Y_{t+1} = 1.5\ Y_t + 3$ with $Y_0 = 2$

   (b) $Y_{t+1} = 0.9\ Y_t + 2$ with $Y_0 = 3$

   (c) $Y_{t+1} = Y_t$ with $Y_0 = 6$

   (d) $Y_{t+1} = 1.1\ Y_t + 4$ with $Y_0 = 1$

2. A savings club collects £15 from each of its members at the beginning of each year. Assuming that the savings increase at 5% per annum compound interest, express the relationship between the values of the savings at the beginning of year $t + 1$ and year t in difference-equation form and hence show that the value of the savings after 20 years is £535.8.

3. Show that the equations

$$\text{demand:} \qquad q_t = 4 - p_t$$

$$\text{supply:} \qquad q_t = 1 + 1.5\ p_{t-1}$$

   with $p_0 = 1$ lead to an 'exploding' cobweb model.

4. Show that the equations

$$\text{demand:} \qquad q_t = 4 - p_t$$

$$\text{supply:} \qquad q_t = 1 + 0.5\ p_{t-1}$$

   with $p_0 = 1$ lead to a stable cobweb model. What is the equilibrium price?

5. Show that the equations

$$\text{demand:} \qquad q_t = 8 - 3\ p_t$$

$$\text{supply:} \qquad q_t = 3\ p_{t-1}$$

   with $p_0 = 2$ lead to an oscillating cobweb model.

6. Solve the Harrod–Domar growth model (see section 10.5) when $a = 0.9$, $g = 0.3$ and $Y_0 = 100$.

7. A similar model to the cobweb arises when there are static supply and demand equations but the new price is affected by the level of inventory (the difference between q^s and q^d) in the previous period. For example,

$$\text{demand:} \qquad q_t^d = 150 - 0.6\ p_t$$

$$\text{supply:} \qquad q_t^s = 0.4\,p_t - 50$$

$$\text{pricing:} \qquad p_{t+1} = p_t - 0.3(q_t^s - q_t^d)$$

Find the solution for p_t and comment on whether it converges.

8. Obtain the solution of the following simple accelerator model:

$$Y_t = C_t + I_t$$

$$I_t = a(Y_t - Y_{t-1}) \qquad (a > 0)$$

$$C_t = bY_{t-1} + c \qquad (b, c > 0)$$

What is the condition for it to be stable?

10.8 Samuelson's multiplier–accelerator model

This assumes that

1. Consumer demand in any time period is a function of consumer income in the previous time period.

   That is $\qquad\qquad C_t = cY_{t-1}$

   where c is a constant which is less than one and is the marginal propensity to consume.

2. Investment in any time period is a function of the difference in income between the two previous time periods.

$$I_t = b(Y_{t-1} - Y_{t-2})$$

   where b is the acceleration coefficient, which is a constant.

3. The national output, or income, Y_t, in any time period is made up of the estimated consumer demand and the estimated investment demand in the same period.

$$Y_t = C_t + I_t$$

Based on these assumptions we have the model

$$C_t = cY_{t-1} \tag{1}$$

$$I_t = b(Y_{t-1} - Y_{t-2}) \tag{2}$$

$$Y_t = C_t + I_t \tag{3}$$

Substituting (1) and (2) into (3) leads to

$$Y_t = cY_{t-1} + b(Y_{t-1} - Y_{t-2})$$

$$= (c + b)Y_{t-1} - bY_{t-2}$$

This is a *second-order difference equation* because it relates the value of income in time period t to the value of income in the two immediately preceding time periods. It is possible to determine the value of income in any time period if we know the value of income in any two adjacent periods and the value of the constants b and c.

For example, let us suppose that

$$b = 0.6 \quad c = 0.9 \quad Y_0 = 100 \quad Y_1 = 120$$

Then
$$Y_2 = (c + b)Y_1 - bY_0 = 1.5(120) - 0.6(100)$$
$$= 180 - 60 = 120$$
$$Y_3 = 1.5(120) - 0.6(120) = 180 - 72 = 108$$

The value of income in each successive time period can be calculated by continuing the above procedure. But this is not necessary, because it is possible to obtain a solution to a second-order difference equation and to use this solution to determine Y_t in any time period as we did in the first-order case.

For convenience the equations and their methods of solution are again discussed under appropriate headings.

10.9 Second-order linear difference equations

HOMOGENEOUS

These can be written in general as $aY_{t+2} + bY_{t+1} + cY_t = 0$ and the solution is of the form $Y_t = \lambda^t$ where λ is a constant whose value is as yet undetermined.

If $Y_t = \lambda^t$, then $Y_{t+1} = \lambda^{t+1}$ and $Y_{t+2} = \lambda^{t+2}$. Substituting these values in the difference equation, we obtain

$$a\lambda^{t+2} + b\lambda^{t+1} + c\lambda^t = 0$$
$$\lambda^t(a\lambda^2 + b\lambda + c) = 0$$

Therefore
$$\lambda_1 = \frac{-b + \sqrt{(b^2 - 4ac)}}{2a}$$
$$\lambda_2 = \frac{-b - \sqrt{(b^2 - 4ac)}}{2a}$$

and the solution is of the form

$$Y_t = k_1\lambda_1^t + k_2\lambda_2^t$$

where k_1 and k_2 are arbitrary constants.

The equation $a\lambda^2 + b\lambda + c = 0$ is known as the *characteristic equation* and is easily formed by substituting λ^2 for Y_{t+2}, λ for Y_{t+1} and 1 for Y_t in the difference equation. When the roots of the characteristic equation are equal, that is $\lambda_1 = \lambda_2 = \lambda$, the solution is

$$Y_t = (k_1 + k_2 t)\lambda^t$$

This contains two arbitrary constants, which is a necessary requirement for it to be the solution to a second-order difference equation (compare differential equations, section 9.5).

When the roots of the characteristic equation are complex, that is $\lambda_1 = a + ib$ and $\lambda_2 = a - ib$, the solution is

$$Y_t = k_1(a + ib)^t + k_2(a - ib)^t$$

The solution is not very useful in this form because it is not possible to analyse the future time path of Y from a combination of real and imaginary numbers. But the solution can be transformed into a more suitable form by making the following substitutions.

Let $\quad\quad\quad\quad a = r\cos\theta \quad\quad$ and $\quad\quad b = r\sin\theta$

Then $\quad\quad\quad\quad a^2 + b^2 = r^2\cos^2\theta + r^2\sin^2\theta$

$$= r^2(\cos^2\theta + \sin^2\theta) = r^2$$

or $\quad\quad\quad\quad\quad\quad r = \sqrt{(a^2 + b^2)}$

and $\quad\quad\quad\quad\quad \theta = \cos^{-1}\dfrac{a}{r} \quad$ or $\quad \sin^{-1}\dfrac{b}{r}$

These can be combined to give

$$\tan\theta = b/a \quad\quad \text{or} \quad\quad \theta = \tan^{-1}(b/a)$$

Here θ is measured in radians (see Appendix A, section A.3). Now we know from the Euler relations (see section 5.16 that for all values of θ,

$$e^{i\theta} = \cos\theta + i\sin\theta$$

Now $(e^{i\theta})^n = e^{in\theta} = \cos n\theta + i\sin n\theta$.

It can be shown, therefore, that

$$(\cos\theta + i\sin\theta)^n = \cos n\theta + i\sin n\theta$$

This is usually referred to as *De Moivre's theorem* and it holds for all values of θ.

Making use of the substitutions $a = r\cos\theta$ and $b = r\sin\theta$

$$(a + ib) = r\cos\theta + ir\sin\theta = r(\cos\theta + i\sin\theta)$$

$$\therefore \qquad (a + ib)^t = [r(\cos \theta + i \sin \theta)]^t$$
$$= r^t(\cos \theta + i \sin \theta)^t$$
$$= r^t(\cos t\theta + i \sin t\theta)$$

Similarly, from the properties of complex numbers it can be shown that

$$(a - ib)^t = r^t(\cos t\theta - i \sin t\theta)$$

The solution to the difference equation which has complex roots to the characteristic equation can now be expressed in terms of trigonometrical functions:

$$Y_t = k_1(a + ib)^t + k_2(a - ib)^t$$

becomes $Y_t = k_1[r^t(\cos t\theta + i \sin t\theta)] + k_2[r^t(\cos t\theta - i \sin t\theta)]$

The right hand side of this equation can be arranged in a more convenient form by collecting together the sine and cosine terms:

$$Y_t = r^t[(k_1 + k_2) \cos t\theta + i(k_1 - k_2) \sin t\theta]$$
$$= r^t(k_3 \cos t\theta + k_4 \sin t\theta)$$

where $k_3 = k_1 + k_2$, $k_4 = i(k_1 - k_2)$, which are both constants.

This result can be compared to the solution obtained for a second-order differential equation.

$$y = e^{ax}(k_3 \cos bx + k_4 \sin bx)$$

It was shown in Chapter 9 that it is possible, and often more useful, to express this in the form $y = Ae^{ax} \cos (bx - \epsilon)$. In a similar way it can be shown, although the mathematics are not discussed here, that the solution to a second-order difference equation can be expressed in the alternative form

$$Y_t = Ar^t \cos (t\theta - \epsilon)$$

where A and ϵ are the two arbitrary constants, whose values are determined in a particular case from two pieces of information which are available about the system. The behaviour of the solution depends mainly on the value of r, since a cosine function is always cyclical. If $r > 1$ or $r < -1$ then Y will tend to oscillate between $\pm\infty$. Otherwise Y will tend to zero.

NON-HOMOGENEOUS

This has the general form

$$aY_{t+2} + bY_{t+1} + cY_t = f(t)$$

It can be shown that the solution to this type of equation is the sum of two parts which must be obtained independently. These are usually called the *transient function* and the *equilibrium*, or *particular solution*.

The transient function is simply the solution of the homogeneous part of the equation formed by letting $f(t) = 0$ and this can be obtained by the method of the previous section.

The particular solution is any solution which satisfies the difference equation and its form depends upon the form of $f(t)$.

1. $f(t)$ is a constant. For example, if $aY_{t+2} + bY_{t+1} + cY_t = K$, we try a constant as the particular solution, that is $Y_t = Z$ for all t and it follows that $Y_{t+2} = Y_{t+1} = Y_t = Z$.

Substituting these values in the difference equation, we have

$$aZ + bZ + cZ = K$$

That is $Z(a + b + c) = K$ or $Z = \dfrac{K}{a + b + c}$

It will be found that this value of Z satisfies the difference equation and is therefore a particular solution for all cases except those in which

$$a + b + c = 0$$

In such cases we must try the solution

$$Y_t = Zt$$

so that

$$Y_{t+1} = Z(t + 1) = Zt + Z$$
$$Y_{t+2} = Z(t + 2) = Zt + 2Z$$

Notice that Zt is the product of Z and t. Substituting these values in the difference equation gives

$$aZ(t + 2) + bZ(t + 1) + cZt = K$$

or

$$aZt + 2aZ + bZt + bZ + cZt = K$$
$$Z(a + b + c)t + Z(b + 2a) = K$$

If this equation is to be valid for all values of t, it is necessary that both of the following conditions hold:

$$Z(a + b + c) = 0 \quad \text{and} \quad Z(b + 2a) = K$$

The second equation gives a value for Z:

$$Z = \frac{K}{b + 2a} \quad \text{and hence} \quad Y_t = Zt = \frac{Kt}{b + 2a}$$

provided that $(b + 2a)$ is not equal to zero.

    2. $f(t)$ is not a constant

        (a) $f(t)$ is a linear function. For example,

$$aY_{t+2} + bY_{t+1} + cY_t = K_1 + K_2t$$

The particular solution is of the form

$$Y_t = Z_0 + Z_1t$$

so that

$$Y_{t+1} = Z_0 + Z_1(t + 1)$$

and

$$Y_{t+2} = Z_0 + Z_1(t + 2)$$

These values are substituted in the difference equation:

$$a[Z_0 + Z_1(t + 2)] + b[Z_0 + Z_1(t + 1)] + c[Z_0 + Z_1t] = K_1 + K_2t$$

This can be rearranged with the constant terms and the terms in t collected separately:

$$(aZ_0 + 2aZ_1 + bZ_0 + bZ_1 + cZ_0) + (aZ_1 + bZ_1 + cZ_1)t = K_1 + K_2t$$

that is $\quad (a + b + c)Z_0 + (2a + b)Z_1 + (a + b + c)Z_1t = K_1 + K_2t$

If this equation is to be true for all values of t, the following two conditions must hold:

$$(a + b + c)Z_0 + (2a + b)Z_1 = K_1 \tag{1}$$

$$(a + b + c)Z_1 = K_2 \tag{2}$$

From (2)

$$Z_1 = \frac{K_2}{a + b + c}$$

and from (1) and (2)

$$Z_0 = \frac{K_1 - (2a + b)Z_1}{a + b + c}$$

$$= \frac{K_1}{a + b + c} - \frac{(2a + b)}{(a + b + c)} \frac{K_2}{(a + b + c)}$$

$$= \frac{K_1}{a + b + c} - \frac{(2a + b)K_2}{(a + b + c)^2}$$

The particular solution can now be expressed in terms of the known constants a, b, c, K_1 and K_2.

$$Y_t = Z_0 + Z_1 t$$

$$= \frac{K_1}{a + b + c} - \frac{(2a + b)K_2}{(a + b + c)^2} + \frac{K_2 t}{a + b + c}$$

$$= \frac{1}{a + b + c}\left[K_1 - K_2\left(\frac{2a + b}{a + b + c} - t\right)\right]$$

(b) $f(t)$ is of the form K^t. For example,

$$aY_{t+2} + bY_{t+1} + cY_t = K^t$$

The particular solution is of the form

$$Y_t = ZK^t$$

so that $$Y_{t+1} = ZK^{t+1}$$

and $$Y_{t+2} = ZK^{t+2}$$

These values are substituted in the difference equations

$$aZK^{t+2} + bZK^{t+1} + cZK^t = K^t$$

$$K^t(aK^2 + bK + c)Z = K^t$$

$$\therefore \qquad Z = \frac{1}{aK^2 + bK + c}$$

and the particular solution is

$$Y_t = ZK^t = \frac{K^t}{aK^2 + bK + c}$$

provided that $aK^2 + bK + c$ is not equal to zero.

If this condition is not satisfied then we must try a particular solution of the form

$$Y_t = ZtK^t$$

If this results in a particular solution with a denominator equal to zero then it is necessary to try the substitution

$$Y_t = Zt^2K^t$$

This procedure can be continued until a solution is obtained.

(c) $f(t)$ is a trigonometrical function. For example,

$$aY_{t+2} + bY_{t+1} + cY_t = K \sin t$$

The particular solution is of the form

$$Y_t = Z_0 \sin t + Z_1 \cos t$$

so that $\qquad\qquad Y_{t+1} = Z_0 \sin (t + 1) + Z_1 \cos (t + 1)$

and $\qquad\qquad Y_{t+2} = Z_0 \sin (t + 2) + Z_1 \cos (t + 2)$

The values of Z_0 and Z_1 can be found by substituting these values in the difference equation and by making use of the following trigonometrical identities.

$$\sin (A + B) \equiv \sin A \cos B + \cos A \sin B$$

$$\cos (A + B) \equiv \cos A \cos B - \sin A \sin B$$

The exercise, which is rather more lengthy than the previous ones, is left for the reader to attempt.

10.10   A consumption–investment model

The following example shows how the above methods are applied in the analysis of the time path of national income Y_t given the assumptions that the consumption and investment functions are

$$C_t = 0.5 \, Y_{t-1} + 0.19 \, Y_{t-2} \tag{1}$$

$$I_t = 10 + 50(1.05)^t \tag{2}$$

$$Y_t = C_t + I_t \tag{3}$$

$$Y_0 = 100 \quad \text{and} \quad Y_1 = 120 \tag{4}$$

A second-order difference equation is obtained by substituting (1) and (2) into (3).

$$Y_t = 0.5 \, Y_{t-1} + 0.19 \, Y_{t-2} + 10 + 50(1.05)^t$$

or $\qquad Y_t - 0.5 \, Y_{t-1} - 0.19 \, Y_{t-2} = 10 + 50(1.05)^t$

The solution to the reduced form is obtained from the characteristic equation

$$\lambda^2 - 0.5\lambda - 0.19 = 0$$

the roots of which are

$$\lambda = \frac{0.5 \pm \sqrt{[(0.5)^2 + 4 \times 0.19]}}{2} = \frac{0.5 \pm 1}{2}$$

∴ $\lambda_1 = 0.75 \qquad \lambda_2 = -0.25$

and the transient solution is

$$Y_t = k_1(0.75)^t + k_2(-0.25)^t$$

The particular solution can be found by applying the methods of section 10.9 to the two terms on the right hand side in turn. For

$$Y_t - 0.5\,Y_{t-1} - 0.19\,Y_{t-2} = 10$$

the particular solution is obtained by trying $Y_t = z$ to give

$$Y_t = \left\{ \frac{1}{1 - 0.5 - 0.19} \right\} 10 = \frac{10}{0.31} = 32.26$$

For the equation

$$Y_t - 0.5\,Y_{t-1} - 0.19\,Y_{t-2} = 50(1.05)^t$$

the particular solution is obtained by trying $Y_t = z(1.05)^t$ to give

$$Y_t = 50 \left\{ \frac{1.05^2}{1.05^2 - 0.5(1.05) - 0.19} \right\} (1.05)^t$$

Hence

$$Y_t = 50 \left\{ \frac{1.05^2}{0.3875} \right\} (1.05)^t = 142.26(1.05)^t$$

The solution can then be written as

$$Y_t = k_1(0.75)^t + k_2(-0.25)^t + 32.26 + 142.26(1.05)^t$$

and the value of the constants, k_1 and k_2, found from the initial conditions.

$$Y_0 = 100$$
$$100 = k_1 + k_2 + 32.26 + 142.26 \qquad (1)$$
$$Y_1 = 120$$
$$120 = 0.75k_1 - 0.25k_2 + 32.26 + 142.26(1.05) \qquad (2)$$

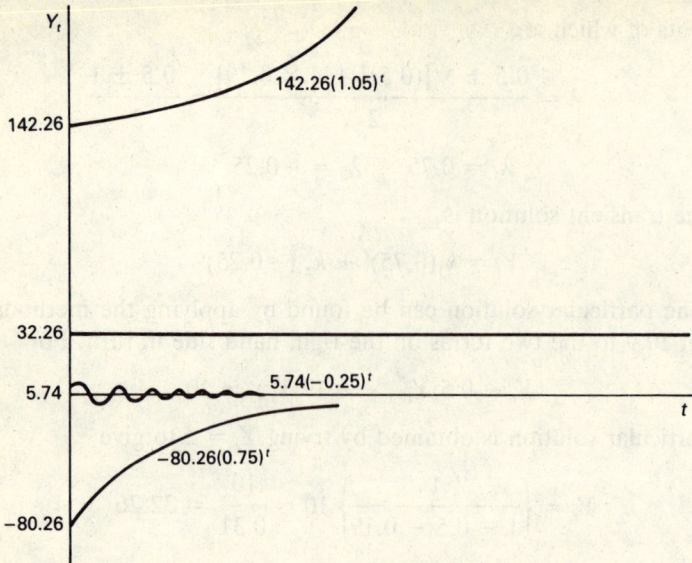

Fig. 10.5

From (1)

$$k_1 + k_2 = 100 - 174.52 = -74.52$$

and from (2)

$$0.75k_1 - 0.25k_2 = 120 - 32.26 - 149.37 = -61.63$$

These two equations can be solved simultaneously with the result

$$k_1 = -80.26 \quad \text{and} \quad k_2 = 5.74$$

The complete solution to the difference equation

$$Y_t - 0.5Y_{t-1} - 0.19Y_{t-2} = 10 + 50(1.05)^t$$

and the initial conditions $Y_0 = 100$, $Y_1 = 120$ is given by

$$Y_t = -80.26(0.75)^t + 5.74(-0.25)^t + 32.26 + 142.26(1.05)^t$$

It is easy to check that this solution satisfies the initial conditions, but not as easy to show that it satisfies the difference equation.

The general shape of the time path of national income can be obtained by considering the function in terms of the four constituent parts. The graphs of these can be sketched as shown in Fig. 10.5, where the term $142.26(1.05)^t$ becomes predominant with time.

10.11 An inflation–unemployment model

In section 9.11 a continuous time model of the interactions between inflation, unemployment and monetary policy was examined based on the expectations-augmented Phillips curve. Here we consider the discrete time version of the same model. Adding time subscripts, the expectations-augmented Phillips curve can be written as

$$p_t = a - bU_t - c + p_t^e \qquad (a, b, c, > 0) \tag{1}$$

and the simple adaptive expectations equation is

$$p_{t+1}^e - p_t^e = k(p_t - p_t^e) \qquad (k > 0) \tag{2}$$

This states that the change in the expectation is a proportion, k, of the error in expectation in the current period, or alternatively, next period's expected inflation is a weighted average of this period's actual and expected inflation, since

$$p_{t+1}^e = kp_t + (1 - k)p_t^e$$

Finally, we assume that the change in unemployment is related to the rate of growth of real money, or

$$U_{t+1} - U_t = -e(m - p_{t+1}) \qquad (e > 0) \tag{3}$$

where m is the nominal rate of growth of money and p is the rate of inflation. We will also assume that m is a constant.

The three equations (1)–(3) explain p, p^e and U. Moving (1) forward one period gives

$$p_{t+1} = a - bU_{t+1} - c + p_{t+1}^e$$

and subtracting (1) from this results in

$$p_{t+1} - p_t = -b(U_{t+1} - U_t) + (p_{t+1}^e - p_t^e)$$

Substituting from (2) and (3) into this gives

$$p_{t+1} - p_t = be(m - p_{t+1}) + k(p_t - p_t^e)$$

or
$$(1 + be)p_{t+1} = (1 + k)p_t + bem - kp_t^e \tag{4}$$

This includes p^e which can be obtained from (1),

$$-p_t^e = a - bU_t - c - p_t$$

and substituting this into (4),

$$(1 + be)p_{t+1} = p_t + bem + ka - bkU_t - ck$$

Finally to eliminate U we first-difference this equation

$$(1 + be)(p_{t+1} - p_t) = (p_t - p_{t-1}) - bk(U_t - U_{t-1})$$

where the difference in m is zero, since m (the growth of nominal money) is assumed to be a constant. Next, we use (3) to eliminate U,

$$(1 + be)(p_{t+1} - p_t) = (p_t - p_{t-1}) + bek(m - p_t)$$

Rearranging,

$$(1 + be)p_{t+1} - (1 + be + 1 - bek)p_t + p_{t-1} = bekm$$

This is a second-order non-homogeneous difference equation which can be put into the more usual form by shifting it forward one period and simplifying the notation:

$$Ap_{t+2} + Bp_{t+1} + p_t = C \qquad (5)$$

where $A = (1 + be)$, $B = -(2 + be[1 - k])$, $C = bekm$. The characteristic equation is

$$A\lambda^2 + B\lambda + 1 = 0$$

and, since the roots depend on the values of the parameters, it is difficult to analyse the behaviour in the general case. Before looking at some examples we notice that the particular or equilibrium solution, found by putting $p_{t+2} = p_{t+1} = p_t = p$ is

$$p = \frac{C}{A + B + 1} = m$$

and so the equilibrium rate of inflation is m, the rate of monetary expansion. This exists for any (non-infinite) values of the parameters.

Taking some examples, first suppose that $k = 1$, so that, from (2), $p_{t+1}^e = p_t$, and the expected rate of inflation next period is this period's actual rate. Here (5) reduces to

$$(1 + be)p_{t+2} - 2p_{t+1} + p_t = bem$$

for which the characteristic equation is

$$(1 + be)\lambda^2 - 2\lambda + 1 = 0$$

and the roots are real if

$$4 > 4(1 + be)$$

which requires either $b < 0$ or $e < 0$. Since these are assumed to be positive then when $k = 1$ the difference equation has complex roots and the solution is expressed in terms of trigonometrical functions (as in section 10.9 above). The time path of inflation (p) will therefore follow a cycle and the value will fluctuate.

For example, if $be = 5$ the characteristic equation is

$$6\lambda^2 - 2\lambda + 1 = 0$$

and the roots are $\lambda = (1 \pm \sqrt{5}i)/6$
so that $a = 1/6$, $b = \sqrt{5}/6$ and $r = \sqrt{(a^2 + b^2)} = 0.4082$. Also,
$\theta = \tan^{-1}(b/a) = \tan^{-1}\sqrt{5} = 65.9° = 1.15$ radians.
Therefore, $p_t = Er^t \cos(t\theta - \epsilon) = E(0.4082^t)\cos(1.15t - \epsilon)$ where E and ϵ are constants. The equilibrium solution, from above, is $p = m$ and so the full solution is

$$p_t = E(0.4082^t)\cos(1.15t - \epsilon) + m$$

For simplicity we will assume that $m = 0$. The values of E and ϵ are given by the initial conditions. Let these be $t = 0$, $p_0 = 0$ and $t = 1$, $p_1 = 1$. The first gives

$$0 = E\cos(-\epsilon)$$

and so either $E = 0$ (which is ignored because it gives $p_t = m$ for all t) or $\epsilon = \pi/2 = 1.5708$.

The second gives

$$1 = E(0.4082)\cos(1.15 - \epsilon)$$

and so $\qquad E = 1/[0.4082 \cos(-0.4208)]$

Since 0.4208 radians $= 24.11°$ then $E = 2.684$.
The full solution is

$$p_t = 2.684(0.4082^t)\cos(1.15t - 1.5708)$$

The first few values are $p_0 = 0$, $p_1 = 1$, $p_2 = 0.33$, $p_3 = -0.06$, $p_4 = -0.07$, $p_5 = -0.03$ and so there is a rapid cyclical convergence towards the equilibrium value of zero.

Next, suppose that $k = 0.4$, so that expected inflation gradually adjusts to actual inflation, and also that $be = 5$. The characteristic equation of (5) is now

$$6\lambda^2 - 5\lambda + 1 = 0$$

and the roots are $\lambda_1 = 0.5$ and $\lambda_2 = 0.33$. Therefore,

$$p_t = F(0.5^t) + G(0.33^t)$$

is the transient solution. As $t \to \infty$, this will tend to zero and so inflation will converge on its equilibrium value.

Finally, suppose that $k = 0.4$ and $be = 2$ so that the characteristic equation of (5) is

$$3\lambda^2 - 3.2\lambda + 1 = 0$$

and the roots are complex. As in the first case above the time path of inflation will follow a cycle.

The expectations-augmented Phillips curve model can also be analysed as a difference equation in unemployment or expected inflation. For unemployment, the equations of the model are

$$p_t = a - bU_t - c + p_t^e \qquad (a, b, c > 0) \qquad (1)$$

$$p_{t+1}^e - p_t^e = k(p_t - p_t^e) \qquad (k > 0) \qquad (2)$$

$$U_{t+1} - U_t = -e(m - p_{t+1}) \qquad (e > 0) \qquad (3)$$

From (1), $p_t^e = p_t - a + bU_t + c$

Substituting for p^e in (2),

$$(p_{t+1} - a + bU_{t+1} + c) - (p_t - a + bU_t + c)$$
$$= kp_t - k(p_t - a + bU_t + c)$$

or $\qquad p_{t+1} - p_t = bU_t - bU_{t+1} + ak - bkU_t - ck \qquad (6)$

Now differencing (3) and then using (6),

$$U_{t+1} - 2U_t + U_{t-1} = e(p_{t+1} - p_t)$$
$$= e[b(1 - k)U_t - bU_{t+1} + ak - ck]$$

or $\qquad (1 + be)U_{t+1} - [2 + be(1 - k)]U_t + U_{t-1} = (a - c)ek$

This can be shifted forward one period to give

$$AU_{t+2} + BU_{t+1} + U_t = C' \qquad (7)$$

where $A = (1 + be)$, $B = -[2 + be(1 - k)]$ and $C' = (a - c)ek$

Comparing this equation with (5) above we see that the coefficients on the left hand side of the difference equation are the same in each case but the constant terms differ. Thus the solutions to the homogeneous equations, that is the transient solutions, will be the same – so that if p is cyclical then U will also be cyclical – but because the constant terms differ the equilibrium solutions will differ. This is to be expected since there is no reason for the equilibrium level of unemployment (which is likely to be affected by labour market institutions and demographic factors) and the equilibrium level of inflation (which will depend on productivity and the rate of increase in nominal money) to be identical. We also note that the rate of monetary expansion, m, does not occur in (7) and so changing m would not affect unemployment. Finally, the equilibrium solution of (7), found by putting $U_t = U_{t+1} = U_{t+2} = Z$, say, is $U = (a - c)/b$ which is also independent of m. This is known as the *natural rate of unemployment* and is unrelated to the rate of inflation. Thus, in the long run, there is no trade-off between unemployment and inflation in this model.

A difference equation in p^e can also be found. Differencing (1),

$$p_t - p_{t-1} = -b(U_t - U_{t-1}) + p_t^e - p_{t-1}^e$$
$$= be(m - p_t) + p_t^e - p_{t-1}^e$$

from using (3). Now (2) can be written

$$kp_t = p_{t+1}^e - (1 - k)p_t^e$$

and so substituting for p in the previous equation and simplifying gives

$$(1 + be)p_{t+1}^e - [2 + be(1 - k)]\, p_t^e + p_{t-1}^e = bekm$$

which has the same coefficients as (5) and (7), and the constant term is the same as in (5). Therefore in equilibrium, $p^e = m$ and the behaviour is as above.

With all mathematical models it is important to realise that the properties follow from the assumptions and that changing the assumptions, in general, changes the behaviour of the model. The challenge is to find simple models which can make a useful contribution to our understanding of the real world.

10.12 Exercises

1. Show that Samuelson's multiplier–accelerator model leads to a characteristic equation with
   (a) complex roots when $c = 0.9$, $b = 0.5$
   (b) repeated roots when $c = 0.96$, $b = 0.64$
   (c) real roots when $c = 0.9$, $b = 0.3$

2. Solve the following difference equations and sketch the time path of Y_t for $t = 0$ to $t = 5$ for (a)–(e).

(a) $Y_{t+2} - Y_{t+1} - 2Y_t = 3$ with $Y_0 = 2$ $Y_1 = 2$

(b) $Y_{t+2} - 3Y_{t+1} + 2Y_t = 4$ with $Y_0 = 1$ $Y_1 = 2$

(c) $Y_{t+2} - 4Y_{t+1} + 4Y_t = 3^t$ with $Y_0 = 3$ $Y_1 = 4$

(d) $Y_{t+2} - 3Y_{t+1} + 2Y_t = 3(2^t)$ with $Y_0 = 1$ $Y_1 = 5$

(e) $4Y_{t+2} + 4Y_{t+1} + Y_t = 2 + 3t$ with $Y_1 = 1$ $Y_2 = 2$

(f) $Y_{t+2} + 4Y_{t+1} + 5Y_t = 4$

(g) $Y_{t+2} - 4Y_{t+1} + 8Y_t = t$.

3. For the equations

$$Y_t = C_t + I_t \tag{1}$$

$$C_t = \alpha Y_t + \beta Y_{t-1} + \delta \tag{2}$$

$$I_t = \gamma(Y_{t-1} - Y_{t-2}) \tag{3}$$

show that substitution from (2) and (3) into (1) produces a second-order difference equation which, if $\delta = 0$, is similar to the equation from Samuelson's multiplier-accelerator model (Section 10.8).
    Solve the equation when

(a) $\alpha = 0.6$, $\beta = 0.2$, $\gamma = 0.3$ and $\delta = 1$

(b) $\alpha = 0.5$, $\beta = 0.2$, $\gamma = 0.2$ and $\delta = 1$

4. In section 10.11 the original Phillips curve was modified to give real wages, $(w - p^e)$, as a function of unemployment. If there is *money illusion* and workers do not get full compensation for inflation, the expectations-augmented Phillips curve can be written

$$p_t = a - bU_t - c + fp_t^e \qquad (a, b, c, f > 0) \tag{1}$$

where in the special case $f = 1$ we have the previous formulation. Using the other two equations of the model,

$$p_{t+1}^e - p_t^e = k(p_t - p_t^e) \qquad (k > 0) \tag{2}$$

$$U_{t+1} - U_t = -e(m - p_{t+1}) \qquad (e > 0) \tag{3}$$

obtain the difference equation in p and compare it with (5) of section 10.11.

5. One variation on the multiplier-accelerator model of section 10.8 is obtained by introducing government expenditure, G, which might be determined by the previous period's income and the previous period's government expenditure. Suppose that the equations of the model are

$$Y_t = C_t + I_t + G_t$$

$$C_t = aY_{t-1}$$

$$I_t = b(Y_t - Y_{t-1})$$

$$G_t = cY_{t-1} + dG_{t-1}$$

Obtain the difference equation in Y and show that (a) if $d = 0$ it reduces to a first-order equation, and (b) if $a = 0.6$, $b = 2$, $c = 0.1$ and $d = 0.5$ and the initial conditions are $Y_0 = 1$, $Y_1 = 1$, the value of Y will eventually approach infinity.

10.13 Revision exercises for Chapter 10 (without answers)

1. Solve the following difference equations and plot the time path of Y_t for $t = 0$ to $t = 5$:

(a) $Y_t = 0.5 Y_{t-1} + 150$ with $Y_0 = 100$

(b) $Y_t = 1.3 Y_{t-1} - 75$ with $Y_0 = 100$

(c) $Y_t = Y_{t-1} + 1$ with $Y_0 = 1$.

2. Examine the behaviour of the following equations as t increases:

$$\text{demand} : q_t = 100 - ap_t$$

$$\text{supply} : q_t = bp_t - 3$$

if (a) $a = 2$, $b = 3$ (b) $a = 2$, $b = 2$ (c) $a = 3$, $b = 2$.

3. The value of sales, S, of a product is found to satisfy

$$S_t = 0.8 S_{t-1} + 5A_t$$

where A is the expenditure on advertising. The company determines its advertising budget according to the rule

$$A_t = 0.1S_{t-1}$$

If $S_0 = 50$, determine the value of sales for $t = 1$ to $t = 5$.

4. Solve the following difference equations and explain how the time path of Y_t will vary as t tends to infinity:

(a) $Y_{t+2} + 2Y_{t+1} = 4$ with $Y_0 = 1$,

(b) $Y_{t+2} + 2Y_{t+1} - 0.5Y_t = 5$ with $Y_0 = 5$, $Y_1 = 6$,

(c) $Y_{t+2} - 2Y_{t+1} + 0.5Y_t = 3$ with $Y_0 = 3$, $Y_1 = 6$,

(d) $Y_{t+2} - 4Y_{t+1} + 6Y_t = t$ with $Y_0 = 1$, $Y_1 = 2$,

(e) $Y_{t+2} - 2Y_{t+1} + Y_t = 5 + 2t$ with $Y_0 = 2$, $Y_1 = 3$.

5. Examine the behaviour of the equilibrium solution of the following supply and demand equations:

$$\text{supply}: q_t = 6p_t - 5 + 4(p_{t-1} - p_{t-2})$$
$$\text{demand}: q_t = 500 - 2p_t.$$

6. Given the simple national-income model

$$Y_t = C_t + I_t + G_t$$
$$C_t = 0.7Y_{t-1} + 2$$
$$I_t = 2(Y_{t-1} - Y_{t-2})$$
$$G_t = 0.2Y_{t-1}$$

and the initial conditions $Y_0 = 100$, $Y_1 = 102$, examine the time path of Y_t for $t = 2$ to $t = 6$.

7. A university gives resources (R) to each department according to the formula

$$R_t = 0.5(R_{t-1} + R_{t-2}) + 0.2S_t$$

where S measures new student registrations. It is also known that $S_t = 1.2S_{t-1}$ with $S_1 = 10$. By first obtaining the solution for S, determine how R moves given the initial conditions that $R_0 = 0$, $R_1 = 10$.

Appendix A
Trigonometric Functions

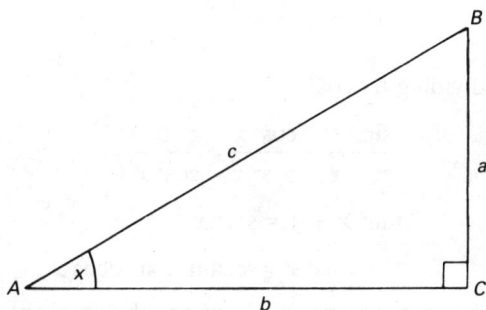

Fig. A.1

A.1 Definitions

Let us consider a right-angled triangle ABC (Fig. A.1). If the side BC is of length a, AC of length b, AB of length c, and x is the angle BAC then we define

$$\text{sine } x = \sin x = \frac{a}{c} \tag{1}$$

$$\text{cosine } x = \cos x = \frac{b}{c} \tag{2}$$

$$\text{tangent } x = \tan x = \frac{a}{b} \tag{3}$$

Values of these functions can be obtained from tables.

For a right-angled triangle,

$$a^2 + b^2 = c^2$$

or, dividing by c^2,

$$\left(\frac{a}{c}\right)^2 + \left(\frac{b}{c}\right)^2 = 1$$

That is

$$(\sin x)^2 + (\cos x)^2 = 1$$

or $$\sin^2 x + \cos^2 x = 1 \tag{4}$$

Therefore, $$\sin^2 x = 1 - \cos^2 x$$
$$\cos^2 x = 1 - \sin^2 x$$

Notice that $$\tan x = \frac{a}{b} = \frac{a/c}{b/c} = \frac{\sin x}{\cos x}$$

From (4), by dividing by $\cos^2 x$,

$$\frac{\sin^2 x}{\cos^2 x} + \frac{\cos^2 x}{\cos^2 x} = \frac{1}{\cos^2 x}$$

or, $$\tan^2 x + 1 = \sec^2 x \tag{5}$$

where $$\sec x = \text{secant } x = 1/\cos x$$

Two other trigonometric functions are the cosecant and cotangent defined by

$$\text{cosecant } x = \text{cosec } x = \frac{1}{\sin x}$$

and $$\text{cotangent } x = \text{cotan } x = \frac{1}{\tan x} = \frac{\cos x}{\sin x}$$

A.2 Compound angles

For any two angles x and y the following relationships can be shown to be true:

$$\sin (x + y) = \sin x \cos y + \cos x \sin y \tag{6}$$
$$\cos (x + y) = \cos x \cos y - \sin x \sin y \tag{7}$$
$$\sin (x - y) = \sin x \cos y - \cos x \sin y \tag{8}$$
$$\cos (x - y) = \cos x \cos y + \sin x \sin y \tag{9}$$

$$\tan (x + y) = \frac{\tan x + \tan y}{1 - \tan x \tan y} \tag{10}$$

$$\tan (x - y) = \frac{\tan x - \tan y}{1 + \tan x \tan y} \tag{11}$$

Using these relationships, we have

(a) when $x = y$, from (6)

$$\sin (x + x) = \sin 2x = 2 \sin x \cos x \tag{12}$$

(b) when $x = y$, from (7),

$$\cos 2x = \cos x \cos x - \sin x \sin x$$
$$= \cos^2 x - \sin^2 x$$
$$= 1 - 2 \sin^2 x = 2 \cos^2 x - 1 \tag{13}$$

since $\cos^2 x + \sin^2 x = 1$. Hence,

$$2 \sin^2 x = 1 - \cos 2x$$

or

$$\sin^2 x = \frac{1 - \cos 2x}{2} \tag{14}$$

and

$$\cos^2 x = \frac{1 + \cos 2x}{2} \tag{15}$$

(c) When $x = y$, from (10)

$$\tan 2x = \frac{2 \tan x}{1 - \tan^2 x} \tag{16}$$

(d)

$$\sin x + \sin y = 2 \sin \left(\frac{x + y}{2}\right) \cos \left(\frac{x - y}{2}\right) \tag{17}$$

$$\sin x - \sin y = 2 \cos \left(\frac{x + y}{2}\right) \sin \left(\frac{x - y}{2}\right) \tag{18}$$

$$\cos x + \cos y = 2 \cos \left(\frac{x + y}{2}\right) \cos \left(\frac{x - y}{2}\right) \tag{19}$$

$$\cos x - \cos y = -2 \sin \left(\frac{x + y}{2}\right) \sin \left(\frac{x - y}{2}\right) \tag{20}$$

The formulae (17) to (20) can be verified by using (6) to (9) to expand the right-hand side of each statement.

A.3 Degrees and radians

In elementary mathematical work it is convenient to measure angles in degrees. In more advanced work, however, it is more convenient to use another measure of angles, known as radians. A *radian* is defined as the angle at the centre of a circle which contains an arc length equal to the radius of the circle (see Fig. A.2).

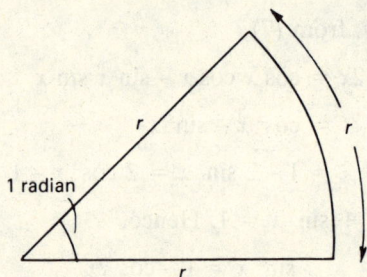

Fig. A.2

If r is the radius of a circle, the circumference is of length $2\pi r$. The angle at the centre of a circle is $360°$, and so

$$360° = \frac{2\pi r}{r} = 2\pi \text{ radians}$$

or π radians $= 180°$

and $\pi/2$ radians $= 90°$

where $\pi = 3.14159$. Now consider the relationship between an angle measured as $D°$ and R radians. Since $360° = 2\pi$ radians,

$$\frac{D}{360} = \frac{R}{2\pi}$$

and so $D = \dfrac{360\,R}{2\pi}$ and $R = \dfrac{2\pi D}{360}$

These convert radians to degrees and degrees to radians. For example,

if $D = 30°$, $R = \dfrac{2\pi(30)}{360} = 0.5236$ radians

Also, if $R = 1.30$ radians, $D = \dfrac{360(1.30)}{2\pi} = 74.48°$

A.4 General angles

From Fig. A.3 and the definitions $\sin x = a/c$, $\cos x = b/c$, $\tan x = a/b$ we can see that varying x causes the point P to mark out a quadrant of a circle from A to B. At the same time the values of a and b vary.

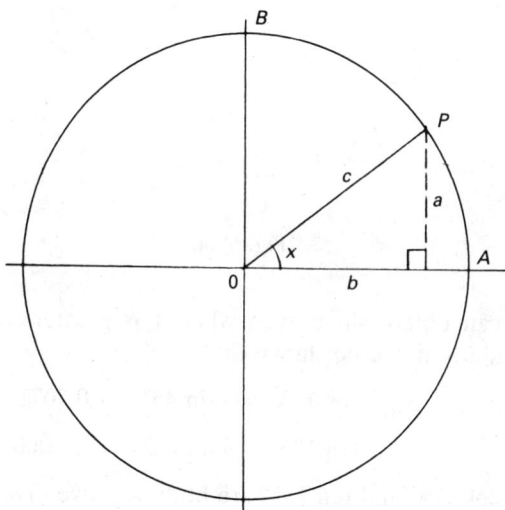

Fig. A.3

If we allow x to become very small and approach zero, a approaches zero, and b approaches c, hence $\sin 0 = 0$, $\cos 0 = 1$, $\tan 0 = 0$.

Similarly if we allow x to approach 90°, a approaches c and b approaches zero. Hence,

$$\sin 90° = \sin (\pi/2) = 1$$

$$\cos 90° = \cos (\pi/2) = 0$$

$$\tan 90° = \tan (\pi/2) = \infty$$

Let us now increase x to 135° or $3\pi/4$ radians (Fig. A.4). We know that, for example, using (6),

$$\sin 135° = \sin (90 + 45)° = \sin 90° \cos 45° + \cos 90° \sin 45°$$

$$= \cos 45° = 0.7071,$$

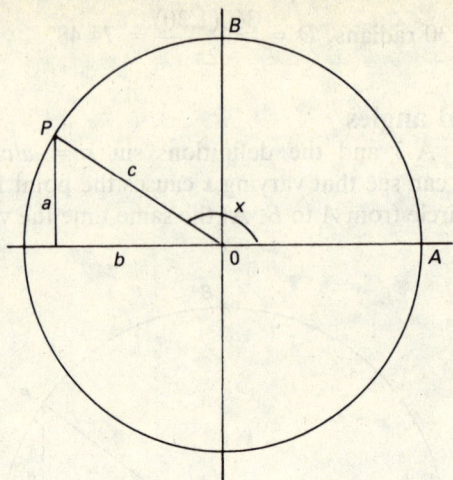

Fig. A.4

that is, we can obtain $\sin x$ even when x is greater than $90°$ or $\pi/2$ radians. Similarly it can be shown that

$$\cos 135° = -\sin 45° = -0.7071$$

$$\tan 135° = -\tan 45° = -1.0000$$

Notice that $\cos 135°$ and $\tan 135°$ are both negative. This is because

$$\cos x = \frac{b}{c} \qquad \tan x = \frac{a}{b}$$

and b is measured in the negative direction from 0, and so is a negative number. Both a and c are positive numbers.

Fig. A.5 shows the angle $x = 225°$. In this case both a and b are negative numbers so that $\sin x$ and $\cos x$ are negative, but $\tan x$ is positive. It can be shown that

$$\sin 225° = -0.7071 \qquad \cos 225° = -0.7071 \qquad \tan 225° = 1.000$$

Similarly, the angle x can be increased to, say, $315°$, and in this case a is negative and both b and c are positive, so that $\sin x$ and $\tan x$ are negative and $\cos x$ is positive. These results can be summarised as in Fig A.6 to indicate which of sine, cosine and tangent are positive.

So far we have considered only angles of up to 2π radians ($360°$),

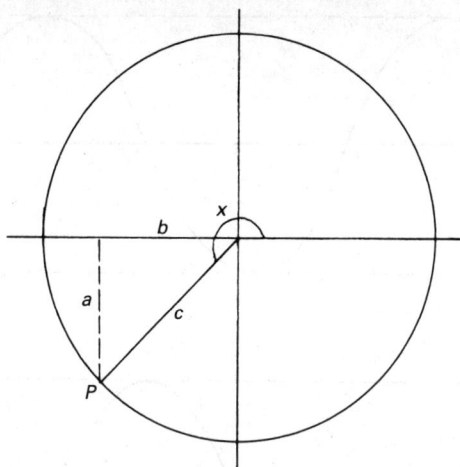

Fig. A.5

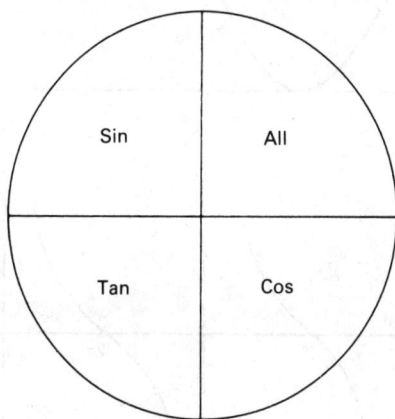

Fig. A.6

but it is possible to allow an angle to exceed this by rotating the point P (Fig. A.5) around the circle. In this way, an angle of, say, $2\pi + x$ has the same sine, cosine, and tangent as x has.

We know that

$$\sin 0 = 0 \qquad \cos 0 = 1 \qquad \tan 0 = 0$$
$$\sin (\pi/2) = 1 \qquad \cos (\pi/2) = 0 \qquad \cos (\pi/2) = \infty$$

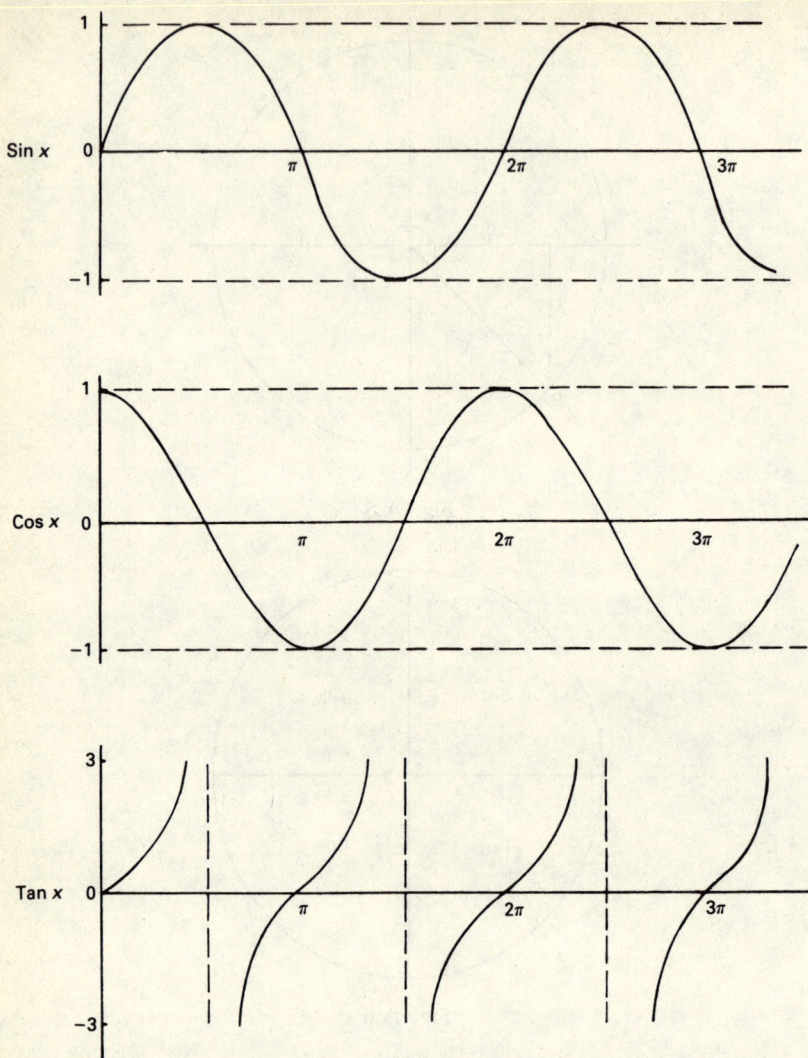

Fig. A.7

By evaluating sin x, cos x, and tan x for other values of x we can draw the graphs of these functions (Fig. A.7).

The graph of cos x is identical to the graph of sin x moved $\pi/2$ to the left. The graph of tan x has asymptotes at $\pi/2$, $3\pi/2$, $5\pi/2$, . . . and so it consists of a series of sections.

A.5 Differentiation

If $y = \sin x$ and we allow x to increase by a small amount δx, and let the corresponding increase in y be δy, then

$$y + \delta y = \sin (x + \delta x)$$

Subtracting, we have

$$\delta y = \sin (x + \delta x) - \sin x$$

Using (6), we obtain

$$\delta y = \sin x \cos \delta x + \cos x \sin \delta x - \sin x$$

$$= \sin x (\cos \delta x - 1) + \cos x \sin \delta x$$

Hence,
$$\frac{\delta y}{\delta x} = \sin x \left(\frac{\cos \delta x - 1}{\delta x} \right) + \frac{\cos x \sin \delta x}{\delta x}$$

Now by MacLaurin's theorem (section 5.15)

$$\cos x = 1 - \frac{x^2}{2!} + \frac{x^4}{4!} - \frac{x^6}{6!} + \cdots$$

and
$$\sin x = x - \frac{x^3}{3!} + \frac{x^5}{5!} - \frac{x^7}{7!} + \cdots$$

That is,
$$\cos \delta x = 1 - \frac{(\delta x)^2}{2!} + \frac{(\delta x)^4}{4!} - \frac{(\delta x)^6}{6!} + \cdots$$

so that
$$\frac{\cos \delta x - 1}{\delta x} = - \frac{\delta x}{2!} + \frac{(\delta x)^3}{4!} - \frac{(\delta x)^5}{6!} + \cdots$$

and
$$\lim_{\delta x \to 0} \left(\frac{\cos \delta x - 1}{\delta x} \right) = 0$$

Similarly,
$$\frac{\sin \delta x}{\delta x} = 1 - \frac{(\delta x)^2}{3!} + \frac{(\delta x)^4}{5!} - \frac{(\delta x)^6}{7!} + \cdots$$

and
$$\lim_{\delta x \to 0} \left(\frac{\sin \delta x}{\delta x} \right) = 1$$

Therefore,
$$\lim_{\delta x \to 0} \left(\frac{\delta y}{\delta x} \right) = \frac{dy}{dx} = \sin x \,(0) + \cos x \,(1) = \cos x$$

Hence, if

$$y = \sin x \qquad \frac{dy}{dx} = \cos x$$

The same method can be used to show that if $y = \cos x$, $dy/dx = -\sin x$.

If $y = \tan x = (\sin x)/(\cos x)$, then, using the rule for differentiating quotients, we have

$$\frac{dy}{dx} = \frac{\cos x \, (\cos x) - \sin x \, (-\sin x)}{(\cos x)^2}$$

$$= \frac{\cos^2 x + \sin^2 x}{\cos^2 x} = \frac{1}{\cos^2 x} = \sec^2 x$$

Similarly it can be shown that if

$$y = \text{cotan } x = \frac{1}{\tan x} \qquad \frac{dy}{dx} = -\text{cosec}^2 x = \frac{-1}{\sin^2 x}$$

$$y = \text{cosec } x = \frac{1}{\sin x} \qquad \frac{dy}{dx} = \frac{-\cos x}{\sin^2 x} = \frac{-\text{cosec } x}{\tan x}$$

$$y = \sec x = \frac{1}{\cos x} \qquad \frac{dy}{dx} = \frac{\sin x}{\cos^2 x} = \tan x \sec x$$

The above results can be generalised for the angle mx (where m is a constant) by use of the function of a function rule.

For example, if $y = \sin mx$, let $\theta = mx$ so that $d\theta/dx = m$.

Then $\qquad\qquad\qquad y = \sin \theta \qquad \dfrac{dy}{d\theta} = \cos \theta$

$$\therefore \qquad\qquad \frac{dy}{dx} = \frac{dy}{d\theta}\frac{d\theta}{dx} = m \cos mx$$

The derivatives of the trigonometric functions are listed in Table A.1.

A.6 Integration

The integrals of some trigonometric functions can be deduced from Table A.1. For example,

$$\int \cos x \, dx = \sin x + c$$

and $\qquad\qquad\qquad \int \sin x \, dx = -\cos x + c$

TABLE A.1

Function	Derivative	Function	Derivative
$\sin x$	$\cos x$	$\sin mx$	$m \cos mx$
$\cos x$	$-\sin x$	$\cos mx$	$-m \sin mx$
$\tan x$	$\sec^2 x$	$\tan mx$	$m \sec^2 mx$
$\cotan x$	$-\cosec^2 x$	$\cotan mx$	$-m \cosec^2 mx$
$\sec x$	$\tan x \sec x$	$\sec mx$	$m \tan mx \sec mx$
$\cosec x$	$\dfrac{-\cosec x}{\tan x}$	$\cosec mx$	$\dfrac{-m \cosec mx}{\tan mx}$

Now
$$\int \tan x \, dx = \int \frac{\sin x}{\cos x} dx = -\int \frac{-\sin x}{\cos x} dx$$

$$= -\log (\cos x) + c$$

The integrals of other trigonometric functions are more difficult but some can be obtained by making the substitutions

$$\sin x = \frac{2 \tan (x/2)}{1 + \tan^2 (x/2)} \qquad \cos x = \frac{1 - \tan^2 (x/2)}{1 + \tan^2 (x/2)}$$

or, writing $t = \tan (x/2)$,

$$\sin x = \frac{2t}{1 + t^2} \qquad \cos x = \frac{1 - t^2}{1 + t^2}$$

These can be shown to be true by using the formulae for compound angles:

$$\frac{2 \tan (x/2)}{1 + \tan^2 (x/2)} = \frac{2[\sin (x/2)]/[\cos (x/2)]}{1 + [\sin^2 (x/2)]/[\cos^2 (x/2)]}$$

$$= \frac{2 \sin (x/2)}{\cos (x/2)[\cos^2 (x/2) + \sin^2 (x/2)]/\cos^2 (x/2)}$$

$$= \frac{2 \sin (x/2) \cos (x/2)}{1} = \sin x, \text{ using (12) and (4)}$$

Similarly,

$$\frac{1 - \tan^2 (x/2)}{1 + \tan^2 (x/2)} = \frac{1 - [\sin^2 (x/2)]/[\cos^2 (x/2)]}{1 + [\sin^2 (x/2)]/[\cos^2 (x/2)]}$$

$$= \frac{\cos^2 (x/2) - \sin^2 (x/2)}{\cos^2 (x/2) + \sin^2 (x/2)} = \cos x$$

using (4) and (13). These can be used to integrate some functions which cannot be integrated by other methods.

For example $\qquad \int \dfrac{dx}{\cos x}$

Let $t = \tan (x/2)$

Then $\qquad \cos x = \dfrac{1 - t^2}{1 + t^2} \qquad \sin x = \dfrac{2t}{1 + t^2}$

$$dx = \frac{2 \, dt}{1 + t^2}$$

Now $\qquad \displaystyle\int \dfrac{dx}{\cos x} = \int \dfrac{(1 + t^2)}{(1 - t^2)} \dfrac{2 \, dt}{(1 + t^2)} = \int \dfrac{2dt}{1 - t^2}$

$$= \int \frac{1}{1 + t} \, dt + \int \frac{1}{1 - t} \, dt$$

$$= \log (1 + t) - \log (1 - t) + C$$

$$= \log \left(\frac{1 + t}{1 - t} \right) + C$$

$$= \log \left(\frac{1 + \tan (x/2)}{1 - \tan (x/2)} \right) + C$$

Some standard integrals are given in Table A.2 (the constants of integration are omitted).

A.7 Inverse functions

If $y = \sin x$ then $x = \sin^{-1} y$ is known as the *inverse sine function*, or, x is an angle whose sine is y. Similarly, $x = \cos^{-1} y$ and $x = \tan^{-1} y$ are the *inverse cosine* and *tangent functions*.

These are useful in integration when trigonometric substitutions can be made. For example,

TABLE A.2

Function	Integral	Function	Integral
sin x	$-\cos x$	sin ax	$-\dfrac{1}{a}\cos ax$
cos x	sin x	cos ax	$\dfrac{1}{a}\sin ax$
tan x	$-\log(\cos x)$	tan ax	$\dfrac{-1}{a}\log(\cos ax)$
cotan x	$\log(\sin x)$	cotan ax	$\dfrac{1}{a}\log(\sin ax)$

$$\int \frac{1}{\sqrt{(a^2 - x^2)}}\,dx$$

Substitute
$$x = a\sin\theta$$
$$dx = a\cos\theta\,d\theta$$

so that
$$\int \frac{1}{\sqrt{(a^2 - x^2)}}\,dx = \int \frac{a\cos\theta\,d\theta}{\sqrt{(a^2 - a^2\sin^2\theta)}} = \int \frac{a\cos\theta\,d\theta}{a\cos\theta}$$
$$= \int 1\,d\theta = \theta + C$$

but if $x = a\sin\theta$, $\theta = \sin^{-1} x/a$, hence
$$\int \frac{dx}{\sqrt{(a^2 - x^2)}} = \sin^{-1}(x/a) + C$$

In the same way,
$$\int \frac{dx}{(a^2 + x^2)} = \frac{1}{a}\tan^{-1}\left(\frac{x}{a}\right) + C$$

where the appropriate substitution is $x = a\tan\theta$.

Appendix B
Set Theory

B.1 Introduction

A *set* is the name given to a group or collection of distinct objects. Each member of a set is called an *element*. For example the set A might be defined as consisting of the three integers 1, 2 and 3, so that

$$A = \{1, 2, 3\},$$

where curly brackets or braces are used around the elements, while the set B might consist of four integers

$$B = \{1, 3, 5, 7\}.$$

To indicate that an element is a member of a set the special symbol $\in$, which is read as 'belongs to' or 'is an element of', is used. Here $1 \in A$ and $3 \in B$. When an element does not belong to a set the symbol $\notin$ is used so that $4 \notin A$ and $2 \notin B$.

The special set which has no elements is called the *null set* or referred to as an *empty set* and is represented by $\emptyset$.

In these examples the sets A and B have been defined by enumerating all of the elements. This is possible when the number of elements in the set is small but in other cases it is more convenient to include a mathematical description, as in

$$P = \{p \mid p > 0\}$$

where the set P is defined as containing those values of p for which p is greater than zero, and here p might be the price of a good. Here the convention of using capital letters for the set and lower case letters for the elements of the set is adopted. Another example is

$$Q = \{q \mid 3 < q < 30\}$$

where Q is those values of q which are greater than 3 and less than 30.

More generally, the name R is commonly given to the set of all real numbers,

$$R = \{ x \mid x \text{ is a real number}\}$$

and the statement 'x is a real number' can be written $x \in R$.

Another useful concept is the idea of the *universal set*, which is all the elements under discussion, and therefore varies with the context. For example, in discussing the properties of the set A the corresponding universal set, S, might be defined in one of the following ways

$$S = \{1, 2, 3, 4, 5, 6, 7, 8\}$$

or $\qquad S = \{-1, 0, 1, 2, 3, 9\}$

or $\qquad S = \{x \mid x \text{ is a real number}\}$

B.2 Combining sets

We now consider relationships between sets. First, two sets are equal only if they contain identical elements. Thus if

$$A = \{1, 2, 3\},$$

and $\qquad E = \{2, 3, 1\}$

then $E = A$, even though the order in which the elements are written differs. Conversely, if at least one element does not occur in both sets the sets are not equal. For example, if

$$B = \{1, 3, 5, 7\}$$

$A \neq B$ since $2 \in A$ and $2 \notin B$. In the special case where all the elements of one set, M say, are also elements in a larger set, N say, then M is said to be a *sub-set* of N, and this is written $M \subset N$ and is read 'M is a sub-set of N' or 'M is contained in N'. An example of a sub-set is where

$$M = \{\text{Monday, Tuesday}\}$$

$$N = \{x \mid x \text{ is a day of the week}\}$$

Another example is where

$$A = \{1, 2, 3\},$$

and $\qquad S = \{1, 2, 3, 4, 5, 6, 7, 8, 9\}$

so that A is a sub-set of S. Notice that every set is a sub-set of its corresponding universal set.

Next, we can combine together two sets to form their *union* (indicated by ∪), which includes those elements belonging to *either or both* sets. For example,

$$A \cup B = \{1, 2, 3\} \;\cup\; \{1, 3, 5, 7\}$$

$$= \{1, 2, 3, 5, 7\}.$$

More formally, $A \cup B = \{x \mid x \in A \text{ or } x \in B\}$

In the same way we can define the *intersection* (indicated by ∩) of two sets as being *only* those elements which occur in *both* sets. Thus,

$$A \cap B = \{1, 2, 3\} \cap \{1, 3, 5, 7\}$$

$$= \{1, 3\}$$

The formal definition is

$$A \cap B = \{x \mid x \in A \text{ and } x \in B\}$$

As another example consider

$$F = \{a, b, c, -1, 0, 6\}$$

and $\qquad\qquad G = \{c, d, 0, 1, 6\},$

then $\qquad\qquad F \cup G = \{a, b, c, d, -1, 0, 1, 6\}$

and $\qquad\qquad F \cap G = \{c, 0, 6\}.$

An interesting special case occurs when two sets have no elements in common, and so are said to be *disjoint*, and their union is the universal set. For example, consider

$$A = \{1, 2, 3\},$$

$$P = \{4, 6, 7, 9\}$$

and $\qquad\qquad S = \{1, 2, 3, 4, 6, 7, 9\}$

Since $\qquad\qquad S = A \cup P$

then P is called the *complement* of A, written $\widetilde{A}$. Thus the union of a set and its complement is the universal set.

B.3   Venn diagrams

The properties of combinations of sets can be illustrated by the use of 'Venn diagrams'. In these, the universal set is represented by a

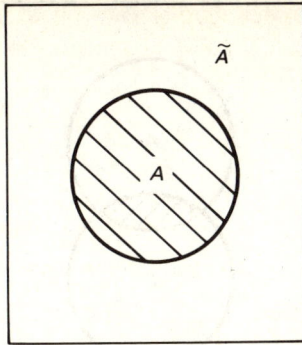

Fig. B.1

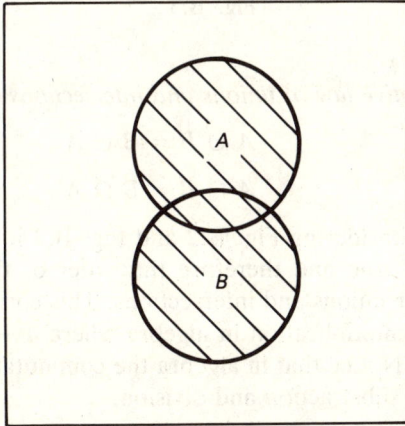

Fig. B.2

rectangle, and sub-sets of this by circles. Thus Fig. B.1 shows the set A, whose elements are within the circle and its complement, $\widetilde{A}$, whose elements are outside the circle but within the rectangle. The union of A and B is shown in Fig. B.2 as the shaded area, which includes all the elements in A or in B or in both. In Fig. B.3 the intersection of A and B is shown as the shaded area and here it includes only those elements in both A and B.

The use of Venn diagrams allows the following statements to be verified for any sets A, B and C.

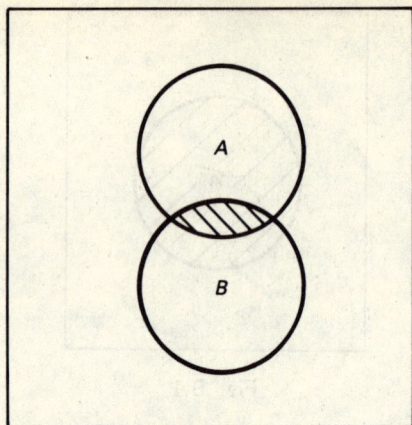

Fig. B.3

(a) *Commutative law of unions and intersections*:

$$A \cup B = B \cup A$$

$$A \cap B = B \cap A$$

By considering Fig. B.2 and Fig. B.3 it can be seen that these are true and therefore the order of the sets does not matter for unions and intersections. This corresponds to addition and multiplication in algebra where $a + b = b + a$ and $ab = ba$. Notice that in algebra the commutative law does not apply for substraction and division.

(b) *Associative law of unions and intersections*:

$$A \cup (B \cup C) = (A \cup B) \cup C = A \cup B \cup C$$

$$A \cap (B \cap C) = (A \cap B) \cap C = A \cap B \cap C$$

These are illustrated in Fig. B.4 and Fig. B.5 and also correspond to addition and multiplication in algebra.

(c) *Distributive law of unions and intersections*:

$$A \cup (B \cap C) = (A \cup B) \cap (A \cup C)$$

$$A \cap (B \cup C) = (A \cap B) \cup (A \cap C)$$

These are shown in Fig. B.6 and Fig. B.7.

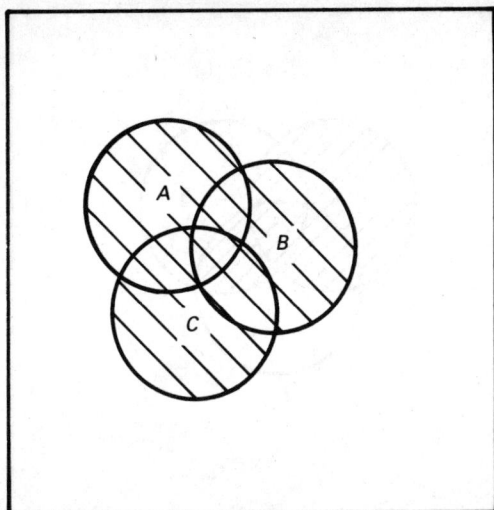

Fig. B.4

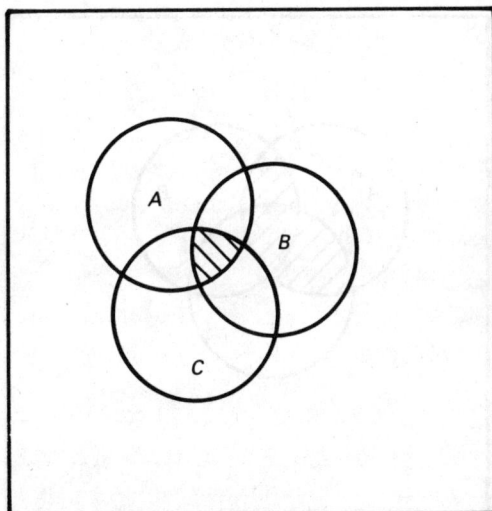

Fig. B.5

The laws of unions and intersections of sets extend to the special sets mentioned earlier, namely the universal set, the null set and the complement of any set.

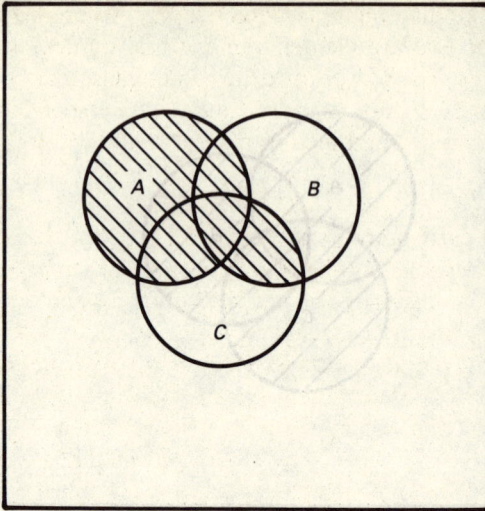

Fig. B.6

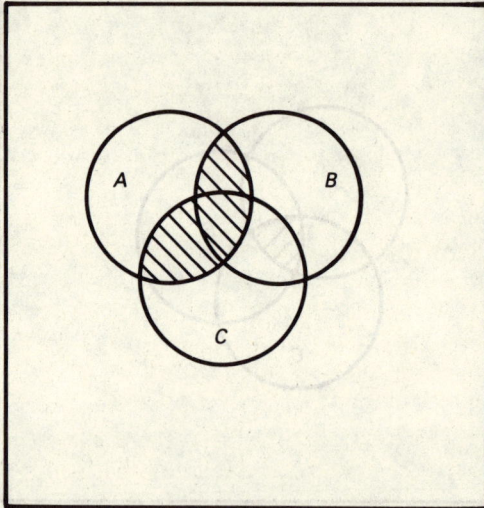

Fig. B.7

B.4 Relations and functions

The concept of a set is essentially one-dimensional since only one variable is considered at a time. We now extend the ideas to combinations of sets. For example, suppose we observe the price, p, of a product and the corresponding quantity demanded, q, where both p and q are real numbers and so are elements of the real number set R. That is, $p \in R$ and $q \in R$. The sets of prices, P, and quantities, Q, are of course related since for each particular price there is an associated quantity and these two sets can be combined into an *ordered set*, say,

$$M = \{(p, q) \mid p \in R, q \in R\}$$

Here the outer curly brackets or braces define the ordered set M, while the round brackets contain the elements of M, which are the sets P and Q. Therefore M consists of *ordered pairs* of prices and quantities and so while two sets, $A = \{1, 2, 3\}$ and $E = \{2, 3, 1\}$ are identical since they contain the same elements, the two ordered pairs from the ordered set M, $(1, 5)$ and $(5, 1)$ are different because the first is $p = 1$ and $q = 5$, while the second is $p = 5$ and $q = 1$.

Any ordered pair of values, say (p, q) or more generally (x, y) define what is called a *relation* between the variables. That is, a given value of x is associated with one or more values of y. For example, the relation defined by the ordered set $\{(x, y) \mid y = x + 2\}$ includes the ordered pairs $(0,2)$, $(-1,1)$ and $(100,102)$. Graphically, this relation can be represented by a straight line. Another example is the relation defined by the ordered set $\{(x, y) \mid y < 2x\}$ which includes the ordered pairs $(3,0)$, $(3,1)$ and $(1,0)$. In this example it is clear that the relation between y and x is not unique since for a given value of x many different values of y can occur. The graphical representation consists of the area below the straight line $y = 2x$.

When there is just one value of y corresponding to each value of x the relation defines y as a function of x or $y = f(x)$. That is, a *function* relating y and x is a set of ordered pairs with the property that for each value of x there is a unique value of y. Alternatively, the way in which y and x are related is said to be single valued. Notice that it is possible for a function to give the same y value for different values of x and an obvious example is $y = x^2$ where $x = -1$ and $x = 1$ both give $y = 1$. In the converse case, for example if $y^2 = x$, for each value of x there are two values of y and here we have a relation that is not a function.

The rule defined by a function, such as $y = x^2$, can also be interpreted as a *mapping* or transformation, since the value of x is mapped or transformed into the value of y. The notation

$$f: x -> y$$

indicates the mapping from x to y with f representing the particular rule, $y = x^2$ here. Also, y is called the *value* of the function and x the *argument*. Returning to the terminology of section 1.1, the variable x is the *independent* variable and y is the *dependent* variable. Two other terms are used in the literature: the *domain* of a function is the set of all permissible x values while the *range* is the set of all the values resulting from the mapping and so is the set of y values. For example suppose

$$y = 5x - 5$$

and x lies between 2 and 20. Then the domain is the set $\{X \mid 2 < X < 20\}$ and the range is the set $\{Y \mid 5 < Y < 95\}$.

Finally, it is sometimes convenient to find the *inverse* relationship for a particular relation. This can generally be found but may not be a function. For example, if

$$y = 5x - 5$$

the inverse relationship, which is a function, is

$$x = 0.2y + 1,$$

while if

$$y = x^2$$

the inverse relationship is

$$x = \sqrt{y}$$

which has two values of x for each y and so is not a function.

Appendix C
Answers to Exercises

Exercises 1.5

1. (a) $Y = 3x + 4$; intercept is 4, slope is 3

   (b) $Y = 3x - 4$; intercept is -4, slope is 3

   (c) $Y = 4 - 3x$; intercept is 4, slope is -3

   (d) $Y = 3x$;   intercept is 0, slope is 3

   (e) $Y = 4$;   intercept is 4, slope is 0.

2. Let $TC = a + bq$
   then from (a) $70 = a + 10b$
   and (b) $120 = a + 20b$

   Solving these equations simultaneously gives the values $a = 20$, $b = 5$

   $\therefore$ $TC = 20 + 5q$

   Fixed cost = 20, variable cost = 5.

3. $TC = 15 + 6q$
   Total cost is £75 for an output of 10 units.

4. (b) Fixed cost is 3, variable cost is 2   (c) 33   (d) 21.

Exercises 1.8

1. $Y = 33 + 2x$

   (a) At the breakeven point Y = total revenue = $13x$

   $\therefore$ $13x = 33 + 2x$

   $\therefore$ $x = 3$

   (b) Net revenue = $pX - Y = 13(15) - 33 - 2(15) = 132$

417

(d) Because the slope is 2, i.e. the variable cost is £2 per unit.

2. (a) $TC = 50 + 5q$

  (b) At breakeven point, $50 + 5q = 10q$ so that $q = 10$

  (c) Net revenue $= 12p - 110$

   (i) $p = 5$, net revenue $= -£50$

   (ii) $p = 10$, net revenue $= £10$

   (iii) $p = 15$, net revenue $= £70$.

3. (a) At equilibrium $q_s = q_d$

   $\therefore 25p - 10 = 200 - 5p$

   $\therefore p = 7$ and $q = 165$

  (b) $20p - 25 = 200 - 5p$

   $\therefore p = 9$ and $q = 155$.

4. (a) $p = 4$, $q = 3$

  (b) $p = \frac{1}{2}$, $q = 0$

  (c) There is no unique solution (only one line)

  (d) The equations are inconsistent (parallel lines).

Exercises   1.12

1. (a) In equilibrium, $q_d = q_s$ or $200 - 4p = -10 + 26p$. Rearranging gives $210 = 30p$ or $p = 7$, hence $q = 172$ and $pq = 1204$.

  (b) Here $p' = p - 5$ so that $q_s = -10 + 26(p - 5) = 26p - 140$. Equating to demand gives $26p - 140 = 200 - 4p$ so $30p = 340$ and $p = 11.3$, $q = 154.7$. Tax revenue $= 5q = 773.3$ and the producer's revenue is $p'q = 974.6$.

  (c) Here $p^r = 0.8p$ so that $q_s = -10 + 26(0.8p) = -10 + 20.8p$. Equating to demand gives $-10 + 20.8p = 200 - 4p$ so $24.8p = 210$ and $p = 8.47$, $q = 166.1$.
  Tax revenue $= 0.2(8.47)(166.1) = 281.4$
  Producer's revenue $= p^r q = 0.8(8.47)(166.1) = 1125.5$.

2. Let the flat-rate tax be t per unit so that $q_s = -40 + 15(p - t)$. Equating to demand, $-40 + 15(p - t) = 300 - 6p$ or $p = (340 +$

15t)/21 and if $t = 1.7$, $p = 17.4$.

For a 10% tax, $q_s = -40 + 15(0.9)p = -40 + 13.5p$

In equilibrium, $300 - 6p = -40 + 13.5p$ or $p = 340/19.5 = 17.4$

3. (a) $Y = 20 + 0.7Y + 4 = 24 + 0.7Y$ or $0.3Y = 24$ so $Y = 80$ and
   $C = 20 + 0.7(80) = 76$

   (b) $Y = 20 + 0.7(Y - 10) + 4 = 17 + 0.7Y$ or $0.3Y = 17$ so
   $Y = 56.7$ and $C = 20 + 0.7(56.7 - 10) = 52.7$

   (c) $Y = 20 + 0.7(0.75)Y + 4 = 24 + 0.525Y$ or $0.475Y = 24$ so
   $Y = 50.5$ and $C = 20 + 0.7(0.75)50.5 = 46.5$

   (d) $Y = 20 + 0.7(0.75Y - 10) + 4 = 17 + 0.525Y$ or $0.475Y = 17$
   so $Y = 35.8$ and $C = 20 + 0.7[(0.75)(35.8) - 10] = 31.8$.

4. (a) IS: $Y = C + I = 15 + 0.8Y + 75 - 100i$ or $Y = 450 - 500i$
   LM: $250 = 250 - 160i + 0.1Y$ or $Y = 1600i$
   Solution is $i = 450/2100 = 0.214$ oe 21.4% and $Y = 342.9$

   (b) IS: $Y = 65 + 0.8Y - 100i$ or $Y = 325 - 500i$
   LM: $Y = 1600i$. Solution is $i = 325/2100 = 0.155$ and $Y = 248$.

   (c) IS: $Y = 450 - 500i$
   LM: $275 = 250 - 160i + 0.1Y$ or $Y = 250 + 1600i$.
   Solution is $i = 200/2100 = 0.095$ and $Y = 402$.

5. (a) IS: $Y = C + I + G = 15 + 0.8Y_d + 75 - 100i + 300$
   and $Y_d = 0.8Y - 5$ so $Y = 386 + 0.64Y - 100i$ and this
   simplifies to $Y = 1072.2 - 277.8i$.

   (b) LM: $950 = 0.2Y + 750 - 260i$ or $Y = 1000 + 1300i$
   Equilibrium requires $1072.2 - 277.8i = 1000 + 1300i$ or
   $i = 72.2/1577.8 = 0.046$ and $Y = 1059.8$.

   (c) IS: $Y = 440 + 0.64Y - 4 - 100i$ or $Y = 1211.1 - 277.8i$
   LM: $Y = 1000 + 1300i$
   Equilibrium has $i = 211.1/1577.8 = 0.134$ and $Y = 1173.9$.
   Here both Y and i are higher than previously.

Exercises   1.14

1. $2x_1 + 2x_2 = 2$

   $3x_1 - x_2 = 1$

$$\therefore \quad x_1 = \frac{\begin{vmatrix} 2 & 2 \\ 1 & -1 \end{vmatrix}}{\begin{vmatrix} 2 & 2 \\ 3 & -1 \end{vmatrix}} = \frac{(-2-2)}{(-2-6)} = \frac{-4}{-8} = \frac{1}{2}$$

and

$$x_2 = \frac{\begin{vmatrix} 2 & 2 \\ 3 & 1 \end{vmatrix}}{\begin{vmatrix} 2 & 2 \\ 3 & -1 \end{vmatrix}} = \frac{2-6}{-8} = \frac{-4}{-8} = \frac{1}{2}$$

2. $x_1 = 1$, $x_2 = 1$.

3. $x_1 - x_2 + x_3 = 4$

$\quad x_1 + x_2 + 3x_3 = 8$

$\quad x_1 + 2x_2 - x_3 = 0$

$$\therefore \quad x_1 = \frac{\begin{vmatrix} 4 & -1 & 1 \\ 8 & 1 & 3 \\ 0 & 2 & -1 \end{vmatrix}}{\begin{vmatrix} 1 & -1 & 1 \\ 1 & 1 & 3 \\ 1 & 2 & -1 \end{vmatrix}} = \frac{4 \begin{vmatrix} 1 & 3 \\ 2 & -1 \end{vmatrix} - (-1) \begin{vmatrix} 8 & 3 \\ 0 & -1 \end{vmatrix} + 1 \begin{vmatrix} 8 & 1 \\ 0 & 2 \end{vmatrix}}{1 \begin{vmatrix} 1 & 3 \\ 2 & -1 \end{vmatrix} - (-1) \begin{vmatrix} 1 & 3 \\ 1 & -1 \end{vmatrix} + 1 \begin{vmatrix} 1 & 1 \\ 1 & 2 \end{vmatrix}}$$

$$= \frac{4(-7) + (-8) + (16)}{(-7) + (-4) + (1)} = \frac{-20}{-10} = 2$$

$$x_2 = \frac{\begin{vmatrix} 1 & 4 & 1 \\ 1 & 8 & 3 \\ 1 & 0 & -1 \end{vmatrix}}{\begin{vmatrix} 1 & -1 & 1 \\ 1 & 1 & 3 \\ 1 & 2 & -1 \end{vmatrix}} = \frac{-8 + 16 - 8}{-10} = 0$$

$$x_3 = \cfrac{\begin{vmatrix} 1 & -1 & 4 \\ 1 & 1 & 8 \\ 1 & 2 & 0 \end{vmatrix}}{\begin{vmatrix} 1 & -1 & 1 \\ 1 & 1 & 3 \\ 1 & 2 & -1 \end{vmatrix}} = \frac{-16 - 8 + 4}{-10} = \frac{-20}{-10} = 2$$

4. $x_1 = 1, \quad x_2 = 1, \quad x_3 = -1.$

5. $x_1 = 2, \quad x_2 = 2, \quad x_3 = 2.$

6. $x_1 = -1, \quad x_2 = -2, \quad x_3 = -3.$

Exercises 1.16

1. (a) $Y = C + I + G = C + I + 20$

$$C = 20 + 0.7(Y - T) = 20 + 0.7(Y - 5 - 0.3Y)$$

$$= 16.5 + 0.49Y$$

$$I = 15 + 0.1Y$$

Arranging in the order $Y, C \ I$:

$$Y - C - I = 20$$

$$-0.49Y + C = 16.5$$

$$-0.1 \ Y + I = 15$$

Using the notation of the text,

$$|A| = \begin{vmatrix} 1 & -1 & -1 \\ -0.49 & 1 & 0 \\ -0.1 & 0 & 1 \end{vmatrix} = \begin{vmatrix} 0.9 & -1 & 0 \\ -0.49 & 1 & 0 \\ -0.1 & 0 & 1 \end{vmatrix}$$

$$= 0.9 - 0.49 = 0.41$$

$$|A_1| = \begin{vmatrix} 20 & -1 & -1 \\ 16.5 & 1 & 0 \\ 15 & 0 & 1 \end{vmatrix} = \begin{vmatrix} 35 & -1 & 0 \\ 16.5 & 1 & 0 \\ 15 & 0 & 1 \end{vmatrix}$$

$$= 35 + 16.5 = 51.5$$

$$|\mathbf{A}_2| = \begin{vmatrix} 1 & 20 & -1 \\ -0.49 & 16.5 & 0 \\ -0.1 & 15 & 1 \end{vmatrix} = \begin{vmatrix} 0.9 & 35 & 0 \\ -0.49 & 16.5 & 0 \\ -0.1 & 15 & 1 \end{vmatrix}$$

$$= 14.85 + 17.15 = 32$$

$$|\mathbf{A}_3| = \begin{vmatrix} 1 & -1 & 20 \\ -0.49 & 1 & 16.5 \\ -0.1 & 0 & 15 \end{vmatrix} = \begin{vmatrix} 0.51 & 0 & 36.5 \\ -0.49 & 1 & 16.5 \\ -0.1 & 0 & 15 \end{vmatrix}$$

$$= 7.65 + 3.65 = 11.3$$

Hence, $Y = \dfrac{51.5}{0.41} = 125.6$, $C = \dfrac{32}{0.41} = 78.0$ and $I = \dfrac{11.3}{0.41} = 27.6$

At this equilibrium, $G = 20$ and $T = 5 + 0.3(125.6) = 42.7$ and so there is a surplus of 22.7.

   1. (b) If $G = T = 5 + 0.3Y$ the national income identity becomes
$Y = C + I + 5 + 0.3Y$ or $0.7Y - C - I = 5$. The other two
equations are unchanged: $-0.49Y + C = 16.5$
$$-0.1\ Y + I = 15$$

$$\text{Here, } |\mathbf{A}| = \begin{vmatrix} 0.7 & -1 & -1 \\ -0.49 & 1 & 0 \\ -0.1 & 0 & 1 \end{vmatrix} = \begin{vmatrix} 0.6 & -1 & 0 \\ -0.49 & 1 & 0 \\ -0.1 & 0 & 1 \end{vmatrix}$$

$$= 0.6 - 0.49 = 0.11$$

$$|\mathbf{A}_1| = \begin{vmatrix} 5 & -1 & -1 \\ 16.5 & 1 & 0 \\ 15 & 0 & 1 \end{vmatrix} = \begin{vmatrix} 20 & -1 & 0 \\ 16.5 & 1 & 0 \\ 15 & 0 & 1 \end{vmatrix}$$

$$= 20 + 16.5 = 36.5$$

$$|\mathbf{A}_2| = \begin{vmatrix} 0.7 & 5 & -1 \\ -0.49 & 16.5 & 0 \\ -0.1 & 15 & 1 \end{vmatrix} = \begin{vmatrix} 0.6 & 20 & 0 \\ -0.49 & 16.5 & 0 \\ -0.1 & 15 & 1 \end{vmatrix}$$

$$= 9.9 + 9.8 = 19.7$$

$$|A_3| = \begin{vmatrix} 0.7 & -1 & 5 \\ -0.49 & 1 & 16.5 \\ -0.1 & 0 & 15 \end{vmatrix} = \begin{vmatrix} 0.21 & 0 & 21.5 \\ -0.49 & 1 & 16.5 \\ -0.1 & 0 & 15 \end{vmatrix}$$

$$= 3.15 + 2.15 = 5.3$$

Hence, $Y = \dfrac{36.5}{0.11} = 331.8$, $C = \dfrac{19.7}{0.11} = 179.1$, $I = \dfrac{5.3}{0.11} = 48.2$

and $T = 5 + 0.3(331.8) = 104.6 = G$ – higher than in (a) above, as are Y, C and I.

2. $Y = C + I + G + (X - M) = C + I_0 + fM + X - M$

   or $Y - C - (f - 1)M = I_0 + X$

   $-bY + C = a$

   $-dY + M = c$

Using the notation from the text,

$$|A| = \begin{vmatrix} 1 & -1 & -f+1 \\ -b & 1 & 0 \\ -d & 0 & 1 \end{vmatrix} = \begin{vmatrix} 1-b & 0 & -f+1 \\ -b & 1 & 0 \\ -d & 0 & 1 \end{vmatrix}$$

$$= 1 - b - df + d$$

$$|A_1| = \begin{vmatrix} I_0 + X & -1 & -f+1 \\ a & 1 & 0 \\ c & 0 & 1 \end{vmatrix} = \begin{vmatrix} I_0 + X + a & 0 & -f+1 \\ a & 1 & 0 \\ -c & 0 & 1 \end{vmatrix}$$

$$= I_0 + X + a + cf - c$$

$$|A_2| = \begin{vmatrix} 1 & I_0 + X & -f+1 \\ -b & a & 0 \\ -d & c & 1 \end{vmatrix}$$

$$= (-f + 1)(ad - bc) + (a + bI_0 + bX)$$

$$|A_3| = \begin{vmatrix} 1 & -1 & I_0 + X \\ -b & 1 & a \\ -d & 0 & c \end{vmatrix} = \begin{vmatrix} 1-b & 0 & a + I_0 + X \\ -b & 1 & a \\ -d & 0 & c \end{vmatrix}$$

$$= (1 - b)c + d(a + I_0 + X)$$

Therefore, $Y = \dfrac{I_0 + X + a + cf - c}{1 - b - df + d}$,

$$C = \dfrac{(-f + 1)(ad - bc) + (a + bI_0 + bX)}{1 - b - df + d}$$

and $M = \dfrac{(1 - b)c + d(a + I_0 + X)}{1 - b - df + d}$

3. For A, $3 - p_A + p_B = p_A - 2$ or $-2p_A + p_B$ $= -5$

For B, $8 - 2p_B + p_C = p_B - 1$ or $-3p_B + p_C$ $= -9$

For C, $6 + 2p_A - p_C = 2p_C - 2$ or $2p_A - 3p_C$ $= -8$

Using the notation from the text,

$$|\mathbf{A}| = \begin{vmatrix} -2 & 1 & 0 \\ 0 & -3 & 1 \\ 2 & 0 & -3 \end{vmatrix} = \begin{vmatrix} 0 & 1 & 0 \\ -6 & -3 & 1 \\ 2 & 0 & -3 \end{vmatrix}$$

$$= -(18 - 2) = -16$$

$$|\mathbf{A}_1| = \begin{vmatrix} -5 & 1 & 0 \\ -9 & -3 & 1 \\ -8 & 0 & -3 \end{vmatrix} = \begin{vmatrix} 0 & 1 & 0 \\ -24 & -3 & 1 \\ -8 & 0 & -3 \end{vmatrix}$$

$$= -(72 + 8) = -80$$

$$|\mathbf{A}_2| = \begin{vmatrix} -2 & -5 & 0 \\ 0 & -9 & 1 \\ 2 & -8 & -3 \end{vmatrix} = \begin{vmatrix} 0 & -13 & -3 \\ 0 & -9 & 1 \\ 2 & -8 & -3 \end{vmatrix}$$

$$= 2(-13 - 27) = -80$$

$$|\mathbf{A}_3| = \begin{vmatrix} -2 & 1 & -5 \\ 0 & -3 & -9 \\ 2 & 0 & -8 \end{vmatrix} = \begin{vmatrix} 0 & 1 & -13 \\ 0 & -3 & -9 \\ 2 & 0 & -8 \end{vmatrix}$$

$$= 2(-9 - 39) = -96$$

Hence, $p_A = \dfrac{-80}{-16} = 5$, $p_B = \dfrac{-80}{-16} = 5$, $p_C = \dfrac{-96}{-16} = 6$

$$q^A = 3, \ q^B = 4, \ q^C = 10$$

4. Equating supply and demand in each market,

$$44 - 2p_1 + p_2 + \ p_3 = -10 + 3p_1 \text{ or } 54 = 5p_1 - p_2 - p_3$$

$$25 + \ p_1 - p_2 + 3p_3 = -15 + 5p_2 \text{ or } 40 = -p_1 + 6p_2 - 3p_3$$

$$40 + 2p_1 + p_2 - 3p_3 = -18 + 2p_3 \text{ or } 58 = -2p_1 - p_2 + 5p_3$$

Using the notation from the text,

$$|\mathbf{A}| = \begin{vmatrix} 5 & -1 & -1 \\ -1 & 6 & -3 \\ -2 & -1 & 5 \end{vmatrix} = \begin{vmatrix} 0 & -1 & 0 \\ 29 & 6 & -9 \\ -7 & -1 & 6 \end{vmatrix}$$

$$= (29 \times 6) - (7 \times 9) = 111$$

$$|\mathbf{A_1}| = \begin{vmatrix} 54 & -1 & -1 \\ 40 & 6 & -3 \\ 58 & -1 & 5 \end{vmatrix} = \begin{vmatrix} -4 & 0 & -6 \\ 40 & 6 & -3 \\ 58 & -1 & 5 \end{vmatrix}$$

$$= -4(30 - 3) - 6(-40 - 348) = 2220$$

$$|\mathbf{A_2}| = \begin{vmatrix} 5 & 54 & -1 \\ -1 & 40 & -3 \\ -2 & 58 & 5 \end{vmatrix} = \begin{vmatrix} 0 & 254 & -16 \\ -1 & 40 & -3 \\ 0 & -22 & 11 \end{vmatrix}$$

$$= 2794 - 352 = 2442$$

$$|\mathbf{A_3}| = \begin{vmatrix} 5 & -1 & 54 \\ -1 & 6 & 40 \\ -2 & -1 & 58 \end{vmatrix} = \begin{vmatrix} 7 & 0 & -4 \\ -13 & 0 & 388 \\ -2 & -1 & 58 \end{vmatrix}$$

$$= 2716 - 52 = 2664$$

Therefore, $p_1 = \dfrac{2220}{111} = 20$, $p_2 = \dfrac{2442}{111} = 22$, $p_3 = \dfrac{2664}{111} = 24$

$$q_1 = 50, q_2 = 95, q_3 = 30.$$

Exercises 2.3

1.
$$A = \begin{bmatrix} 2 & 4 \\ 1 & 3 \end{bmatrix} \qquad B = \begin{bmatrix} 2 & 0 \\ -1 & 1 \end{bmatrix} \qquad C = \begin{bmatrix} 3 \\ 1 \end{bmatrix}$$

(a) $A + B = \begin{bmatrix} 2+2 & 4+0 \\ 1+(-1) & 3+1 \end{bmatrix} = \begin{bmatrix} 4 & 4 \\ 0 & 4 \end{bmatrix}$

(b) $A - 2B = \begin{bmatrix} 2-(2\times 2) & 4-(2\times 0) \\ 1-(2\times -1) & 3-(2\times 1) \end{bmatrix} = \begin{bmatrix} -2 & 4 \\ 3 & 1 \end{bmatrix}$

(c) $AB = \begin{bmatrix} 2 & 4 \\ 1 & 3 \end{bmatrix}\begin{bmatrix} 2 & 0 \\ -1 & 1 \end{bmatrix} = \begin{bmatrix} (2\times 2)+(4\times -1) & (2\times 0)+(4\times 1) \\ (1\times 2)+(3\times -1) & (1\times 0)+(3\times 1) \end{bmatrix}$

$$= \begin{bmatrix} 4-4 & 0+4 \\ 2-3 & 0+3 \end{bmatrix} = \begin{bmatrix} 0 & 4 \\ -1 & 3 \end{bmatrix}$$

(d) $AC = \begin{bmatrix} 2 & 4 \\ 1 & 3 \end{bmatrix}\begin{bmatrix} 3 \\ 1 \end{bmatrix} = \begin{bmatrix} (2\times 3)+(4\times 1) \\ (1\times 3)+(3\times 1) \end{bmatrix} = \begin{bmatrix} 10 \\ 6 \end{bmatrix}$

$(2\times 2)\ (2\times 1)$ $\qquad\qquad\qquad\qquad\qquad (2\times 1)$

(e) $CA = \begin{bmatrix} 3 \\ 1 \end{bmatrix}\begin{bmatrix} 2 & 4 \\ 1 & 3 \end{bmatrix}$ It is not possible to form the product matrix **CA**

$(2\times 1)\ (2\times 2)$

(f) $\mathbf{B} = \begin{bmatrix} 2 & 0 \\ -1 & 1 \end{bmatrix}$ $\mathbf{B}' = \begin{bmatrix} 2 & -1 \\ 0 & 1 \end{bmatrix}$

$\mathbf{AB}' = \begin{bmatrix} 2 & 4 \\ 1 & 3 \end{bmatrix} \begin{bmatrix} 2 & -1 \\ 0 & 1 \end{bmatrix}$

$= \begin{bmatrix} (2 \times 2) + (4 \times 0) & (2 \times -1) + (4 \times 1) \\ (1 \times 2) + (3 \times 0) & (1 \times -1) + (3 \times 1) \end{bmatrix}$

$= \begin{bmatrix} 4 & 2 \\ 2 & 2 \end{bmatrix}$

(g) $\mathbf{C} = \begin{bmatrix} 3 \\ 1 \end{bmatrix}$ $\mathbf{C}' = [3 \quad 1]$

$\mathbf{C}'\mathbf{A} = [3 \quad 1] \begin{bmatrix} 2 & 4 \\ 1 & 3 \end{bmatrix}$

$(1 \times 2) \ (2 \times 2)$

$= [(3 \times 2) + (1 \times 1) \quad (3 \times 4) + (1 \times 3)] = [7 \quad 15]$
(1×2)

(h) $\mathbf{BA} = \begin{bmatrix} 2 & 0 \\ -1 & 1 \end{bmatrix} \begin{bmatrix} 2 & 4 \\ 1 & 3 \end{bmatrix}$

$= \begin{bmatrix} (2 \times 2) + (0 \times 1) & (2 \times 4) + (0 \times 3) \\ (-1 \times 2) + (1 \times 1) & (-1 \times 4) + (1 \times 3) \end{bmatrix} = \begin{bmatrix} 4 & 8 \\ -1 & -1 \end{bmatrix}$

This is not equal to **AB** (obtained earlier in (c)).

2.
$$\mathbf{A} = \begin{bmatrix} 1 & 0 & 2 \\ 1 & 1 & 3 \\ 0 & 2 & -1 \end{bmatrix} \qquad \mathbf{B} = \begin{bmatrix} 2 & 1 \\ 1 & -1 \\ 2 & 2 \end{bmatrix} \qquad \mathbf{C} = [1 \quad 0 \quad 2]$$

(a) $\mathbf{AB} = \begin{bmatrix} 1 & 0 & 2 \\ 1 & 1 & 3 \\ 0 & 2 & -1 \end{bmatrix} \begin{bmatrix} 2 & 1 \\ 1 & -1 \\ 2 & 2 \end{bmatrix}$

$$= \begin{bmatrix} (1\times2) + (0\times1) + (2\times2) & (1\times1) + (0\times-1) + (2\times2) \\ (1\times2) + (1\times1) + (3\times2) & (1\times1) + (1\times-1) + (3\times2) \\ (0\times2) + (2\times1) + (-1\times2) & (0\times1) + (2\times-1) + (-1\times2) \end{bmatrix}$$

$$= \begin{bmatrix} 6 & 5 \\ 9 & 6 \\ 0 & -4 \end{bmatrix}$$

(b) $\mathbf{A'B} = \begin{bmatrix} 1 & 1 & 0 \\ 0 & 1 & 2 \\ 2 & 3 & -1 \end{bmatrix} \begin{bmatrix} 2 & 1 \\ 1 & -1 \\ 2 & 2 \end{bmatrix}$

             (3 × 3)   (3 × 2)

$$= \begin{bmatrix} (1\times2) + (1\times1) + (0\times2) & (1\times1) + (1\times-1) + (0\times2) \\ (0\times2) + (1\times1) + (2\times2) & (0\times1) + (1\times-1) + (2\times2) \\ (2\times2) + (3\times1) + (-1\times2) & (2\times1) + (3\times-1) + (-1\times2) \end{bmatrix}$$

$$= \begin{bmatrix} 3 & 0 \\ 5 & 3 \\ 5 & -3 \end{bmatrix}$$

          (3 × 2)

(c) $\mathbf{A}$ $\mathbf{C}$ does not exist

    (3 × 3) (1 × 3)

(d) $\mathbf{AC'} = \begin{bmatrix} 1 & 0 & 2 \\ 1 & 1 & 3 \\ 0 & 2 & -1 \end{bmatrix} \begin{bmatrix} 1 \\ 0 \\ 2 \end{bmatrix}$

$$= \begin{bmatrix} (1 \times 1) + (0 \times 0) + (2 \times 2) \\ (1 \times 1) + (1 \times 0) + (3 \times 2) \\ (0 \times 1) + (2 \times 0) + (-1 \times 2) \end{bmatrix} = \begin{bmatrix} 5 \\ 7 \\ -2 \end{bmatrix}$$

(e) $\mathbf{CB} = \begin{bmatrix} 1 & 0 & 2 \end{bmatrix} \begin{bmatrix} 2 & 1 \\ 1 & -1 \\ 2 & 2 \end{bmatrix}$

$\qquad (1 \times 3) \quad (3 \times 2)$

$= [(1 \times 2) + (0 \times 1) + (2 \times 2) \quad (1 \times 1) + (0 \times -1) + (2 \times 2)]$

$= [6 \quad 5]$

$\qquad (1 \times 2)$

Exercises 2.5

1. $\mathbf{A} = \begin{bmatrix} 2 & 1 \\ 1 & 3 \end{bmatrix}$

$\therefore \quad \mathbf{C} = \begin{bmatrix} 3 & -1 \\ -1 & 2 \end{bmatrix} \qquad \text{and} \qquad |\mathbf{A}| = 5$

$\therefore \quad \mathbf{A}^{-1} = \frac{1}{5} \begin{bmatrix} 3 & -1 \\ -1 & 2 \end{bmatrix} = \begin{bmatrix} \frac{3}{5} & -\frac{1}{5} \\ -\frac{1}{5} & \frac{2}{5} \end{bmatrix}$

The solution to the equations is given by

$\mathbf{X} = \mathbf{A}^{-1}\mathbf{b} = \begin{bmatrix} \frac{3}{5} & -\frac{1}{5} \\ -\frac{1}{5} & \frac{2}{5} \end{bmatrix} \begin{bmatrix} 4 \\ 7 \end{bmatrix}$

$= \begin{bmatrix} (\frac{3}{5} \times 4) + (-\frac{1}{5} \times 7) \\ (-\frac{1}{5} \times 4) + (\frac{2}{5} \times 7) \end{bmatrix} = \begin{bmatrix} \frac{12}{5} - \frac{7}{5} \\ -\frac{4}{5} + \frac{14}{5} \end{bmatrix} = \begin{bmatrix} 1 \\ 2 \end{bmatrix}$

$\therefore \quad x_1 = 1, \qquad x_2 = 2$

2. (a) $x_1 = 58/23, \qquad x_2 = -14/23$

   (b) $x_1 = 0, \qquad x_2 = -1, \qquad x_3 = 3$

   (c) $x_1 = 2, \qquad x_2 = 0, \qquad x_3 = 1$

Exercises 2.7

1. Including the unit matrix with **A** gives

$$\left[\begin{array}{cc|cc} 2 & -3 & 1 & 0 \\ 1 & 1 & 0 & 1 \end{array}\right]$$

Subtract row 2 from row 1 to give 1 in the (1,1) position,

$$\left[\begin{array}{cc|cc} 1 & -4 & 1 & -1 \\ 1 & 1 & 0 & 1 \end{array}\right]$$

Substract row 1 from row 2 to give 0 in the (2,1) position,

$$\left[\begin{array}{cc|cc} 1 & -4 & 1 & -1 \\ 0 & 5 & -1 & 2 \end{array}\right]$$

Divide row 2 by 5, to give a 1 in the (2,2) position, and add four times the new row 2, to give a 0 in the (1,2) position,

$$\left[\begin{array}{cc|cc} 1 & 0 & 0.2 & 0.6 \\ 0 & 1 & -0.2 & 0.4 \end{array}\right]$$

Hence, $\mathbf{A}^{-1} = \left[\begin{array}{cc} 0.2 & 0.6 \\ -0.2 & 0.4 \end{array}\right]$ and $\mathbf{A}^{-1}\mathbf{A} = \mathbf{I}$

For **B**, $\left[\begin{array}{cc|cc} 0 & 2 & 1 & 0 \\ -1 & -1 & 0 & 1 \end{array}\right]$

To get 1 in the (1,1) position, subtract row 2 from row 1,

$$\left[\begin{array}{cc|cc} 1 & 3 & 1 & -1 \\ -1 & -1 & 0 & 1 \end{array}\right]$$

Add row 1 to row 2

$$\left[\begin{array}{cc|cc} 1 & 3 & 1 & -1 \\ 0 & 2 & 1 & 0 \end{array}\right]$$

Subtract 1.5 times row 2 from row 1, divide row 2 by 2

$$\begin{bmatrix} 1 & 0 & | & -0.5 & -1 \\ 0 & 1 & | & 0.5 & 0 \end{bmatrix}$$

Hence, $\mathbf{B}^{-1} = \begin{bmatrix} -0.5 & -1 \\ 0.5 & 0 \end{bmatrix}$ and $\mathbf{B}^{-1}\mathbf{B} = \mathbf{I}$

For **C**, $\begin{bmatrix} 3 & -4 & | & 1 & 0 \\ 1 & 4 & | & 0 & 1 \end{bmatrix}$

Divide row 1 by 3 and subtract the new row 1 from row 2

$$\begin{bmatrix} 1 & -4/3 & | & 1/3 & 0 \\ 0 & 16/3 & | & -1/3 & 1 \end{bmatrix}$$

Add 0.25 times row 2 to row 1, and multiply row 2 by 3/16

$$\begin{bmatrix} 1 & 0 & | & 1/4 & 1/4 \\ 0 & 1 & | & -1/16 & 3/16 \end{bmatrix}$$

Hence, $\mathbf{C}^{-1} = \begin{bmatrix} 1/4 & 1/4 \\ -1/16 & 3/16 \end{bmatrix}$ and $\mathbf{C}^{-1}\mathbf{C} = \mathbf{I}$

For **D**, $\begin{bmatrix} 1 & -1 & 0 & | & 1 & 0 & 0 \\ -2 & 2 & 3 & | & 0 & 1 & 0 \\ 3 & 0 & 2 & | & 0 & 0 & 1 \end{bmatrix}$

Add twice row 1 to row 2, and subtract three times row 1 from row 3,

$$\begin{bmatrix} 1 & -1 & 0 & | & 1 & 0 & 0 \\ 0 & 0 & 3 & | & 2 & 1 & 0 \\ 0 & 3 & 2 & | & -3 & 0 & 1 \end{bmatrix}$$

Adding one-third of row three to row 1 and row 2,

$$\begin{bmatrix} 1 & 0 & 2/3 & | & 0 & 0 & 1/3 \\ 0 & 1 & 11/3 & | & 1 & 1 & 1/3 \\ 0 & 3 & 2 & | & -3 & 0 & 1 \end{bmatrix}$$

Subtract three times row 2 from row 3

$$\begin{bmatrix} 1 & 0 & 2/3 & 0 & 0 & 1/3 \\ 0 & 1 & 11/3 & 1 & 1 & 1/3 \\ 0 & 0 & -9 & -6 & -3 & 0 \end{bmatrix}$$

Divide row 3 by -9 and combine with row 1 and row 2

$$\begin{bmatrix} 1 & 0 & 0 & -4/9 & -2/9 & 1/3 \\ 0 & 1 & 0 & -13/9 & -2/9 & 1/3 \\ 0 & 0 & 1 & 2/3 & 1/3 & 0 \end{bmatrix}$$

Hence, $\mathbf{D}^{-1} = \begin{bmatrix} -4/9 & -2/9 & 1/3 \\ -13/9 & -2/9 & 1/3 \\ 2/3 & 1/3 & 0 \end{bmatrix}$ and $\mathbf{D}^{-1}\,\mathbf{D} = \mathbf{I}$

For, $\mathbf{E}$, $\begin{bmatrix} 2 & 0 & 1 & 1 & 0 & 0 \\ 2 & 3 & -2 & 0 & 1 & 0 \\ -1 & 1 & -1 & 0 & 0 & 1 \end{bmatrix}$

Divide row 1 by 2, subtract twice the new row 1 from row 2, and add the new row 1 to row 3,

$$\begin{bmatrix} 1 & 0 & 0.5 & 0.5 & 0 & 0 \\ 0 & 3 & -3 & -1 & 1 & 0 \\ 0 & 1 & -0.5 & 0.5 & 0 & 1 \end{bmatrix}$$

Divide row 2 by 3 and subtract the new row 2 from row 3,

$$\begin{bmatrix} 1 & 0 & 0.5 & 0.5 & 0 & 0 \\ 0 & 1 & -1 & -1/3 & 1/3 & 0 \\ 0 & 0 & 0.5 & 5/6 & -1/3 & 1 \end{bmatrix}$$

Subtract row 3 from row 1, add twice row 3 to row 2, and multiply row 3 by 2,

$$\begin{bmatrix} 1 & 0 & 0 & -1/3 & 1/3 & -1 \\ 0 & 1 & 0 & 4/3 & -1/3 & 2 \\ 0 & 0 & 1 & 5/3 & -2/3 & 2 \end{bmatrix}$$

Hence, $\mathbf{E}^{-1} = \begin{bmatrix} -1/3 & 1/3 & -1 \\ 4/3 & -1/3 & 2 \\ 5/3 & -2/3 & 2 \end{bmatrix}$ and $\mathbf{E}^{-1}\,\mathbf{E} = \mathbf{I}$

For **F**,
$$\left[\begin{array}{ccc|ccc} 3 & -2 & 3 & 1 & 0 & 0 \\ -1 & 0 & 3 & 0 & 1 & 0 \\ 2 & 1 & -1 & 0 & 0 & 1 \end{array}\right]$$

Divide row 1 by 3, add the new row 1 to row 2 and subtract twice the new row 1 from row 3,

$$\left[\begin{array}{ccc|ccc} 1 & -2/3 & 1 & 1/3 & 0 & 0 \\ 0 & -2/3 & 4 & 1/3 & 1 & 0 \\ 0 & 7/3 & -3 & -2/3 & 0 & 1 \end{array}\right]$$

Subtract row 2 from row 1, multiply row 2 by $-3/2$, and subtract 7/3 times the new row 2 from row 3

$$\left[\begin{array}{ccc|ccc} 1 & 0 & -3 & 0 & -1 & 0 \\ 0 & 1 & -6 & -0.5 & -1.5 & 0 \\ 0 & 0 & 11 & 0.5 & 3.5 & 1 \end{array}\right]$$

Divide row 3 by 11, add three times the new row 3 to row 1, and add six times the new row 3 to row 2

$$\left[\begin{array}{ccc|ccc} 1 & 0 & 0 & 3/22 & -1/22 & 3/11 \\ 0 & 1 & 0 & -5/22 & 9/22 & 6/11 \\ 0 & 0 & 1 & 1/22 & 7/22 & 1/11 \end{array}\right]$$

Hence, $\mathbf{F}^{-1} = \left[\begin{array}{ccc} 3/22 & -1/22 & 3/11 \\ -5/22 & 9/22 & 6/11 \\ 1/22 & 7/22 & 1/11 \end{array}\right]$ and $\mathbf{F}^{-1}\,\mathbf{F} = \mathbf{I}$

For **G**,
$$\left[\begin{array}{ccc|ccc} -1 & 0 & 2 & 1 & 0 & 0 \\ 1 & 2 & 1 & 0 & 1 & 0 \\ 2 & 1 & 3 & 0 & 0 & 1 \end{array}\right]$$

Add row 1 to row 2, add twice row 1 to row 3, and multiply row 1 by -1,

$$\left[\begin{array}{ccc|ccc} 1 & 0 & -2 & -1 & 0 & 0 \\ 0 & 2 & 3 & 1 & 1 & 0 \\ 0 & 1 & 7 & 2 & 0 & 1 \end{array}\right]$$

Divide row 2 by 2 and subtract the new row 2 from row 3

$$\left[\begin{array}{ccc|ccc} 1 & 0 & -2 & -1 & 0 & 0 \\ 0 & 1 & 1.5 & 0.5 & 0.5 & 0 \\ 0 & 0 & 5.5 & 1.5 & -0.5 & 1 \end{array}\right]$$

Divide row 3 by 5.5, add twice the new row 3 to row 1 and subtract 1.5 times the new row 3 from row 2

$$\left[\begin{array}{ccc|ccc} 1 & 0 & 0 & -5/11 & -2/11 & 4/11 \\ 0 & 1 & 0 & 1/11 & 7/11 & -3/11 \\ 0 & 0 & 1 & 3/11 & -1/11 & 2/11 \end{array}\right]$$

Hence $\mathbf{G}^{-1} = \begin{bmatrix} -5/11 & -2/11 & 4/11 \\ 1/11 & 7/11 & -3/11 \\ 3/11 & -1/11 & 2/11 \end{bmatrix}$ and $\mathbf{G}^{-1}\,\mathbf{G} = \mathbf{I}$

2. As explained in the text it is not necessary to calculate the inverse matrix in order to get the solution but it is included in these answers for completeness.

(a)
$$\left[\begin{array}{cc|cc|c} 4 & -2 & 1 & 0 & 6 \\ 2 & 1 & 0 & 1 & 5 \end{array}\right]$$

Divide row 1 by 4 and subtract twice the new row 1 from row 2

$$\left[\begin{array}{cc|cc|c} 1 & -0.5 & 0.25 & 0 & 1.5 \\ 0 & 2 & -0.5 & 1 & 2 \end{array}\right]$$

Divide row 2 by 2 and add 0.5 times the new row 2 to row 1

$$\left[\begin{array}{cc|cc|c} 1 & 0 & 0.125 & 0.25 & 2 \\ 0 & 1 & -0.25 & 0.5 & 1 \end{array}\right]$$ and so $x = 2$, $y = 1$

(b)
$$\left[\begin{array}{ccc|ccc|c} 3 & -1 & 2 & 1 & 0 & 0 & 7 \\ 1 & 3 & 1 & 0 & 1 & 0 & 3 \\ 2 & 1 & 1 & 0 & 0 & 1 & 4 \end{array}\right]$$

Divide row 1 by 3, subtract the new row 1 from row 2, and subtract twice the new row 1 from row 3

$$\left[\begin{array}{ccc|ccc|c} 1 & -1/3 & 2/3 & 1/3 & 0 & 0 & 7/3 \\ 0 & 10/3 & 1/3 & -1/3 & 1 & 0 & 2/3 \\ 0 & 5/3 & -1/3 & -2/3 & 0 & 1 & -2/3 \end{array}\right]$$

Multiply row 2 by 3/10, add one-third of the new row 2 to row 1, and subtract 5/3 times the new row 2 from row 3.

$$\begin{bmatrix} 1 & 0 & 0.7 & 0.3 & 0.1 & 0 & 2.4 \\ 0 & 1 & 0.1 & -0.1 & 0.3 & 0 & 0.2 \\ 0 & 0 & -0.5 & -0.5 & -0.5 & 1 & -1 \end{bmatrix}$$

Multiply row 3 by -2, subtract 0.7 times the new row 3 from row 1, and subtract 0.1 times the new row 3 from row 2

$$\begin{bmatrix} 1 & 0 & 0 & -0.4 & -0.6 & 1.4 & 1 \\ 0 & 1 & 0 & -0.2 & 0.2 & 0.2 & 0 \\ 0 & 0 & 1 & 1 & 1 & -2 & 2 \end{bmatrix} \text{ and } x = 1, y = 0, z = 2$$

(c)
$$\begin{bmatrix} 1 & 1 & 1 & 1 & 0 & 0 & 2 \\ 4 & 3 & 2 & 0 & 1 & 0 & 7 \\ 1 & -1 & -1 & 0 & 0 & 1 & 0 \end{bmatrix}$$

Subtract four times row 1 from row 2 and subtract row 1 from row 3,

$$\begin{bmatrix} 1 & 1 & 1 & 1 & 0 & 0 & 2 \\ 0 & -1 & -2 & -4 & 1 & 0 & -1 \\ 0 & -2 & -2 & -1 & 0 & 1 & -2 \end{bmatrix}$$

Add row 2 to row 1, multiply row 2 by -1, add twice the new row 2 to row 3,

$$\begin{bmatrix} 1 & 0 & -1 & -3 & 1 & 0 & 1 \\ 0 & 1 & 2 & 4 & -1 & 0 & 1 \\ 0 & 0 & 2 & 7 & -2 & 1 & 0 \end{bmatrix}$$

Subtract row 3 from row 2, add half of row 3 to row 1, and multiply row 3 by 0.5,

$$\begin{bmatrix} 1 & 0 & 0 & 0.5 & 0 & 0.5 & 1 \\ 0 & 1 & 0 & -3 & 1 & -1 & 1 \\ 0 & 0 & 1 & 3.5 & -1 & 0.5 & 0 \end{bmatrix} \text{ and } x = 1, y = 1, z = 0$$

(d)
$$\begin{bmatrix} 2 & 1 & 1 & 1 & 1 & 0 & 0 & 0 & 3 \\ 1 & -1 & -1 & 1 & 0 & 1 & 0 & 0 & 3 \\ 1 & 2 & 3 & -1 & 0 & 0 & 1 & 0 & 2 \\ 3 & 3 & -1 & 2 & 0 & 0 & 0 & 1 & 4 \end{bmatrix}$$

Divide row 1 by 2, subtract the new row 1 from row 2 and from row 3, and subtract three times the new row 1 from row 4

$$\begin{bmatrix} 1 & 0.5 & 0.5 & 0.5 & 0.5 & 0 & 0 & 0 & 1.5 \\ 0 & -1.5 & -1.5 & 0.5 & -0.5 & 1 & 0 & 0 & 1.5 \\ 0 & 1.5 & 2.5 & -1.5 & -0.5 & 0 & 1 & 0 & 0.5 \\ 0 & 1.5 & -2.5 & 0.5 & -1.5 & 0 & 0 & 1 & -0.5 \end{bmatrix}$$

Add row 2 to row 3 and row 4, divide row 2 by -1.5 and subtract half of the new row 2 from row 1

$$\begin{bmatrix} 1 & 0 & 0 & 2/3 & 1/3 & 1/3 & 0 & 0 & 2 \\ 0 & 1 & 1 & -1/3 & 1/3 & -2/3 & 0 & 0 & -1 \\ 0 & 0 & 1 & -1 & -1 & 1 & 1 & 0 & 2 \\ 0 & 0 & -4 & 1 & -2 & 1 & 0 & 1 & 1 \end{bmatrix}$$

Subtract row 3 from row 2, add four times row 3 to row 4,

$$\begin{bmatrix} 1 & 0 & 0 & 2/3 & 1/3 & 1/3 & 0 & 0 & 2 \\ 0 & 1 & 0 & 2/3 & 4/3 & -5/3 & -1 & 0 & -3 \\ 0 & 0 & 1 & -1 & -1 & 1 & 1 & 0 & 2 \\ 0 & 0 & 0 & -3 & -6 & 5 & 4 & 1 & 9 \end{bmatrix}$$

Divide row 4 by -3, add the new row 4 to row 3, subtract 2/3 times the new row 4 from row 1 and from row 2

$$\begin{bmatrix} 1 & 0 & 0 & 0 & -1 & 13/9 & 8/9 & 2/9 & 4 \\ 0 & 1 & 0 & 0 & 0 & -5/9 & -1/9 & 2/9 & -1 \\ 0 & 0 & 1 & 0 & 1 & -2/3 & -1/3 & -1/3 & -1 \\ 0 & 0 & 0 & 1 & 2 & -5/3 & -4/3 & -1/3 & -3 \end{bmatrix} \quad \begin{aligned} \text{and } w &= 4 \\ x &= -1 \\ y &= -1 \\ z &= -3 \end{aligned}$$

Exercises 2.9

1. (a) Here, $\begin{vmatrix} 2 & -1 \\ 1 & 1 \end{vmatrix} = 2 - (-1) = 3$ and so rank $= 2$

By Gaussian elimination,

$$\begin{bmatrix} 2 & -1 & 1 & 0 \\ 1 & 1 & 0 & 1 \end{bmatrix}$$

Subtract row 2 from row 1 and subtract the new row 1 from row 2

$$\begin{bmatrix} 1 & -2 & 1 & -1 \\ 0 & 3 & -1 & 2 \end{bmatrix}$$

Divide row 2 by 3 and add twice the new row 2 to row 1; this will give the unit matrix and so the rank is 2.

(b) $\begin{vmatrix} 1 & 1 & 3 \\ 2 & 1 & 1 \\ 1 & -2 & -1 \end{vmatrix} = \begin{vmatrix} 1 & 1 \\ -2 & -1 \end{vmatrix} - \begin{vmatrix} 2 & 1 \\ 1 & -1 \end{vmatrix}$

$$+ 3 \begin{vmatrix} 2 & 1 \\ 1 & -1 \end{vmatrix} = 1 + 3 - 9 = -5$$

and so the rank is 3. By Gaussian elimination,

$$\left[\begin{array}{ccc|ccc} 1 & 1 & 3 & 1 & 0 & 0 \\ 2 & 1 & 1 & 0 & 1 & 0 \\ 1 & -2 & -1 & 0 & 0 & 1 \end{array} \right]$$

Subtract twice row 1 from row 2 and row 1 from row 3

$$\left[\begin{array}{ccc|ccc} 1 & 1 & 3 & 1 & 0 & 0 \\ 0 & -1 & -5 & -2 & 1 & 0 \\ 0 & -3 & -4 & -1 & 0 & 1 \end{array} \right]$$

Add row 2 to row 1 and subtract three times row 2 from row 3, and multiply row 2 by -1

$$\left[\begin{array}{ccc|ccc} 1 & 0 & -2 & -1 & 1 & 0 \\ 0 & 1 & 5 & 2 & -1 & 0 \\ 0 & 0 & 11 & 5 & -3 & 1 \end{array} \right]$$

It can be seen that further operations will result in the unit matrix and so the rank is 3.

(c) The determinant can be simplified by adding twice column 2 to column 1, subtracting column 2 from column 3 and expanding by row 2

$$\begin{vmatrix} 1 & 2 & -4 \\ 2 & -1 & -1 \\ 3 & 1 & -5 \end{vmatrix} = \begin{vmatrix} 5 & 2 & -6 \\ 0 & -1 & 0 \\ 5 & 1 & -6 \end{vmatrix} = - \begin{vmatrix} 5 & -6 \\ 5 & -6 \end{vmatrix} = 0 \text{ so the rank} < 3.$$

Next, try to find a non-zero (2×2) determinant. Here,

$$\begin{vmatrix} 1 & 2 \\ 2 & -1 \end{vmatrix} = -1 - 4 = -5 \text{ and so the rank is 2.}$$

By Gaussian elimination,

$$\left[\begin{array}{ccc|ccc} 1 & 2 & -4 & 1 & 0 & 0 \\ 2 & -1 & -1 & 0 & 1 & 0 \\ 3 & 1 & -5 & 0 & 0 & 1 \end{array} \right]$$

Subtract twice row 1 from row 2 and subtract three times row 1 from row 3

$$\begin{bmatrix} 1 & 2 & -4 & 1 & 0 & 0 \\ 0 & -5 & 7 & -2 & 1 & 0 \\ 0 & -5 & 7 & -3 & 0 & 1 \end{bmatrix}$$

Subtracting row 2 from row 3 will give 0, 0, 0 on row 3 and further row operations will give a (2×2) unit matrix and so the rank is 2.

(d) Expanding the determinant by the first row,

$$\begin{vmatrix} 1 & 0 & 0 \\ 1 & 1 & 1 \\ 0 & 1 & 1 \end{vmatrix} = \begin{vmatrix} 1 & 1 \\ 1 & 1 \end{vmatrix} = 0 \text{ and rank} < 3$$

Searching for a non-zero (2×2) determinant, $\begin{vmatrix} 1 & 0 \\ 0 & 1 \end{vmatrix} = 1$

and the rank $= 2$

By Gaussian elimination,

$$\begin{bmatrix} 1 & 0 & 0 & 1 & 0 & 0 \\ 1 & 1 & 1 & 0 & 1 & 0 \\ 0 & 1 & 1 & 0 & 0 & 1 \end{bmatrix}$$

Subtract row 1 from row 2 and subtract the new row 2 from row 3,

$$\begin{bmatrix} 1 & 0 & 0 & 1 & 0 & 0 \\ 0 & 1 & 1 & -1 & 1 & 0 \\ 0 & 0 & 0 & 1 & -1 & 1 \end{bmatrix} \text{ and the rank is 2.}$$

(e) The determinant can be simplified by subtracting column 1 from column 2 and from column 3, and then expanding by row 1

$$\begin{vmatrix} 1 & 1 & 1 \\ 1 & 0 & 2 \\ 2 & 2 & -1 \end{vmatrix} = \begin{vmatrix} 1 & 0 & 0 \\ 1 & -1 & 1 \\ 2 & 0 & -3 \end{vmatrix} = 3 \text{ and rank} = 3$$

By Gaussian elimination,

$$\begin{bmatrix} 1 & 1 & 1 & 1 & 0 & 0 \\ 1 & 0 & 2 & 0 & 1 & 0 \\ 2 & 2 & -1 & 0 & 0 & 1 \end{bmatrix}$$

Subtract row 1 from row 2 and subtract twice row 1 from row 3

$$\begin{bmatrix} 1 & 1 & 1 & | & 1 & 0 & 0 \\ 0 & -1 & 1 & | & -1 & 1 & 0 \\ 0 & 0 & -3 & | & -2 & 0 & 1 \end{bmatrix}$$

It can be seen that further row operations will result in the unit matrix and so the rank is 3.

(f) Here the matrix is (3 × 4) and so the maximum rank is 3. Taking the first 3 columns, the determinant is

$$\begin{vmatrix} 1 & 0 & 0 \\ 0 & 2 & 2 \\ 2 & 1 & 2 \end{vmatrix} = \begin{vmatrix} 2 & 2 \\ 1 & 2 \end{vmatrix} = 2 \text{ and rank} = 3$$

By Gaussian elimination,

$$\begin{bmatrix} 1 & 0 & 0 & -1 & | & 1 & 0 & 0 & 0 \\ 0 & 2 & 2 & 1 & | & 0 & 1 & 0 & 0 \\ 2 & 1 & 3 & 1 & | & 0 & 0 & 1 & 0 \end{bmatrix}$$

Subtract twice row 1 from row 3 and divide row 2 by 2

$$\begin{bmatrix} 1 & 0 & 0 & -1 & | & 1 & 0 & 0 & 0 \\ 0 & 1 & 1 & 0.5 & | & 0 & 0.5 & 0 & 0 \\ 0 & 1 & 3 & 3 & | & -2 & 0 & 1 & 0 \end{bmatrix}$$

Subtracting row 2 from row 3 will give 0   0   2 on row 3 and so a unit matrix can be formed and the rank is 3.

(g) The maximum rank is 3 and taking the first three columns, adding column 1 to column 3

$$\begin{vmatrix} 1 & 0 & -1 \\ 1 & 1 & -1 \\ 0 & 1 & -1 \end{vmatrix} = \begin{vmatrix} 1 & 0 & 0 \\ 1 & 1 & 0 \\ 0 & 1 & -1 \end{vmatrix}$$

$$= \begin{vmatrix} 1 & 0 \\ 1 & -1 \end{vmatrix} = -1 \text{ so rank} = 3$$

By Gaussian elimination,

$$\begin{bmatrix} 1 & 0 & -1 & -1 & | & 1 & 0 & 0 & 0 \\ 1 & 1 & -1 & -1 & | & 0 & 1 & 0 & 0 \\ 0 & 1 & -1 & -1 & | & 0 & 0 & 1 & 0 \end{bmatrix}$$

Subtracting row 1 from row 2 and the new row 2 from row 3

$$
\left[\begin{array}{rrrr|rrrr}
1 & 0 & -1 & -1 & 1 & 0 & 0 & 0 \\
0 & 1 & 0 & 0 & -1 & 1 & 0 & 0 \\
0 & 0 & -1 & -1 & 1 & -1 & 1 & 0
\end{array}\right]
$$

and further row operations will give a (3×3) unit matrix and so the rank is 3.

2. In each case there is a unique solution if the determinant has a non-zero value.

(a)

$$
\begin{vmatrix}
1 & 3 & -2 \\
1 & -2 & 1 \\
2 & -2 & 3
\end{vmatrix}
=
\begin{vmatrix}
1 & 3 & -2 \\
0 & -5 & 3 \\
0 & -8 & 7
\end{vmatrix}
=
\begin{vmatrix}
-5 & 3 \\
-8 & 7
\end{vmatrix}
= -11
$$

where row 1 was subtracted from row 2 and twice row 1 from row 3, and there is a unique solution.

(b)

$$
\begin{vmatrix}
1 & 2 & 1 \\
2 & -1 & 2 \\
1 & 1 & 1
\end{vmatrix}
=
\begin{vmatrix}
5 & 2 & 5 \\
0 & -1 & 0 \\
3 & 1 & 3
\end{vmatrix}
= -
\begin{vmatrix}
5 & 5 \\
3 & 3
\end{vmatrix}
= 0
$$

where twice column 2 was added to column 1 and column 3, and as the determinant is zero there is no unique solution.

(c)

$$
\begin{vmatrix}
1 & 3 & -1 & 2 \\
1 & -1 & 2 & 1 \\
1 & 1 & 1 & -2 \\
1 & 1 & 0 & 5
\end{vmatrix}
=
\begin{vmatrix}
1 & 2 & -1 & -3 \\
1 & -2 & 2 & -4 \\
1 & 0 & 1 & -7 \\
1 & 0 & 0 & 0
\end{vmatrix}
$$

$$
= -
\begin{vmatrix}
2 & -1 & -3 \\
-2 & 2 & -4 \\
0 & 1 & -7
\end{vmatrix}
$$

where column 1 was subtracted from column 2 and five times column 1 was subtracted from column 4. Adding row 2 to row 1 gives a determinant with two identical rows and so it has a value zero. There is no unique solution.

Exercises 2.11

1. $\mathbf{A} = \begin{bmatrix} 2 & 1 & 2 \\ 1 & 0 & 1 \end{bmatrix}$ so $\mathbf{A}' = \begin{bmatrix} 2 & 1 \\ 1 & 0 \\ 2 & 1 \end{bmatrix}$ and $(\mathbf{A}')' = \begin{bmatrix} 2 & 1 & 2 \\ 1 & 0 & 1 \end{bmatrix} = \mathbf{A}$

$$\mathbf{A} + \mathbf{B} = \begin{bmatrix} 2 & 1 & 2 \\ 1 & 0 & 1 \end{bmatrix} + \begin{bmatrix} 3 & -1 & 1 \\ 2 & -2 & 0 \end{bmatrix} = \begin{bmatrix} 5 & 0 & 3 \\ 3 & -2 & 1 \end{bmatrix}$$

so that
$$(\mathbf{A} + \mathbf{B})' = \begin{bmatrix} 5 & 3 \\ 0 & -2 \\ 3 & 1 \end{bmatrix} = \begin{bmatrix} 2 & 1 \\ 1 & 0 \\ 2 & 1 \end{bmatrix} + \begin{bmatrix} 3 & 2 \\ -1 & -2 \\ 1 & 0 \end{bmatrix} = \mathbf{A}' + \mathbf{B}'$$

Now
$$\mathbf{BC} = \begin{bmatrix} 3 & -1 & 1 \\ 2 & -2 & 0 \end{bmatrix} \begin{bmatrix} -1 & 0 \\ 2 & 1 \\ -1 & 0 \end{bmatrix} = \begin{bmatrix} -6 & -1 \\ -6 & -2 \end{bmatrix}$$

and
$$(\mathbf{BC})' = \begin{bmatrix} -6 & -6 \\ -1 & -2 \end{bmatrix}$$

Also, $\mathbf{C}'\mathbf{B}' = \begin{bmatrix} -1 & 2 & -1 \\ 0 & 1 & 0 \end{bmatrix} \begin{bmatrix} 3 & 2 \\ -1 & -2 \\ 1 & 0 \end{bmatrix} = \begin{bmatrix} -6 & -6 \\ -1 & -2 \end{bmatrix} = (\mathbf{BC})'.$

2. Require $\mathbf{A}^2 = \mathbf{A}$ for $\mathbf{A}$ to be idempotent

$$\mathbf{A}^2 = \begin{bmatrix} 1 & 0 & 0 \\ 0 & 1 & 0 \\ 0 & 0 & 0 \end{bmatrix} \begin{bmatrix} 1 & 0 & 0 \\ 0 & 1 & 0 \\ 0 & 0 & 0 \end{bmatrix} = \begin{bmatrix} 1 & 0 & 0 \\ 0 & 1 & 0 \\ 0 & 0 & 0 \end{bmatrix} = \mathbf{A} - \text{idempotent}$$

$$\mathbf{B}^2 = \begin{bmatrix} 0.25 & 0.25 & 0.25 & 0.25 \\ 0.25 & 0.25 & 0.25 & 0.25 \\ 0.25 & 0.25 & 0.25 & 0.25 \\ 0.25 & 0.25 & 0.25 & 0.25 \end{bmatrix} \begin{bmatrix} 0.25 & 0.25 & 0.25 & 0.25 \\ 0.25 & 0.25 & 0.25 & 0.25 \\ 0.25 & 0.25 & 0.25 & 0.25 \\ 0.25 & 0.25 & 0.25 & 0.25 \end{bmatrix}$$

$$= \begin{bmatrix} 0.25 & 0.25 & 0.25 & 0.25 \\ 0.25 & 0.25 & 0.25 & 0.25 \\ 0.25 & 0.25 & 0.25 & 0.25 \\ 0.25 & 0.25 & 0.25 & 0.25 \end{bmatrix} = \mathbf{B} - \text{idempotent}$$

$$\mathbf{C}^2 = \begin{bmatrix} a & 0 \\ 0 & a \end{bmatrix} \begin{bmatrix} a & 0 \\ 0 & a \end{bmatrix} = \begin{bmatrix} a^2 & 0 \\ 0 & a^2 \end{bmatrix} = \mathbf{C} \text{ only if } a = 0 \text{ or } a = 1.$$

3. $\mathbf{A}' = \begin{bmatrix} 2 & 1 \\ -3 & 1 \end{bmatrix}$ and $(\mathbf{A}')^{-1} = \begin{bmatrix} 0.2 & -0.2 \\ 0.6 & 0.4 \end{bmatrix}$

Now $\mathbf{A}^{-1} = \begin{bmatrix} 0.2 & 0.6 \\ -0.2 & 0.4 \end{bmatrix}$ and so $(\mathbf{A}^{-1})' = \begin{bmatrix} 0.2 & -0.2 \\ 0.6 & 0.4 \end{bmatrix} = (\mathbf{A}')^{-1}$

$\mathbf{AB} = \begin{bmatrix} 2 & -3 \\ 1 & 1 \end{bmatrix} \begin{bmatrix} 2 & 2 \\ -1 & 1 \end{bmatrix} = \begin{bmatrix} 7 & 1 \\ 1 & 3 \end{bmatrix}$ and $(\mathbf{AB})^{-1} = \dfrac{1}{20} \begin{bmatrix} 3 & -1 \\ -1 & 7 \end{bmatrix}$

Also,
$$\mathbf{B}^{-1} = \begin{bmatrix} 0.25 & -0.5 \\ 0.25 & 0.5 \end{bmatrix}$$

and so
$$\mathbf{B}^{-1}\mathbf{A}^{-1} = \frac{1}{20}\begin{bmatrix} 3 & -1 \\ -1 & 7 \end{bmatrix} = (\mathbf{AB})^{-1}.$$

4. Since $\mathbf{A} = \begin{bmatrix} 3 & 1 \\ 1 & 3 \end{bmatrix}$ then $|\mathbf{A} - r\mathbf{I}| = \begin{vmatrix} 3-r & 1 \\ 1 & 3-r \end{vmatrix}$

$$= (3-r)(3-r) - 1$$

and setting this to zero gives $9 - 6r + r^2 - 1 = 0$. Solving gives $r = 4$ and $r = 2$.

For $r = 4$ the characteristic vector is from $[\mathbf{A} - 4\mathbf{I}]\mathbf{x} = 0$,

or $\begin{bmatrix} 3-4 & 1 \\ 1 & 3-4 \end{bmatrix}\begin{bmatrix} x_1 \\ x_2 \end{bmatrix} = \begin{bmatrix} 0 \\ 0 \end{bmatrix}$ and $\begin{array}{l} -x_1 + x_2 = 0 \text{ so } x_1 = x_2 \\ x_1 - x_2 = 0. \end{array}$

Normalising by setting $x_1^2 + x_2^2 = 1$ gives $x_1 = 1/\sqrt{2} = x_2$. For $r = 2$ the characteristic vector is from $[\mathbf{A} - 2\mathbf{I}]\mathbf{x} = 0$,

or $\begin{bmatrix} 3-2 & 1 \\ 1 & 3-2 \end{bmatrix}\begin{bmatrix} x_1 \\ x_2 \end{bmatrix} = \begin{bmatrix} 0 \\ 0 \end{bmatrix}$ and $\begin{array}{l} x_1 + x_2 = 0 \text{ so } x_1 = -x_2 \\ x_1 + x_2 = 0. \end{array}$

Normalising by setting $x_1^2 + x_2^2 = 1$ gives $x_1 = 1/\sqrt{2}$ and $x_2 = -1/\sqrt{2}$.

Since $\mathbf{B} = \begin{bmatrix} 2 & -1 \\ 0 & 1 \end{bmatrix}$ then $|\mathbf{B} - r\mathbf{I}| = \begin{vmatrix} 2-r & -1 \\ 0 & 1-r \end{vmatrix}$

$$= (2-r)(1-r)$$

and setting this to zero gives $r = 2$ and $r = 1$.

For $r = 2$ the characteristic vector is from $[\mathbf{B} - 2\mathbf{I}]\mathbf{x} = 0$,

or $\begin{bmatrix} 2-2 & -1 \\ 0 & 1-2 \end{bmatrix}\begin{bmatrix} x_1 \\ x_2 \end{bmatrix} = \begin{bmatrix} 0 \\ 0 \end{bmatrix}$ and $\begin{array}{l} -x_2 = 0 \\ -x_2 = 0. \end{array}$

Normalising by setting $x_1^2 + x_2^2 = 1$ gives $x_1 = 1$, $x_2 = 0$.

For $r = 1$ the characteristic vector is from $[\mathbf{B} - 1\mathbf{I}]\mathbf{x} = 0$,

or $\begin{bmatrix} 2-1 & -1 \\ 0 & 1-1 \end{bmatrix}\begin{bmatrix} x_1 \\ x_2 \end{bmatrix} = \begin{bmatrix} 0 \\ 0 \end{bmatrix}$ and $x_1 - x_2 = 0$ so $x_1 = x_2$

Normalising by setting $x_1^2 + x_2^2 = 1$ gives $x_1 = 1/\sqrt{2}$ and $x_2 = 1/\sqrt{2}$.

Exercises 2.14

1. Level of final demand which can be met by Industry 1 is

   $1500 - (200 + 300) = 1000$, and by Industry 2 is

   $$2500 - (500 + 100) = 1900.$$

2. The matrix of technological coefficients is

   $$\begin{bmatrix} \frac{200}{1500} & \frac{300}{2500} \\ \frac{500}{1500} & \frac{100}{2500} \end{bmatrix} = \begin{bmatrix} \frac{2}{15} & \frac{3}{25} \\ \frac{1}{3} & \frac{1}{25} \end{bmatrix}$$

   and the new situation is as shown in the table.

	Input to Industry 1	Input to Industry 2	Level of output
Industry 1	$\frac{2}{15} \times 2{,}000$	$\frac{3}{25} \times 2{,}500$	2,000
Industry 2	$\frac{1}{3} \times 2{,}000$	$\frac{1}{25} \times 2{,}500$	2,500

   Therefore final demand which can be met by Industry 1 is equal to $2{,}000 - (\frac{2}{15} \times 2{,}000 + \frac{3}{25} \times 2{,}500) = 1{,}433\frac{1}{3}$ and by Industry 2 to $2{,}500 - (\frac{1}{3} \times 2{,}000 + \frac{1}{25} \times 2{,}500) = 1{,}733\frac{1}{3}$.

3. (a)

   $$\begin{bmatrix} 1 - 0.2 & -0.4 \\ -0.3 & 1 - 0.2 \end{bmatrix} \begin{bmatrix} X_1 \\ X_2 \end{bmatrix} = \begin{bmatrix} 1000 \\ 2000 \end{bmatrix}$$

   which can be solved using the inverse matrix to give $X_1 = 3077$ and $X_2 = 3654$ approx.

   (b)

   $$\begin{bmatrix} 0.9 & -0.6 \\ -0.4 & 0.9 \end{bmatrix} \begin{bmatrix} X_1 \\ X_2 \end{bmatrix} = \begin{bmatrix} 1000 \\ 2000 \end{bmatrix}$$

   $X_1 = 3684$ and $X_2 = 3860$ approx.

4. (a) Here $\mathbf{A} = \begin{bmatrix} 100/1000 & 200/2000 \\ 500/1000 & 100/2000 \end{bmatrix} = \begin{bmatrix} 0.1 & 0.10 \\ 0.5 & 0.05 \end{bmatrix}$

   Therefore, $\mathbf{I} - \mathbf{A} = \begin{bmatrix} 0.9 & -0.10 \\ -0.5 & 0.95 \end{bmatrix}$ and

   $$(\mathbf{I} - \mathbf{A})^{-1} = \begin{bmatrix} 1.180 & 0.124 \\ 0.621 & 1.118 \end{bmatrix}$$

by row operations. The multipliers are $1.180 + 0.621 = 1.801$ and $0.124 + 1.118 = 1.242$.

If the final demand for exports becomes 500 for agriculture and 900 for industry, the total final demands are 800 and 1500 respectively. Therefore,

$$\mathbf{X} = (\mathbf{I} - \mathbf{A})^{-1} \mathbf{C} = \begin{bmatrix} 1.180 & 0.124 \\ 0.621 & 1.118 \end{bmatrix} \begin{bmatrix} 800 \\ 1500 \end{bmatrix} = \begin{bmatrix} 1130.0 \\ 2173.8 \end{bmatrix}$$

instead of 1000 and 2000.

(b) Here

$$\mathbf{A} = \begin{bmatrix} 0.1 & 0.10 & 300/900 \\ 0.5 & 0.05 & 600/900 \\ 0.1 & 0.40 & 0/900 \end{bmatrix} = \begin{bmatrix} 0.1 & 0.10 & 0.33 \\ 0.5 & 0.05 & 0.67 \\ 0.1 & 0.40 & 0.00 \end{bmatrix}$$

Therefore,

$$\mathbf{I} - \mathbf{A} = \begin{bmatrix} 0.9 & -0.10 & -0.33 \\ -0.5 & 0.95 & -0.67 \\ -0.1 & -0.40 & 1.00 \end{bmatrix}$$

and $(\mathbf{I} - \mathbf{A})^{-1}$ can be found by row operations:

$$\begin{bmatrix} 0.9 & -0.10 & -0.33 & | & 1 & 0 & 0 \\ -0.5 & 0.95 & -0.67 & | & 0 & 1 & 0 \\ -0.1 & -0.40 & 1.00 & | & 0 & 0 & 1 \end{bmatrix}$$

Divide row 1 by 0.9 and combine the new row 1 with rows 2 and 3:

$$\begin{bmatrix} 1 & -0.111 & -0.370 & | & 1.111 & 0 & 0 \\ 0 & 0.895 & -0.855 & | & 0.556 & 1 & 0 \\ 0 & -0.411 & 0.963 & | & 0.111 & 0 & 1 \end{bmatrix}$$

Divide row 2 by 0.895 and combine the new row 2 with rows 1 and 3:

$$\begin{bmatrix} 1 & 0 & -0.476 & | & 1.180 & 0.124 & 0 \\ 0 & 1 & -0.955 & | & 0.621 & 1.117 & 0 \\ 0 & 0 & 0.570 & | & 0.366 & 0.459 & 1 \end{bmatrix}$$

Divide row 3 by 0.570 and combine the new row 3 with row 1 and row 2:

$$\begin{bmatrix} 1 & 0 & 0 & | & 1.486 & 0.507 & 0.835 \\ 0 & 1 & 0 & | & 1.234 & 1.886 & 1.675 \\ 0 & 0 & 1 & | & 0.642 & 0.805 & 1.754 \end{bmatrix}$$

(Check that $(\mathbf{I} - \mathbf{A})(\mathbf{I} - \mathbf{A})^{-1} = \mathbf{I}$, apart from rounding errors)

The multipliers, from adding the columns, are 3.362, 3.198 and 4.264, compared to 1.801 and 1.242 previously. If the final demand for exports becomes 500 for agriculture and 900 for industry, these are the total final demands (since households are an industry). Therefore,

$$\mathbf{X} = (\mathbf{I} - \mathbf{A})^{-1}\mathbf{C} = \begin{bmatrix} 1.486 & 0.507 & 0.835 \\ 1.234 & 1.886 & 1.675 \\ 0.642 & 0.805 & 1.754 \end{bmatrix} \begin{bmatrix} 500 \\ 900 \\ 0 \end{bmatrix} = \begin{bmatrix} 1199.3 \\ 2314.4 \\ 1045.5 \end{bmatrix}$$

instead of the original values of 1000, 2000, 900 and the values in (a) of 1130 and 2173 (and 900 for households).

5. (a) The technology matrix is obtained by dividing each column by the gross output of the sector:

$$\mathbf{A} = \begin{bmatrix} 11/41 & 19/240 & 1/185 \\ 5/41 & 89/240 & 40/185 \\ 5/41 & 37/240 & 37/185 \end{bmatrix} = \begin{bmatrix} 0.268 & 0.079 & 0.005 \\ 0.122 & 0.371 & 0.216 \\ 0.122 & 0.154 & 0.200 \end{bmatrix}$$

Hence, $\mathbf{I} - \mathbf{A} = \begin{bmatrix} 0.732 & -0.079 & -0.005 \\ -0.122 & 0.629 & -0.216 \\ -0.122 & -0.154 & 0.800 \end{bmatrix}$

and $(\mathbf{I} - \mathbf{A})$ multiplied by the given matrix is approximately $\mathbf{I}$.

(b) Here, $\mathbf{X} = (\mathbf{I} - \mathbf{A})^{-1}\mathbf{C} = \begin{bmatrix} 1.409 & 0.192 & 0.062 \\ 0.372 & 1.753 & 0.476 \\ 0.286 & 0.367 & 1.351 \end{bmatrix} \begin{bmatrix} 15 \\ 120 \\ 130 \end{bmatrix}$

$$= \begin{bmatrix} 52.2 \\ 277.8 \\ 224.0 \end{bmatrix}$$

(c) Combining the industry and services sectors gives:

Output from	Input to Agriculture	Other	Final demand	Total output
Agriculture	11	20	10	41
Other	10	203	212	425

The technology matrix is $\mathbf{A} = \begin{bmatrix} 11/41 & 20/425 \\ 10/41 & 203/425 \end{bmatrix}$

$$= \begin{bmatrix} 0.268 & 0.047 \\ 0.244 & 0.478 \end{bmatrix}$$

and so $\qquad$ $I - A = \begin{bmatrix} 0.732 & -0.047 \\ -0.244 & 0.522 \end{bmatrix}$

By row operations $\quad (I - A)^{-1} = \begin{bmatrix} 1.408 & 0.127 \\ 0.658 & 1.975 \end{bmatrix}$

Hence, $\qquad X = \begin{bmatrix} 1.408 & 0.127 \\ 0.658 & 1.975 \end{bmatrix} \begin{bmatrix} 15 \\ 250 \end{bmatrix} = \begin{bmatrix} 52.9 \\ 503.6 \end{bmatrix}$

These values compare with 52.2 and 277.8 + 224.0 = 501.8 and so, apart from rounding errors (which could be reduced by working to more places after the decimal point in both A and $(I - A)^{-1}$), the values for agriculture and the aggregate are the same. Thus, for an economist interested only in agriculture it does not matter whether we have one or two other sectors and, by implication, the results for agriculture would be the same if there were many other sectors.

Also, the multipliers for agriculture are 2.066 in both cases, while for industry the multiplier is 2.312 and for services, 1.889, and for the combined sector 2.102, the average of the two values.

Exercises 3.2

1. (a) Let x = output and TC = total cost, and since the cost function is quadratic, let $TC = a + bx + cx^2$.

   Using the three pairs of values given:

   $$4 = a \qquad\qquad\qquad\qquad\qquad\qquad (1)$$

   $$14 = a + 2b + 4c \qquad\qquad\qquad\quad (2)$$

   $$58 = a + 6b + 36c \qquad\qquad\qquad\quad (3)$$

   Substituting from (1) into (2) and (3) and simplifying:

   $$10 = 2b + 4c \qquad\qquad\qquad\qquad (4)$$

   $$54 = 6b + 36c \qquad\qquad\qquad\qquad (5)$$

   Subtract three times (4) from (5)

   $$24 = 24c \qquad \text{or } c = 1$$

   Putting $a = 4$ and $c = 1$ in (2) gives $b = 3$ and so

   $$TC = 4 + 3x + x^2$$

x	0	2	4	6	8	10
TC	4	14	32	58	92	134

(b) $TC = 10 + 2x + 0.5x^2$

x	0	2	4	6	8	10
TC	10	16	26	40	58	80

(c) $TC = 25 + 0.5x + 0.25x^2$

x	0	2	4	6	8	10
TC	25	27	31	37	45	55

(d) $TC = 20 + 0.8x + 0.02x^2$

x	0	1	3	5	7	9	10
TC	20	20.82	22.58	24.50	26.58	28.82	30.00

(e) $TC = 20 + 0.5x^2$

x	0	2	4	6	8	10
TC	20	22	28	38	52	70

2.

	x	0	1	2	3	4	5	6	7	8	9	10
(a)	y	100	111	124	139	156	175	196	219	244	271	300
(b)	y	100	91	84	79	76	75	76	79	84	91	100
(c)	y	100	109	116	121	124	125	124	121	116	109	100
(d)	y	100	89	76	61	44	25	4	−19	−44	−71	−100

Exercises 3.4

1. Formula is $x = \{-b \pm \sqrt{(b^2 - 4ac)}\}/2a$

    (a) $a = 1$, $b = -7$, $c = 12$ so $x = 4$ or 3

(b) $a = 1$, $b = 1$, $c = -2$ so $x = 1$ or -2

(c) $a = 2$, $b = 7$, $c = 3$ so $x = -0.5$ or -3

(d) $a = 1$, $b = -2$, $c = 1$ so $x = 1$ (repeated root)

(e) $a = 1$, $b = 0$, $c = -1$ so $x = 1$ or -1

(f) $a = 1$, $b = 0$, $c = 1$ so $x = i$ or $-i$

(g) $a = 1$, $b = -4$, $c = 5$ so $x = 2 + i$ or $2 - i$

(h) $a = 1$, $b = 2$, $c = 2$ so $x = -1 + i$ or $-1 - i$.

2. (a) $y = 0$ when $x^2 - 10x + 25 = 0$ or $x = 5$. Drawing up a table of values:

x	0	3	5	7	10
y	25	4	0	4	25

(b) $y = 0$ when $-x^2 + 4x - 3 = 0$ or $x = 1$ or $x = 3$:

x	0	1	2	3	4
y	-3	0	1	0	-3

Exercises 3.6

1. (a) $y = 0$ when $x = 2$ or 6:

x	0	1	2	3	4	5	6
y	12	5	0	-3	-4	-3	0

(b)

x	0	1	2	3	4	5	6
y	600	418	216	0	-224	-450	-672

Here $x = 3$ is a root and the cubic $= (x - 3)(x^2 - 10x - 200)$. The roots of this quadratic are $x = 20$ and $x = -10$, which are outside the range of interest and can be ignored.

(c) Setting $y = 0$ gives $x = 1.5$, $x = -1$ and $x = 5$ so these are included in the table of values:

x	-2	-1	0	1	1.5	2	3	4	5	6
y	-49	0	15	8	0	-9	-24	-25	0	63

(d) Setting $y = 0$ gives $x = 0$, and so $32 - 2x^2 = 0$ resulting in $x = 4$ or -4.

x	-4	-3	-2	-1	0	1	2	3	4
y	0	-42	-48	-30	0	30	48	42	0

(e) When $y = 0$, $x = 0$:

x	0	1	2	3
y	0	1	32	243

(f) $y = 20/x$ so as $x \to 0$, $y \to \infty$

x	1	5	10	20	100
y	20	4	2	1	0.2

(g) $y = (20/x) + 1$. As $x \to 0$, $y \to \infty$, and as $x \to \infty$, $y \to 1$:

x	1	5	10	20	100
y	21	5	3	2	1.2

2. $Q = KL$ so that $K = Q/L$

L	10	20	50	80	100
Q_1/L	2	1	0.4	0.25	0.2
Q_2/L	5	2.5	1	0.625	0.5
Q_3/L	10	5	2	1.25	1

3. $AC = TC/x$. Here $x > 0$ and negative values can be ignored:

	x	1	10	20	50	100	200
(a)	AC	100	10	5	2	1	0.5
(b)	AC	105	15	10	7	6	5.5
(c)	AC	106	25	30	57	106	205.5

Exercises 3.10

1. The breakeven point has $TC = TR$.

   (a) $x^2 - 6x + 9 = 0$ has $x = 3$ as the repeated root and this is the breakeven point

   (b) $x^2 - 9x + 18 = 0$ has $x = 6$ and $x = 3$ as the breakeven points

   (c) $2x^2 - 3x + 10 = 0$ has no real roots and so there is no breakeven point.

2. (a) $TC = 25 - 6x + x^2$
   (b) 305
   (d) total revenue equals total cost when

   $$10x + 15 = 25 - 6x + x^2$$

   that is when $x^2 - 16x + 10 = 0$
   The roots of this quadratic are given by

   $$x = \frac{16 \pm \sqrt{(16^2 - 4 \times 10)}}{2}$$

   $$= \tfrac{1}{2}(16 \pm \sqrt{(256 - 40)}) = \tfrac{1}{2}(16 \pm \sqrt{(216)})$$

   $$= 8 \pm 7.35 = 15.35 \quad \text{or} \quad 0.65.$$

   These are the two levels of output at which total revenue equals total cost.

4. The equations are satisfied by the values $x = 1$ and $x = 15$.

5.
$$y = 20 + 3x + x^2$$
$$y = 5x + b$$

$\therefore$
$$20 + 3x + x^2 = 5x + b$$

or
$$x^2 - 2x + (20 - b) = 0$$

$$x = \frac{2 \pm \sqrt{[4 - 4(20 - b)]}}{2}$$

$$= 1 \pm \sqrt{[1 - (20 - b)]}$$

$$= 1 \pm \sqrt{(b - 19)}$$

when $b = 20$ $x = 1 \pm 1 = 2$ or 0 (2 real solutions)

$$b = 19 \quad x = 1 \quad \text{(the solutions are coincident)}$$
$$b = 18 \quad x = 1 \pm \sqrt{(-1)} \quad \text{(2 complex solutions)}$$

6.
$$2p^2 - 3p - 40 = 250 - 4p - p^2$$

$\therefore$
$$3p^2 + p - 290 = 0$$

$$p = \frac{-1 \pm \sqrt{(1 + 12 \times 290)}}{6}$$

$$= \frac{-1 \pm \sqrt{3481}}{6} = \frac{-1 \pm 59}{6}$$

$$= \frac{58}{6} \quad \text{or} \quad \frac{-60}{6}$$

In this example we take the positive root, $p = 9\frac{2}{3}$, and at this value $q = 117\frac{8}{9}$.

7. (a) $p = 5$ $q = 10$ (b) $p = 4$ $q = 4$
   (c) $p = 2$ $q = 3$

8. Net revenue $NR = TR - TC$ and $TR = pq$. Since TC is expressed in terms of q rearrange demand to give $p = (200 - q_D)/2$ so $NR = q_D (200 - q_D)/2 - 20 - 5q_D$.

At the breakeven points, NR is zero when $q_D = 0.21$ or 189.8. The corresponding prices are 99.9 and 5.1:

q_D	0	10	20	50	100	150	200
NR	-20	880	1680	3480	4480	2980	-1020

(maximum NR is when q_D is 95 and $NR = 4492.5$)

Exercises 3.12

1. (a) When $x = 19$, $TC = 72$ and when $x = 21$, $TC = 78$ and extra cost is $78 - 72 = 6$
   (b) When $x = 29$, $TC = 102$ and when $x = 31$, $TC = 133$ and extra cost is $133 - 102 = 31$
   (c) When $x = 49$, $TC = 187$ and when $x = 51$, $TC = 223$ and extra cost is $223 - 187 = 36$.

2. Let x = number of cans sold
   Revenue $= 30x$ for $0 < x \leqslant 100$ and
             $= 30 (0.85)x$ for $x > 100$

If $x = 99$, cost $= 30 \, (99) = 2970$p

If $x = 101$, cost $= 30 \, (0.85) \, 101 = 2575.5 -$ cheaper than for 99!

New system: Revenue $= 30x$ for $0 < x < 101$

Revenue $= 100 \, (30) + (x - 100) \, (0.80) \, 30$ for $x > 100$

If $x = 120$, old revenue $= 30 \, (0.85) \, 120 = 3060$,

new revenue $= 100 \, (30) + (20) \, (0.80) \, 30 = 3480$.

Exercises 4.3

1. (a) 19, 225   (b) 950, 12,750
   (c) -210, $-2,250$ (d) 82, 1,290

2. £300   3. £11,300, £67,900

4. Arithmetic progression with $20 = a + (n - 1)d = 10 + 15d$. Hence $d = \frac{2}{3}$ and rate of interest is $\frac{2}{3}/10 = 0.067$ or 6.7 per cent.

Exercises 4.5

1. (a) 2,430; $S_{10} = \dfrac{10(1 - 3^{10})}{-2} = 295240$

   (b) $\frac{1}{3}$; $S_{10} = \dfrac{81[1 - (\frac{1}{3})^{10}]}{1 - \frac{1}{3}} = 121.5$ approximately

   $$S_\infty = \dfrac{81}{1 - (\frac{1}{3})} = 243/2 = 121\tfrac{1}{2}$$

   (c) -64; $S_{10} = \dfrac{2[1 - (-2)^{10}]}{3} = -682$

   (d) 32; $S_{10} = \dfrac{-1024[1 - (-\frac{1}{2})^{10}]}{1 - (-\frac{1}{2})} = -682$ approximately

   $$S_\infty = \dfrac{-1024}{1 - (-\frac{1}{2})} = -682\tfrac{2}{3}$$

   (e) 0.00001; $S_{10} = \dfrac{1 \, [1 - (0.1)^{10}]}{1 - 0.1} = 1.111$ approximately

   $$S_\infty = \dfrac{1}{1 - 0.1} = \tfrac{10}{9}$$

2. (a) $300(1.05)^{15} = 623.7$ (b) $300(1.10)^{15} = 1253.2$

3. $S_{25} = \dfrac{150(1 - 1.06^{25})}{(1 - 1.06)} = 8229.7$

4. $4(1 + r)^4 = 5$

$\therefore$ $(1 + r)^4 = 1.25$

$\therefore$ $1 + r = 1.06$ approximately

and implied rate of compound interest is 6 per cent.

Exercises 4.7

1. (a) Present value of £800 received in 10 years time is given by £800/$(1 + i)^{10}$; this is equal to £300.

$\therefore$ $(1 + i)^{10} = 800/300 = \frac{8}{3}$

We therefore require the discount factors which make

$$\frac{1}{(1 + i)^{10}} = \frac{3}{8} = 0.375$$

From Table 4.5 it can be seen that 0.386 is the factor used to discount sums of money received 10 years hence when the interest rate is equal to 10 per cent. Therefore the implied rate of compound interest is slightly greater than 10 per cent.

(b) The present value of the £800 when the interest rate is 16 per cent is given by £800 $\times$ 0.227 = £181.6

2. (a) £1,000 $\times$ 0.270 = £270   (b) £1,000 $\times$ 0.463 = £463

3. (a) £250 $\times$ 0.621 = £155.25

(b) £155.25/0.751 = £206.72

4. Present value of $A = 100 \times 0.909 + 200 \times 0.826$

$+ 300 \times 0.751 = 481.4$

Present value of $B = 150 \times 0.909 + 300 \times 0.826$

$+ 100 \times 0.751 = 459.3$

$\therefore$ Project A has the greater present value if the discount rate is 10 per cent.

5. Let r = internal rate of return; then

$$\frac{(100-120)}{(1+r)} + \frac{(110-120)}{(1+r)^2} + \frac{(160-100)}{(1+r)^3} = 0$$

$$\frac{-20(1+r)^2 - 10(1+r) + 60}{(1+r)^3} = 0$$

$$\therefore \qquad 2(1+r)^2 + (1+r) - 6 = 0$$

$$2 + 4r + 2r^2 + 1 + r - 6 = 0$$

$$2r^2 + 5r - 3 = 0$$

$$r = \frac{-5 \pm \sqrt{(25+24)}}{4} = \frac{-5 \pm 7}{4}$$

Taking the positive root, $r = 0.5$, or 50 per cent.

Exercises 4.9

1. $PV = \dfrac{100}{0.08}\left[1 - \dfrac{1}{1.08^{20}}\right] = 981$.

2. $PV = A/i = 25/0.04 = 625$.

3. $A = 2500\,(1 - 1.09)/(1 - 1.09^{15}) = 85.15$.

4. (a) $PV = \dfrac{60}{0.06}\left[1 - \dfrac{1}{1.06^{10}}\right] = 441.6$

   (b) $PV = 60/0.06 = 1000$.

5. For a mortgage, $A = \dfrac{i\,(1+i)^n\,p}{(1+i)^n - 1}$

   where A = annual payment, i = interest rate, p = amount borrowed.

   Here $\qquad A = \dfrac{0.12\,(1.12)^{20}\,70000}{(1.12)^{20} - 1} = 9371.51$ per year

   With $n = 30$, $A = 8690.05$.

6. Value, $V = A\{(1+i)^n - 1\}/i = 300\{1.09^{10} - 1\}/0.09 = 4557.88$
   With $n = 20$, $V = 15348.04$.

7. Using the same formula as in Question 6, $V = 250\,\{1.08^n - 1\}/$
   0.08 and (a) $V = 18276.49$ (b) $V = 43079.20$.

Exercises 4.13

1. (a) Simple interest of 6% per annum produces   £45
   Compound interest of 5% produces   £41
   Compound interest of 4% paid twice per
    year produces   £33
   $\therefore$ most profitable investment is that returning simple interest
   at 6%.

   (b) £90, £94, £73; most profitable investment is that returning
   compound interest of 5% per annum.

2. $50(1.02)^{12} = 63.4$

3. (a) $1 + 5x + 10x^2 + 10x^3 + 5x^4 + x^5$

   (b) $8 - 12x + 6x^2 - x^3$

   (c) $81 + 108\left(\dfrac{1}{x}\right) + 54\left(\dfrac{1}{x}\right)^2 + 12\left(\dfrac{1}{x}\right)^3 + \left(\dfrac{1}{x}\right)^4$

   (d) $64 - 192x^2 + 240x^4 - 160x^6 + 60x^8 - 12x^{10} + x^{12}$

4. (a) $e^{0.1} = 1 + 0.1 + \dfrac{(0.1)^2}{2} + \dfrac{(0.1)^3}{6} + \cdots$

   $\phantom{e^{0.1}} = 1 + 0.1 + 0.005 + 0.00017 + \cdots$

   $\phantom{e^{0.1}} = 1.10517$ approx. (from tables $e^{0.1} = 1.1052$)

   (b) $e^{0.5} = 1 + 0.5 + \dfrac{(0.5)^2}{2} + \dfrac{(0.5)^3}{6} + \cdots$

   $\phantom{e^{0.5}} = 1.6458$ approx. (from tables $e^{0.5} = 1.6487$)

   If we include the next term in the expansion we obtain a closer
   approximation with $e^{0.5} = 1.6484$.

   (c) $e^2 = 1 + 2 + \frac{4}{2} + \frac{8}{6} + \frac{16}{24} + \frac{32}{120} + \frac{64}{720} + \cdots$

   $ = 7.35556$ approx. (from tables $e^2 = 7.3891$)

   $\therefore$ in this case we need to include more terms to obtain a
   good approximation.

   (d) $e^{-1} = 1 - 1 + \frac{1}{2} - \frac{1}{6} + \frac{1}{24} - \frac{1}{120} + \frac{1}{720} - \cdots$

   $\phantom{e^{-1}} = 0.368$ approx. (from tables $e^{-1} = 0.3679$).

5. Let P_t be the population in millions after t years so that $P_t = 2e^{0.04t}$.

When $t = 5$, $P_5 = 2e^{0.2} = 2(1.2214) = 2.44$ million. When $t = 25$, $P_{25} = 2e^{1.0} = 2(2.7183) = 5.44$ million.

If growth rate is 2%, $P_t = 2e^{0.02t}$ and $P_5 = 2e^{0.1} = 2(1.1052) = 2.21$ instead of 2.44, while $P_{25} = 2e^{0.5} = 2(1.6487) = 3.30$ instead of 5.44.

6. Let Y_t be the value of national income in after t years so that $Y_t = 125e^{0.03t}$. When $t = 30$, $Y_{30} = 125e^{0.9} = 125(2.4596) = 307.5$.

7. Let S_t be the number of firms surviving in year t then $S_t = 150000e^{-0.15t}$ and for $t = 4$, $S_4 = 150000e^{-0.6} = 150000(0.5488) = 82320$. For $t = 20$, $S_{20} = 150000e^{-3.0} = 150000(0.04979) = 7469$.

8. Let r be the required continuous rate of interest, then $e^r = (1 + 0.06/2)^2 = 1.03^2 = 1.0609$. Taking natural logarithms, $r = 0.0591$ or 5.91%.

Exercises 4.15

1. These use $\log_e \dfrac{(1 + x)}{(1 - x)} = 2\{ x + \dfrac{x^3}{3} + \dfrac{x^5}{5} + \cdots\}$

   and $x = (a - 1)/(a + 1)$. Also, $\log_{10} x = \log_e x / \log_e 10$. For $a = 0.1$, $x = -0.9/1.1 = -0.8181818$ so all the terms in the series are negative. The series is $2\{-0.8181818 - 0.1825695 - 0.0733296 - 0.0350631 - 0.0182560 - 0.0099990 - 0.0056638 - 0.0032859 - 0.0019409 - \ldots\} = 2\{-1.1482896\} = -2.2965792$ compared with -2.302585 in the tables. Using $\log_e 10 = 2.3025851$, $\log_{10} 0.1 = -2.2965792/2.3025851 = -0.9974$ while the accurate value is -1.0.

   For $a = 100$, $x = 99/101 = 0.980198$ and the series is $2\{0.980198 + 0.3139209 + 0.1809669 + 0.1241934 + 0.0928072 + 0.0729557 + 0.0593111 + 0.0493874 + 0.0418684 + \cdots\} = 2\{1.915609\} = 3.831218$. The accurate value is 4.60517 and the error is because of the slow convergence of the series which occurs because x is close to 1. $\log_{10} 100 = 3.831218/2.3025851 = 1.6639$ instead of the accurate value of 2.0.

2. Let P_t be the population in year t. Here, $P_t = 2.4e^{0.05t}$.

   (a) Require solution of $3 = 2.4e^{0.05t}$ or $1.25 = e^{0.05t}$. Taking natural logarithms, $\log_e 1.25 = 0.05t$ or $0.22314 = 0.05t$ so $t = 4.46$ years.

   (b) Require solution of $5 = 2.4e^{0.05t}$ or $2.0833 = e^{0.05t}$. Taking natural logarithms, $0.73397 = 0.05t$ so $t = 14.68$ years.

3. Let S_t be the number of firms surviving in year t. Here, S_t = $100000e^{-0.12t}$. Require the solution of $50000 = 100000e^{-0.12t}$ or $0.5 = e^{-0.12t}$. Taking natural logarithms, $-0.69315 = -0.12t$ so $t = 5.78$ years. But this is with this year taking the value 1, so it will be in 4.78 years from now.

4. Pareto's law: $n = N/x^{1.5}$. Here $N = 55000000$

   (a) $x = 10000$ so $n = 55000000/10000^{1.5}$
   Taking logarithms to the base 10, $\log_{10}n = 7.74036 - \{1.5(4.00)\} = 7.74036 - 6 = 1.74036$ and antilog is 55.0 and 55 are expected to have incomes over £10,000. This is rather low and implies that the power to which x is raised should be smaller than 1.5.

   (b) $x = 100000$ so $n = 55000000/100000^{1.5}$
   Taking logarithms to the base 10, $\log_{10}n = 7.74036 - \{1.5(5)\}$ $= 0.24036$ and antilog is 1.7. Therefore 2 are expected to have incomes over £100 000. Again this suggests the power to which x is raised should be lower.

Exercises 5.3

1. $y = 3x - 4$, $dy/dx = 3$

2. $y = x^2 - 3x + 3$, $dy/dx = 2x - 3$

3. $12x^3 - 3x^2 + 2x + 25$ 4. $18x^2 - x^3 + x^2 + 2/x^3$

5. $(1/x - 3x + 2)(4x + 3) + (2x^2 + 3x - 1)(-1/x^2 - 3)$

6. $(3x + 2/x^2)(2) + (2x + 4)(3 - 4/x^3)$

7. $\dfrac{(x^2 + 3x + 2)(8x) - (4x^2 + 4)(2x + 3)}{(x^2 + 3x + 2)^2}$

8. $5(2x + 3)^4 . 2$ 9. $\dfrac{1}{2x + 3} . 2$

10. $\log(2x^2 + 3x - 5) + \dfrac{x(4x + 3)}{2x^2 + 3x - 5}$

11. $6e^{2x} - 12e^{-3x} + 2xe^x + x^2e^x$

12. $2\cos 2x - 15\sin 5x - \dfrac{3}{\cos^2 3x}$

Exercises 5.6

1. (a) $TC = 4q^3 + 2q^2 - 25q$

Marginal cost $(MC) = \dfrac{d(TC)}{dq} = 12q^2 + 4q - 25$

Average cost $(AC) = \dfrac{TC}{q} = 4q^2 + 2q - 25$

(b) $MC = (q^3 - 3q)5 + (16 + 5q)(3q^2 - 3)$

$AC = (q^2 - 3)(16 + 5q)$

(c) $MC = e^{2q}(6) + 12qe^{2q}$

$AC = \dfrac{25}{q} + 6e^{2q}$

(d) $MC = (q^2 - q)\left(\dfrac{3}{q}\right) + (3 \log q + 5)(2q - 1)$

$AC = (3 \log q + 5)(q - 1)$

2. Elasticity of demand $= \dfrac{dq}{dp} \dfrac{p}{q}$

(a) $p + 2q = 50$ $\therefore$ $dq/dp = -\frac{1}{2}$

and when $p = 10$, $q = 20$

$\therefore$ elasticity of demand $= -(\frac{1}{2})(\frac{10}{20}) = -\frac{1}{4}$

(b) -2 (c) -6

3. Marginal revenue (MR) is found by expressing total revenue (TR) in terms of q and finding $d(TR)/dq$.

(a) $TR = pq = 200q - 3q^2$ $d(TR)/dq = 200 - 6q$

(b) $TR = pq = 100$ $d(TR)/dq = 0$

(c) $TR = pq = 250q - qe^{4q}$ $d(TR)/dq = 250 - e^{4q}(1 + 4q)$.

4. (a) Marginal cost, $MC = dTC/dq = 4q + 20$

(b) Total revenue, $TR = pq = 0.25(260 - q)q$

Marginal revenue, $MR = d(TR)/dq = 65 - 0.5q$

(c) Profit or net revenue, $NR = TR - TC$

$$= (65q - 0.25q^2) - (2q^2 + 20q + 300)$$

$$= 45q - 2.25q^2 - 300$$

Marginal profits $= d(NR)/dq = 45 - 4.5q$

(d) $MR = MC$ when $65 - 0.5q = 4q + 20$ or $45 = 4.5q$ so $q = 10$.

5. Profit or net revenue is $NR = TR - TC = pq - TC$

$$= 177q - q^3 - 30 - 2q - 2q^2 = 175q - q^3 - 30 - 2q^2$$

Marginal profit is $d(NR)/dq = 175 - 3q^2 - 4q$ and this is zero when $3q^2 + 4q - 175 = 0$ or $q = 7$ or -8.33 and so the answer is $q = 7$.

6. Average propensity to consume:

$$APC = C/Y = a/Y + b + cY$$

Marginal propensity to consume: $dC/dY = b + 2cY$.

Exercises 5.9

1. (a) $y = 3x^2 - 120x + 30$

$$\frac{dy}{dx} = 6x - 120$$

For a stationary value $\dfrac{dy}{dx} = 0$

so $6x - 120 = 0$ and $x = 20$

$$\frac{d^2y}{dx^2} = 6 \text{ which is positive}$$

∴ $x = 20$ is a minimum value and $y = -1170$

(b) $y = 16 - 8x - x^2$

$$\frac{dy}{dx} = -8 - 2x \qquad \frac{d^2y}{dx^2} = -2$$

$dy/dx = 0$ when $x = -4$ and there is a maximum value at this point with $y = 32$

(c) $y = x^4$

$$\frac{dy}{dx} = 4x^3 \qquad \frac{d^2y}{dx^2} = 12x^2$$

There is a stationary value when $4x^3 = 0$, that is $x = 0$. This is in fact a minimum, with $y = 0$, and the curve is symmetrical about the y-axis. (Draw the curve for $x = -3$ to $x = +3$.)

(d) $y = (x - 5)^3$

$$\frac{dy}{dx} = 3(x - 5)^2 \qquad \frac{d^2y}{dx^2} = 6(x - 5)$$

There is a stationary value when

$$3(x - 5)^2 = 0 \text{ that is when } x = 5$$

at this point d^2y/dx^2 is also equal to 0 and it is necessary to sketch the curve in order to show it is a point of inflexion.

2. $C = 500 + 4q + \frac{1}{2}q^2$

$$\therefore \quad \text{average cost} = \frac{500}{q} + 4 + \frac{q}{2}$$

$$\frac{d(AC)}{dq} = \frac{-500}{q^2} + \frac{1}{2} \quad \text{and} \quad \frac{d^2(AC)}{dq^2} = \frac{1000}{q^3}$$

There is a stationary point when $500/q^2 = \frac{1}{2}$

that is $\quad q^2 = 1000 \quad$ and $\quad q = \pm 31.6$

There is a minimum when $q = 31.6$ because $d^2AC/dq^2 > 0$ for this value.

3. Demand function is given by $q + 2p = 10$

Total revenue $= pq = p(10 - 2p)$

$$\therefore \quad \frac{d(TR)}{dp} = 10 - 4p, \qquad \frac{d^2(TR)}{dp^2} = -4$$

This has a stationary value when $p = \frac{10}{4} = \frac{5}{2}$

The second-order derivative is negative and there is therefore a maximum when $p = \frac{5}{2}$.

$$q = 10 - 2p = 10 - 5 = 5$$

$$\text{Elasticity of demand} = \frac{dq}{dp}\frac{p}{q} = \frac{(-2)(\frac{5}{2})}{5} = -1$$

4. Net revenue $= pq - TC = q(240 - q) - (q^3 + 20q^2 + 20)$

$$\therefore \quad \frac{d(NR)}{dq} = 240 - 2q - 3q^2 - 40q = -3q^2 - 42q + 240$$

This has a stationary value when $q^2 + 14q - 80 = 0$, that is when $q = 4.36$ or -18.35

$$\frac{d^2(NR)}{dq^2} = -6q - 42$$

This is positive when $q = -18.35$ and negative when $q = 4.36$.

$\therefore$ net revenue is maximised when $q = 4.36$ and $p = 235.6$.

5. (a) Total revenue, $TR = pq = (50 - 0.5q)q = 50q - 0.5q^2$
   Maximum is when $d(TR)/dq = 50 - q = 0$ so $q = 50$, provided that the second-order derivative is negative. Here $d^2(TR)/dq^2 = -1$ so $q = 50$ maximises total revenue.
   (b) Average cost, $AC = TC/q = 256/q + 2 + 2q$. Stationary values occur when $d(AC)/dq = -256/q^2 + 2 = 0$ so $q = 11.3$ or -11.3 The second-order condition for a minimum is that $d^2(AC)/dq^2$ is positive. Since $d^2(AC)/dq^2 = 512/q^3$ which is positive when $q = 11.3$, this is the value of q which minimises average cost.
   (c) Profit or net revenue is $NR = TR - TC = 50q - 0.5q^2 - (256 + 2q + 2q^2) = 48q - 2.5q^2 - 256$ and the stationary value is when $d(NR)/dq = 48 - 5q = 0$ or $q = 9.6$. Here $d^2(NR)/dq^2 = -5$ and so $q = 9.6$ maximises profit.
   Notice that to maximise total revenue, $q = 50$, to minimise average costs, $q = 11.3$ and to maximise profits, $q = 9.6$. It is the last of these that the monopolist should adopt.

Exercises 5.13

1. With price discrimination the firm will maximise profits in each market.
   For its own workers, profit or net revenue is $NR_1 =$ total revenue $-$ total costs $= P_1Q_1 - TC = (500 - 10Q_1)Q_1 - 1000 - 15Q_1 - 15Q_2 - 15Q_3$ and the maximum is when $d(NR_1)/dQ_1 = 500 - 20Q_1 - 15 = 0$ or $Q_1 = 24.25$. That this maximises profit is shown by $d^2(NR_1)/dQ_1^2 = -20$. Here $P_1 = 257.5$.

For the domestic market, $NR_2 = (100000 - 5Q_2)Q_2 - 1000 - 15Q_1 - 15Q_2 - 15Q_3$ and $d(NR_2)/dQ_2 = 100000 - 10Q_2 - 15$ which is zero when $Q_2 = 9998.5$. The second-order derivative is negative and so this maximises profit. Here $P_2 = 50007.5$.

For the overseas market, $NR_3 = (62500 - 2.5Q_3)Q_3 - 1000 - 15Q_1 - 15Q_2 - 15Q_3$ and $d(NR_3)/dQ_3 = 62500 - 5Q_3 - 15$ which is zero when $Q_3 = 12497$. The second-order derivative is negative and so this maximises profit. Here $P_3 = 31257.5$.

The total profit under price discrimination is $P_1Q_1 + P_2Q_2 + P_3Q_3 - TC = 6244.375 + 499999988.8 + 390624977.5 - (1000 + 15(24.25 + 9998.8 + 12497) = 890292414.4$. Notice the wide divergence of the prices in these three markets and that the total sales, $Q = Q_1 + Q_2 + Q_3 = 22519.75$.

With no price discrimination all the prices are P, say, and the three demand functions are combined by adding $Q_1 + Q_2 + Q_3 = (50 - 0.1P) + (20000 - 0.2P) + (25000 - 0.4P) = 45050 - 0.7P = Q$. The profit or net revenue is given by

$$NR = PQ - TC = (64357.14 - 1.4286Q)Q - 1000 - 15Q.$$

The stationary value occurs when $d(NR)/dQ = 0$ or $64357 - 2.8572Q - 15 = 0$ so $Q = 22519.25$ and the second-order derivative is negative, indicating a maximum value. Here $P = 32186.00$ and $NR = 724465791.8$ which is lower than the price-discrimination one of 890292414.4. Therefore it pays the monopolist to follow a policy of price discrimination. Also, notice that the value of Q is (apart from rounding errors) the same in both cases with linear demand functions. See question 2 below for a demonstration of this.

2. With price discrimination, maximising profits in the first market requires the maximisation of $NR_1 = p_1q_1 - TC = (a - bq_1)q_1 - f - g(q_1 + q_2)$ which occurs when $d(NR_1)/dq_1 = 0$ or $a - 2bq_1 - g = 0$ so $q_1 = (a - g)/2b$. The second-order derivative is $-2b$ which is negative if $b > 0$ (which requires the demand curve to be downward sloping) and so NR_1 is maximised. From the demand function $p_1 = (a + g)/2$.

For the second market, $NR_2 = (c - eq_2)q_2 - f - g(q_1 + q_2)$ and when $d(NR_2)/dq_2 = 0$ then $c - 2eq_2 - g = 0$ and $q_2 = (c - g)/2e$. The second-order condition proves that this gives a maximum of NR_2. Here $p_2 = (c + g)/2$.

The quantity sold with price discrimination is

$$q = q_1 + q_2 = \frac{(a - g)}{2b} + \frac{(c - g)}{2e} = \frac{ae - eg + bc - bg}{2be}$$

With a common price p, the total demand is

$$q = q_1 + q_2 = \frac{(a - p)}{b} + \frac{(c - p)}{e} = \frac{(ae + bc) - (b + e)p}{be} \quad \text{or}$$

$$p = \frac{(ae + bc) - beq}{(b + e)} = A - Bq \text{ where } A = \frac{(ae + bc)}{(b + e)},$$

$B = be/(b + e)$ and $NR = pq - TC = (A - Bq)q - f - gq$

Differentiating gives

$$\frac{dNR}{dq} = A - 2Bq - g \text{ which is zero when } q = \frac{(A - g)}{2B}$$

The second-order derivative is $-2B$ which is negative (since b and e are positive) so that the stationary value is a maximum. Here $p = (A + g)/2$.

Therefore the quantity sold without price discrimination will be

$$q = \frac{(A - g)}{2B} = \frac{ae + bc - bg - eg}{2be}$$

which is the same value as with price discrimination.

3. (a) With a flat-rate tax, demand is $p + 2q = 250$

supply is $(p - t) - 4q = 100$

and tax revenue is $T = tq$ where t is the flat-rate tax. Equilibrium is when $q = 0.5(250 - p) = 0.25(p - t - 100)$ or $500 - 2p = p - t - 100$ so $p = (600 + t)/3$ and $q = 25 - (t/6)$. Hence, $T = tq = 25t - (t^2/6)$ and so $dT/dt = 25 - t/3$ and this is zero when $t = 75$ and $q = 12.5$, $p = 225$. This is a maximum since the second-order derivative, $d^2T/dt^2 = -1/3$ is negative. The maximum tax revenue is $T = tq = 12.5(75) = 937.5$.

(b) With tax of $r\%$ the supply function is $p(1 - r) - 4q = 100$ and demand is $p + 2q = 250$. In equilibrium, $q = 0.5(250 - p) = 0.25[p(1 - r) - 100]$ or $500 - 2p = p - rp - 100$ so that $p = 600/(3 - r)$, $q = 125 - [300/(3 - r)] = (75 - 125r)/(3 - r)$. The tax revenue is

$$T = rpq = \frac{r(600)[75 - 125r]}{(3 - r)(3 - r)} = \frac{600[75r - 125r^2]}{(3 - r)^2}$$

So,

$$\frac{dT}{dr} = \frac{600\{(3 - r)^2(75 - 250r) - (75r - 125r^2)(2)(-1)(3 - r)\}}{(3 - r)^4}$$

Setting to zero: $(3 - r)(75 - 250r) = -2(75r - 125r^2)$ or $225 - 75r - 750r + 250r^2 = 250r^2 - 150r$ or $225 = 675r$ and $r = 1/3$. To check for a maximum the second-order derivative is required. Now, simplifying,

$$\frac{dT}{dr} = \frac{600\{225 - 75r - 750r + 250r^2 + 150r - 250r^2\}}{(3 - r)^3}$$

$$= \frac{600\{225 - 675r\}}{(3 - r)^3} = \frac{135000\{1 - 3r\}}{(3 - r)^3}$$

$$\frac{d^2T}{dr^2} = \frac{135000\{(3 - r)^3(-3) - (1 - 3r)3(-1)(3 - r)^2}{(3 - r)^6}$$

When $r = 1/3$, the sign of this is negative and so the maximum is when $r = 1/3$, $p = 225$, $q = 12.5$ and the tax revenue is $T = rpq = 937.5$. This is the same as for the flat-rate tax case in (a).

4. (a) In equilibrium, $110 - 3q^2 = 10 + q^2$ or $100 = 4q^2$ and $q = 5$ (taking the positive root), $p = 35$.

   (b) With a flat-rate tax of t per unit supply is now $p - t = 10 + q^2$ and demand is $p = 110 - 3q^2$. The new equilibrium is when $10 + t + q^2 = 110 - 3q^2$ or $q^2 = 25 - (t/4)$ and $p = 110 - 3q^2 = 35 + 3(t/4)$. Tax revenue is $T = tq = t(25 - (t/4))^{0.5}$ and so

$$\frac{dT}{dt} = (25 - (t/4))^{0.5} + \frac{t(0.5)(-0.25)}{(25 - (t/4))^{0.5}}$$

Setting to zero, $(25 - (t/4)) = 0.125t$ and $t = 66.67$. For this value $q = 2.89$ and $p = 85$. To check for a maximum, the second-order derivative is needed. It is convenient here to use $q^2 = 25 - (t/4)$

and to note that differentiating this gives

$$2q\frac{dq}{dt} = \frac{-1}{4}$$

so that
$$\frac{dq}{dt} = \frac{-1}{8q} = q'$$

Since
$$\frac{dT}{dt} = q - \frac{0.125t}{q}$$

$$\frac{d^2T}{dt^2} = \frac{dq}{dt} - \frac{\{q - tq'\}}{8q^2} = \frac{-1}{8q} - \frac{1}{8q} - \frac{t}{64q^3}$$

which is negative. The maximum is when $t = 66.67$ and $T = tq = 192.68$.

5. Here $N = 20000$ = annual demand, $S = 2000$ = set-up cost Q = batch size (to be determined), $i = 5$ = inventory cost per unit per year. The number of batches per year is N/Q. The total annual cost, C is given by the sum of total set-up costs times number of batches and the inventory cost

$$C = \frac{SN}{Q} + \frac{iQ}{2} = \frac{40000000}{Q} + \frac{5Q}{2} \text{ and } \frac{dC}{dQ} = \frac{-40000000}{Q^2} + \frac{5}{2}$$

Setting to zero gives $Q = 4000$ and the second-order derivative is positive, indicating this is a minimum. The total annual cost is £20,000.

6. Let $R = 50000$ be the rate of production. As before, $i = 5$, $N = 20000$, $S = 2000$. The maximum inventory is now $Q(1 - N/R)$ (see the text, p. 204) and the total annual cost is

$$C = 0.5iQ(1 - \frac{N}{R}) + \frac{SN}{Q} = 1.5Q + \frac{40000000}{Q}$$

and $\dfrac{dC}{dQ} = 1.5 - \dfrac{40000000}{Q^2}$ which is zero when $Q = 5164$.

The second-order derivative is positive and so $Q = 5164$ give the production run size which minimises the total annual cost. The total cost is 15492, which is below the value in question 5 since there is a saving in inventory costs when production is spread over time, rather than being instantaneous.

7. Here, annual demand, $N = 3000$ boxes, inventory cost $i = 3$, set-up cost $S = 500$ and the annual production rate is $R = 9000$ boxes. The average inventory size is $0.5Q(1 - N/R) = Q/3$ and so the total annual cost is

$$C = iQ/3 + SN/Q = Q + 1500000/Q$$

$$\frac{dC}{dQ} = 1 - \frac{1500000}{Q^2} \text{ and this is zero when } Q = 1224.7.$$

As the second-order derivative is positive this is the minimising value and the total cost is 2449.6.

Exercises 5.19

1. Taylor's theorem states

$$f(a + x) = f(a) + xf'(a) + \frac{x^2}{2!}f''(a) + \cdots$$

Let $f(a + x) = (1 + x)^4$

$f(a) = 1$

$f'(a + x) = 4(1 + x)^3 \quad \text{and} \quad f'(a) = 4$

$f''(a + x) = 12(1 + x)^2 \quad \text{and} \quad f''(a) = 12$

$f'''(a + x) = 24(1 + x) \quad \text{and} \quad f'''(a) = 24$

$f^4(a + x) = 24 \quad\quad\quad\quad \text{and} \quad f^4(a) = 24$

$$\therefore \quad (1 + x)^4 = 1 + x(4) + \frac{x^2}{2}(12) + \frac{x^3}{6}(24) + \frac{x^4}{24}(24)$$

$$= 1 + 4x + 6x^2 + 4x^3 + x^4$$

and $1.5^4 = (1 + \frac{1}{2})^4 = 1 + 4 \times \frac{1}{2} + 6 \times \frac{1}{4} + 4 \times \frac{1}{8} + \frac{1}{16}$

$$= 1 + 2 + 1.5 + 0.5 + 0.0625 = 5.0625$$

2. MacLaurin's theorem states

$$f(x) = f(0) + xf'(0) + \frac{x^2}{2!}f''(0) + \cdots$$

Let $f(x) = (1 + x)^{-1}$

then $f'(x) = -(1 + x)^{-2}$

$f''(x) = 2(1 + x)^{-3}$

$f'''(x) = -6(1 + x)^{-4}$

$$\vdots$$

Therefore $(1 + x)^{-1} = 1 - x + x^2 - x^3 + \cdots$

3. (a) $f(x) = x^2 + 4.55x - 8.70$

   $f'(x) = 2(x) + 4.55$

   Using as a starting point $a = 1.5$

   $f(1.5) = 2.25 + 6.825 - 8.70 = 0.375$

   $f'(1.5) = 3.0 + 4.55 = 7.55$

   $\therefore \quad h_1 = -\dfrac{f(a)}{f'(a)} = -\dfrac{0.375}{7.55} = -0.0497$

   $\therefore$ new approximate root is $1.5 - 0.05 = 1.45$ for which

   $$f(1.45) = 2.1 + 6.6 - 8.7 = 0$$

   This result shows that 1.45 is a good approximation to one of the values of x which satisfies the equation.

   (b) $f(x) = x^3 - 2.34x^2 + 2x - 4.68$

   $f'(x) = 3x^2 - 4.68x + 2$

   Using as a starting point $a = 2.25$

   $f(a) = 11.39 - 11.85 + 4.50 - 4.68 = -0.64$

   $f'(a) = 15.19 - 10.53 + 2 = 6.66$

   $\therefore \quad h_1 = -\dfrac{f(a)}{f'(a)} = \dfrac{0.64}{6.66} = 0.096$

   $\therefore$ new approximate root is $2.25 + 0.096 = 2.346$

   for which $f(2.346) = 0.045$. A closer approximation can be obtained by repeating this process.

   (c) $x = 4.1$ for which $f(4.1) = 0$.

4. $PV_1 = \dfrac{70}{R} + \dfrac{80}{R^2} + \dfrac{100}{R^3} + \dfrac{100}{R^4}$

   $PV_2 = \dfrac{60}{R} + \dfrac{90}{R^2} + \dfrac{80}{R^3} + \dfrac{130}{R^4}$

   these are equal when $PV_1 - PV_2 = 0$ that is

   $$\dfrac{10}{R} - \dfrac{10}{R^2} + \dfrac{20}{R^3} - \dfrac{30}{R^4} = 0$$

or $$R^3 - R^2 + 2R - 3 = 0$$

$$f(R) = R^3 - R^2 + 2R - 3$$

$$f'(R) = 3R^2 - 2R + 2$$

Using Newton's method with $a = 1$ as a starting point, we obtain

$$f(a) = 1 - 1 + 2 - 3 = -1$$

$$f'(a) = 3 - 2 + 2 = 3$$

$$\therefore \qquad h_1 = -\frac{f(a)}{f'(a)} = \tfrac{1}{3} = 0.33$$

$\therefore$ approximate root is $1 + 0.33 = 1.33$; using this as a new starting point, we have

$$f(1.33) = 0.2437$$

$$f'(1.33) = 4.6467$$

$$\therefore \qquad h_2 = -\frac{f(1.33)}{f'(1.33)} = -\frac{0.2437}{4.6467} = -0.052$$

Try $R = 1.33 - 0.052 = 1.278$ as the next root: $f(1.278) = 0.0101$ and so the rate of interest is approximately 27.8 per cent. A more accurate value can be obtained by repeating the process.

5. In each case the places where the function cuts the axes are required and also where the stationary values are.

(a) $y = x^2 + x + 12$. This cuts the y-axis when $x = 0$, $y = 12$, and the x-axis when $0 = x^2 + x + 12$ or

$$x = \frac{-1 \sqrt{(1 - 48)}}{2} \text{ so there are no real roots}$$

The stationary values are when

$$\frac{dy}{dx} = 2x + 1 = 0 \text{ or } x = \frac{-1}{2}$$

$$\frac{d^2y}{dx^2} = 2$$

and so there is a local minimum when $x = -0.5$, $y = 11.75$. As extra values to help to position the graph, $x = -3$, $y = 18$, $x = 3$, $y = 24$, and the table of values is:

x	-3	-0.5	0	3
y	18	11.75	12	24
		min		

The graph is shown in Fig. C.1.

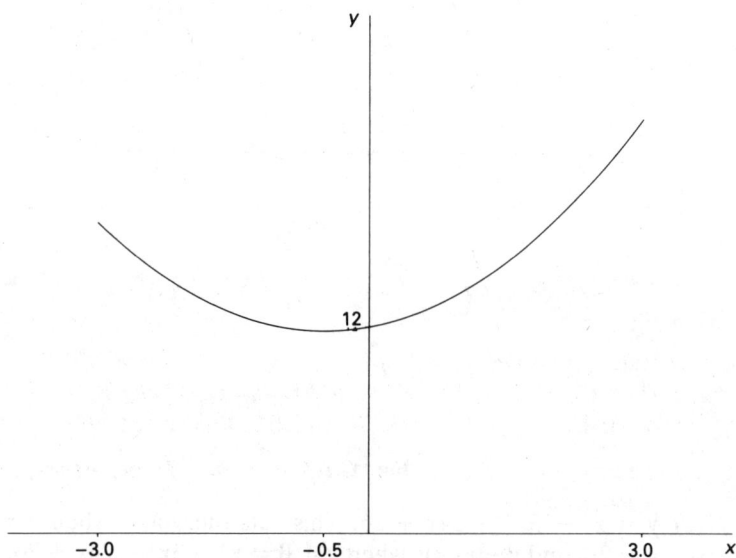

Fig. C.1

(b) $y = x^3 - x^2$. This has $x = 0$, $y = 0$, and also since $y = x^2(x - 1)$, when $x = 1$, $y = 0$

Here, $\dfrac{dy}{dx} = 3x^2 - 2x$ and $\dfrac{d^2y}{dx^2} = 6x - 2$

The stationary values are when $3x^2 - 2x = 0$ and so $x = 0$ or $x = 2/3$. The second-order derivative is negative for $x = 0$ (a local maximum with $y = 0$) and positive for $x = 2/3$ (a local minimum with $y = -0.148$). For extra values take $x = -1$, $y = -2$ and $x = 2$, $y = 4$. The table of values is:

x	-1	0	2/3	1	2
y	-2	0	-0.148	0	4
		max	min		

The graph is shown in Fig. C.2.

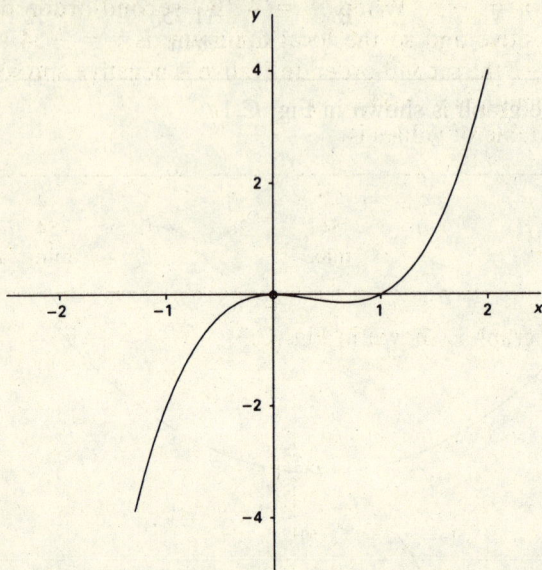

Fig. C.2

(c) $y = x^3 - 3x^2 - 24x + 26$. This cuts the y-axis when $x = 0$, $y = 26$, and the x-axis when $y = 0 = x^3 - 3x^2 - 24x + 26$. To find an approximate root, since the cubic has some terms with negative signs, positive values of x are worth trying. For $x = 1$, $y = 0$ so that $(x - 1)$ is a factor and dividing into the cubic:

$$
\begin{array}{r}
x^2 - 2x - 26 \\
x - 1 \overline{\smash{\big)}\ x^3 - 3x^2 - 24x + 26} \\
\underline{x^3 - x^2} \\
-2x^2 - 24x \\
\underline{-2x^2 + 2x} \\
-26x + 26 \\
-26x + 26
\end{array}
$$

and so $y = (x - 1)(x^2 - 2x - 26)$ and if $y = 0$, $x^2 - 2x - 26 = 0$ so $x = 6.2$ or $x = -4.2$

Here, $\dfrac{dy}{dx} = 3x^2 - 6x - 24$ and $\dfrac{d^2y}{dx^2} = 6x - 6$

The stationary values are when $3x^2 - 6x - 24 = 0$ and so $x = 4$ and $x = -2$. When $x = 4$ the second-order derivative is positive and so the local minimum is $y = -54$ and when $x = -2$ the second-order derivative is negative and so the local maximum is $y = 54$,
The table of values is:

x	−4.2	−2	0	1	4	6.2
y	0	54	26	0	−54	0
		max			min	

The graph is shown in Fig. C.3.

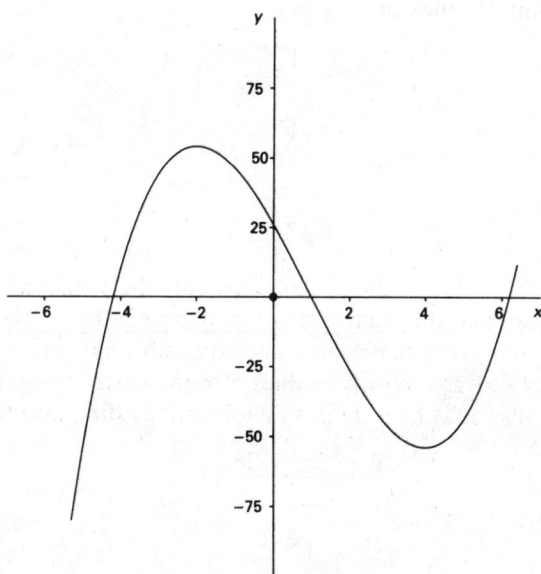

Fig. C.3

(d) $y = x^5$. This goes through the origin ($x = 0$, $y = 0$) and does not cut the axes anywhere else. In attempting to find stationary points, we get

$$\frac{dy}{dx} = 5x^4 \text{ and} \frac{d^2y}{dx^2} = 20x^3$$

and when

$$\frac{dy}{dx} = 0, \; 5x^4 = 0 \text{ so } x = 0$$

but for this value the second-order derivative is also zero. This means that the general rules do not give a clear indication of whether at $x = 0$ there is a maximum, a minimum or a point of inflexion. We therefore work out a table of values centred on $x = 0$ to obtain the general shape of the curve:

x	-3	-2	-1	0	1	2	3
y	-243	-32	-1	0	1	32	243

The graph is shown in Fig. C.4 and it can be seen that there is a point of inflexion at $x = 0$.

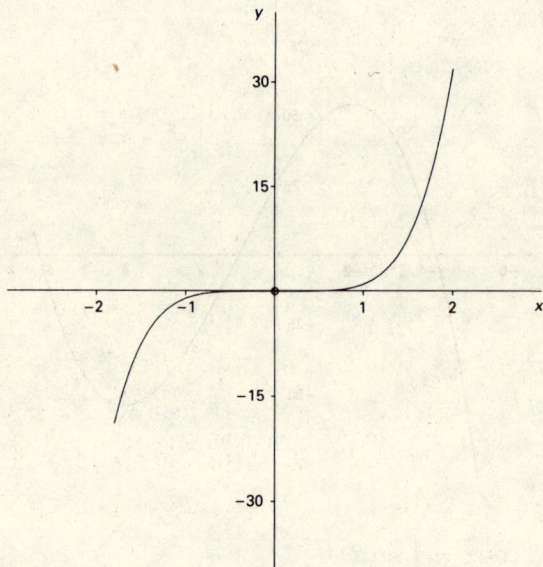

Fig. C.4

Exercises 6.2

1. (a) $\dfrac{dTC}{dq} = 1 \quad \therefore \quad TC = q + C$

   when $q = 0, \quad TC = 600$

$$\therefore \quad TC = 600 + q$$

(b) $TC = 50 + q$

(c) $TC = x^2 + 3x + C$

$$100 = 25 + 15 + C$$

$$\therefore \quad C = 60 \qquad \text{and} \qquad TC = x^2 + 3x + 60.$$

2. Given $C = a + bY + cY^2$ then $mpc = \dfrac{dC}{dY} = b + 2cY$

Since $mpc = 0.1 + 0.02Y$, $b = 0.1$ and $c = 0.01$

The consumption function is $C = a + 0.1Y + 0.01Y^2$

When $Y = 0$, $C = 0$ and so $a = 0$, giving $C = 0.1Y + 0.01Y^2$

3. Here $I(t) = \dfrac{dK}{dt} = a + 2bt + 3ct^2$

and $K = \int I(t)dt = at + bt^2 + ct^3 + d$

Since $K = 0$ when $t = 0$, $d = 0$ and $K = at + bt^2 + ct^3$

Exercises 6.4

1. (a) $TC = x^2 + 3x + 10$

(b) $TC = \frac{1}{3}x^3 + x^2 + 4x + 90$

2. (a) $\frac{3}{2}x^2 - 4x + C$ (b) $\frac{2}{3}x^3 - 2x^2 + 3x + C$

(c) $\frac{1}{2}x^6 + 4/x + x^2 + C$ (d) $-1/x^2 - 4 \log x + C$

(e) $\log(3x - 5) + C$ (f) $\log(2x^2 + x - 4) + C$

(g) $\frac{1}{4}\log(x^4 - 2x^2 + 2) + C$ (h) $2 \log(x^2 + 3x + 1)$

(i) $\frac{2}{3}e^{3x} - e^{-x} + C$ (j) $2e^x - \frac{4}{3}e^{3x} + C$

(k) $-2 \cos x - 4 \sin x$

3. Given $MR = a - 2bq$, $TR = \int MR\, dq = aq - bq^2 + c$

When $q = 0$, $TR = 0$ so that $c = 0$ and $TR = aq - bq^2$

Now $TR = pq$ and so $p = TR/q = a - bq$ is the demand function.

4. Given $e^d = -0.3\dfrac{p}{q}$ and since $e^d = \dfrac{pdq}{qdp}$ then $\dfrac{dq}{dp} = -0.3$

Integrating, $q = -0.3p + c$

When $p = 10$, $q = 100 = -30 + c$ and so $c = 130$

The demand curve is $q = 130 - 0.3p$

5. Given $mpc = 0.5 + Y^{-1} - Y^{-2} = dC/dY$

Integrating, $C = 0.5Y + \log Y + Y^{-1} + A$

When $Y = 1$, $C = 1 = 0.5 + 0 + 1 + A$ so $A = -0.5$ and the consumption function is $C = 0.5Y + \log Y + Y^{-1} - 0.5$.

Exercises 6.6

1. (a) $\displaystyle\int_{x_1=0}^{x_2=3} y\, dx = \int_0^3 2x\, dx = [x^2]_0^3 = 9$

   (b) $[x^2]_3^6 = 36 - 9 = 27$

   (c) $[x^2]_0^6 = 36$

   (d) $[\tfrac{5}{2}x^2]_6^8 = \tfrac{5}{2}[64 - 36] = 70$

   (e) $[2x + \tfrac{3}{2}x^2]_0^3 = 6 + \tfrac{27}{2} = 19\tfrac{1}{2}$

2. Since $MC = 100 - 4q + q^2$, $TC = \displaystyle\int_{20}^{30} MC dq$

   $= [100q - 2q^2 + (q^3/3)]_{20}^{30}$

   $= (3000 - 1800 + 9000) - (2000 - 800 + 2666.7)$

   $= 10200 - 3866.7 = 6333.3.$

3. $TR = \displaystyle\int_{15}^{20} MR\, dq = [200q - 3q^2]_{15}^{20}$

   $= (4000 - 1200) - (3000 - 675) = 475.$

4. $P_n = \displaystyle\int_0^{10} 1000e^{-0.05t}\, dt = 1000[e^{-0.05t}/(-0.05)]_0^{10}$

   $= -20000\{e^{-0.5} - 1\} = -20000\{0.6065 - 1\} = 7870.$

Exercises 6.8

1. (a) $y = 2x$ crosses the x-axis at $x = 0$.

   $\therefore$ area $= 2\displaystyle\int_0^1 2x\, dx = 2[x^2]_0^1 = 2$

   (b) $y = 2 + 3x$ crosses the x-axis at $x = -\tfrac{2}{3}$

$$\therefore \quad \text{area} = - \int_{-3}^{-2/3} (2 + 3x)dx + \int_{-2/3}^{2} (2 + 3x)dx$$

$$= - [2x + \tfrac{3}{2} x^2]_{-3}^{-2/3} + [2x + \tfrac{3}{2} x^2]_{-2/3}^{2}$$

$$= - [-\tfrac{4}{3} + \tfrac{2}{3} - (-6 + 13.5)] + [4 + 6 - (-\tfrac{4}{3} + \tfrac{2}{3})]$$

$$= 8.16 + 10.67 = 18.83$$

(c) Area $= \int_{-2}^{3} x^2 \, dx = [\tfrac{1}{3} x^3]_{-2}^{3} = \tfrac{8}{3} + 9 = 11\tfrac{2}{3}$ \qquad (as the curve does not cross the x-axis)

(d) $y = 1 + x + x^2$ does not cross the x-axis

$$\therefore \qquad\qquad \text{area} = \int_{-4}^{+1} (1 + x + x^2)dx$$

$$= [x + \tfrac{1}{2} x^2 + \tfrac{1}{3} x^3]_{-4}^{+1} = 115/6$$

2. $E(t) = 55 - 16t + t^2$. This cuts the t-axis when $E(t) = 0$ or $t = 5$ or 11. The required area is evaluated in two parts: from $t = 0$ to $t = 5$ and from $t = 5$ to $t = 10$

$$\int_{0}^{5} E(t)dt = [55t - 8t^2 + 0.3333t^3]_{0}^{5}$$

$$= (275 - 200 + 41.67) - (0) = 116.67$$

$$\int_{5}^{10} E(t)dt = [55t - 8t^2 + 0.3333t^3]_{5}^{10}$$

$$= (550 - 800 + 333.33) - (116.67) = -33.33$$

Total absolute error $= 116.67 + 33.33 = 150$.

Exercises 6.11

1. When $p = 40$, $q = 10$. Consumer's surplus $= \int_{0}^{10} pdq - pq$

$$= \int_{0}^{10} (100 - 6q)dq - 400 = [100q - 3q^2]_{0}^{10} - 400$$

$$= (1000 - 300) - (0) - 400 = 300$$

2. Equilibrium is when $60 - q - q^2 = 10 + 4q$ so $q = 5$ or $q = -10$. For $q = 5$, $p = 30$:

area under demand curve is $\int_{0}^{5} pdq = [60q - 0.5q^2 - 0.3333q^3]_{0}^{5}$

$$= (300 - 12.5 - 41.67) - (0) = 245.83$$

consumer's surplus $= 245.83 - 150 = 95.83$

area under supply curve $= \int_0^5 (10 + 4q)dq = [10q + 2q^2]_0^5 = 100$

producer's surplus $= pq - 100 = 150 - 100 = 50$

With a tax of 2 per unit supply is now $p - 2 = 10 + 4q$ or $p = 12 + 4q$ (see section 5.11). The new equilibrium is when $12 + 4q = 60 - q - q^2$ or $q = 4.865$, $p = 31.468$.

area under demand curve is

$$[60q - 0.5q^2 - 0.3333q^3]_0^{4.865} = (291.9 - 11.834 - 38.382)$$
$$= 241.684$$

and the consumer's surplus $= 241.684 - (4.865)(31.468) = 88.59$, which is smaller than the previous value of 95.83

area under the supply curve is now $\int_0^{4.865} (12 + q)dq = [12q + 0.5q^2]_0^{4.865} = (58.38 + 11.834) = 70.21$

producer's surplus is $pq - 70.21 = 153.09 - 70.21 = 82.88$, an increase on the previous value.

3. $MR = 200 - 3q$ and $MC = 2q$ so profit $= \int_0^{30} (MR-MC)dq = \int_0^{30} (200 - 5q)dq = [200q - 2.5q^2]_0^{30} = (6000 - 2250) - 0 = 3750$

Maximum profit is when $MR = MC$ or $200 - 3q = 2q$ and $q = 40$ (assuming that the second-order condition is satisfied, which can easily be checked)

$$\int_0^{40} (MR-MC)dq = [200q - 2.5q^2]_0^{40} = 4000.$$

4. $S(t) - C(t) = 10 - 5t + 30t^2 - 4t^3$. Total net revenue $= \int_0^4 (S(t) - C(t)) dt = \int_0^4 (10 - 5t + 30t^2 - 4t^3) dt = [10t - 2.5t^2 + 10t^3 - t^4]_0^4 = (40 - 40 + 640 - 256) - (0) = 384$

If the campaign is extended from 6 to 8 months the extra net revenue is

$$\int_6^8 (S(t) - C(t)) dt = [10t - 2.5t^2 + 10t^3 - t^4]_6^8 = (944) - (834)$$
$$= 110$$

and so it is worthwhile. But this is not optimal since $S(7) - C(7) = 73 > 0$ while $S(8) - C(8) = -158 < 0$. That is, the optimal length is the value of t for which $S(t) = C(t)$, and is approx. 7.4 months.

5. $MR = MC$ when $600 - 10t - 3t^2 = 50 + 15t$ or $3t^2 + 25t - 550 = 0$

so $t = 10$ years. Total net revenue is

$$\int_0^{10} (MR - MC)\,dt = \int_0^{10} (550 - 25t - 3t^2)\,dt = [550t - 12.5t^2 -$$
$$t^3]_0^{10} = (5500 - 1250 - 1000) - (0) = 3250$$

Total return at time T is $R(T) = \int_0^T (MR - MC)\,dt + S(T)$

$$= 550T - 12.5T^2 - T^3 + 5000 - 300T$$

$$= 250T - 12.5T^2 - T^3 + 5000$$

$\dfrac{dR}{dT} = 250 - 25T - 3T^2$ which is zero when $T = 5.87$

The maximum value is $R(5.87) = 5834.53$.

Exercises 6.13

1. (a) Exact: $\int_1^3 4x^3\,dx = [x^4]_1^3 = 81 - 1 = 80$

   (b) Simpson's rule: Taking $n = 4$ and $c = 0.5$ the table of values is:

x	1	1.5	2	2.5	3
y	4	13.5	32	62.5	108

   The approximate area $= c[y_0 + 4y_1 + 2y_2 + 4y_3 + y_4]/3$
   $= 0.5\,[4 + 54 + 64 + 250 + 108]/3 = 80$.

2. Here $n = 4$ and $c = (b - a)/n = 1$. Total cost $= \int_{10}^{14} MC\,dq$
   $\approx c[MC_0 + 4MC_1 + 2MC_2 + 4MC_3 + MC_4]/3$
   $\approx [3 + 16 + 18 + 40 + 15]/3 = 30.67$.

3. (a) $n = 2$, $c = (b - a)/n = 4$. The table of values is:

x	0	4	8
y	0	256	4096

   area $\approx c[y_0 + 4y_1 + y_2]3 = 4[0 + 1024 + 4096]/3 = 6826.7$.

   (b) $n = 8$, $c = 1$. The table of values is:

x	0	1	2	3	4	5	6	7	8
y	0	1	16	81	256	625	1296	2401	4096

$$\text{area} \approx c[y_0 + 4y_1 + 2y_2 + 4y_3 + 2y_4 + 4y_5 + 2y_6 + 4y_7 + y_8]/3$$
$$\approx [0 + 4 + 32 + 324 + 512 + 2500 + 2592 + 9604 + 4096]/3 = 6554.7.$$

(c) Exact: $\int_0^8 x^4\,dx = [0.2x^5]_0^8 = 0.2(32768) = 6553.6$

Here Simpson's rule with $n = 8$ is close to the exact value.

4. Require $\int_1^4 x^{-1}dx$. Here $f(x) = x^{-1}$, $f'(x) = -x^{-2}$, $f''(x) = 2x^{-3}$, $f'''(x) = -6x^{-4}$ and the integral of Taylor's series is:

$$\int_1^4 f(x)\,dx = [f(a)x + \frac{(x-a)^2 f'(a)}{2!} + \frac{(x-a)^3}{3!}f''(a) + \cdots]_1^4$$

(a) For $\quad a = 1, f(1) = 1, f'(1) = -1, f''(1) = 2, f'''(1) = -6$

$$\int_1^4 f(x)\,dx = [x - \frac{(x-1)^2}{2} + \frac{(x-1)^3}{3} - \frac{(x-1)^4}{4} + \cdots]_1^4$$
$$\approx (4 - 4.5 + 9 - 20.25) - (1)$$
$$\approx -12.75$$

This poor result occurs because the Taylor's series for $a = 1$ does not converge.

(b) For $\quad a = 3, f(3) = 0.3333, f'(3) = -0.1111,$

$$f''(3) = 0.0741, f'''(3) = -0.0741$$

$$\int_1^4 f(x) = [0.3333x - 0.0556\,(x-3)^2 + 0.0124\,(x-3)^3 - 0.0031\,(x-3)^4 + \cdots]_1^4$$
$$\approx (1.333 - 0.0556 + 0.0124 - 0.0031)$$
$$- (0.3333 - 0.2224 - 0.0992 - 0.0496)$$
$$\approx 1.287 - (-0.0379) = 1.3249.$$

(c) From section 6.12, the value when $a = 2.5$ is 1.344 and the exact value is 1.3863. For $a = 1$ the result -12.75 is very inaccurate, while with $a = 3$, 1.3249 is worse than when $a = 2.5$. This suggests that the value of a is important and should be in the middle of the range of x-values. Also, with Taylor's series it may be necessary to check whether the expansion converges and include more than just the first three derivatives.

5. (a) Exact: $\int_0^4 e^x dx = [e^x]_0^4 = (e^4) - (e^0) = 54.598 - 1 = 53.598$.

(b) Simpson's rule with $n = 4$: With $c = 1$ the table of values is:

x	0	1	2	3	4
y	1	2.718	7.389	20.086	54.598

area $\approx c[y_0 + 4y_1 + 2y_2 + 4y_3 + y_4]/3$

$\approx [1 + 10.872 + 14.778 + 80.344 + 54.598]/3 = 53.864$.

(c) By Taylor's series: $f(x) = e^x = f'(x) = f''(x) = f'''(x)$ etc. Taking $a = 2$, $f(2) = 7.389 = f'(2) = f''(2) = f'''(2)$ etc. and this can be taken outside the brackets:

$$\int_0^4 f(x)\,dx = [f(a)x + \frac{(x-a)^2}{2!}f'(a) + \frac{(x-a)^3}{3!}f''(a) + \cdots]_0^4$$

$$\approx 7.389\,[x + \frac{(x-2)^2}{2!} + \frac{(x-2)^3}{3!} + \frac{(x-2)^4}{4!} + \frac{(x-2)^5}{5!}]_0^4$$

$$\approx 7.389\,\{(4 + 2 + 1.333 + 0.6667 + 0.2667)$$

$$- (2 - 1.3333 + 0.6667 - 0.2667)\}$$

$$\approx 7.389\,(8.2667 - 1.0667) = 53.2$$

Here Simpson's rule gives a better approximation than does the Taylor's series result, but adding extra terms to the Taylor's series expansion will improve its accuracy.

Exercises 6.15

1. Let $(5x^2 + 4) = u$ and $e^{3x}\,dx = dv$

Then $\int (5x^2 + 4)e^{3x}\,dx = (5x^2 + 4)\tfrac{1}{3}e^{3x} - \int \frac{10xe^{3x}}{3}\,dx$

and $\int 10xe^{3x}dx = 10x(\tfrac{1}{3}e^{3x}) - \int \frac{10e^{3x}}{3}\,dx$

and $\int 10e^{3x}dx = 10(\tfrac{1}{3}e^{3x})$

$$\therefore \quad \int (5x^2 + 4)\, e^{3x} dx = e^{3x} \left[\frac{5x^2 + 4}{3} - \frac{10x}{9} + \frac{10}{27} \right] + C.$$

2. Let $\log x = u$ and $x^2 dx = dv$

   Then $\int x^2 \log x\, dx = \int \log x\, x^2 dx$

$$= \log x (\tfrac{1}{3} x^3) - \int \tfrac{1}{3} x^3 \frac{1}{x}\, dx$$

$$= \tfrac{1}{3} x^3 \log x - \tfrac{1}{3} \int x^2 dx$$

$$= \tfrac{1}{3} [x^3 \log x - \tfrac{1}{3} x^3] + C$$

$$= \tfrac{1}{3} x^3 [\log x - \tfrac{1}{3}] + C.$$

3. Let $\quad \dfrac{3}{(x+1)(x-1)} = \dfrac{A}{(x+1)} + \dfrac{B}{(x-1)}$

   Then $\qquad\qquad 3 = A(x-1) + B(x+1)$

   and $\qquad\qquad 3 = -A + B$

   $\qquad\qquad\qquad 0 = A + B$

   $\therefore A = -\tfrac{3}{2} \quad B = \tfrac{3}{2}$

$$\therefore \int \frac{3}{(x+1)(x-1)}\, dx = \int \frac{-\tfrac{3}{2}}{(x+1)}\, dx + \int \frac{\tfrac{3}{2}}{(x-1)}\, dx$$

$$= -\tfrac{3}{2} \log (x+1) + \tfrac{3}{2} \log (x-1) + C$$

$$= \tfrac{3}{2} [\log (x-1) - \log (x+1)] + C$$

$$= \tfrac{3}{2} \log \frac{(x-1)}{(x+1)} + C.$$

4. Let $\quad \dfrac{x}{(x-2)(x^2-3)} = \dfrac{A}{(x-2)} + \dfrac{Bx + C}{(x^2-3)}$

   then $\qquad\qquad x = A(x^2 - 3) + (Bx + C)(x - 2)$

   $\qquad\qquad\qquad = Ax^2 - 3A + Bx^2 + Cx - 2Bx - 2C$

   $\therefore \qquad\qquad A + B = 0$

   $\qquad\qquad -2B + C = 1$

   $\qquad\qquad -3A - 2C = 0$

and $\quad A = 2, \quad B = -2, \quad C = -3$

$$\therefore \int \frac{x}{(x-2)(x^2-3)} \, dx = \int \frac{2}{(x-2)} \, dx - \int \frac{2x+3}{(x^2-3)} \, dx$$

$$= 2 \log (x-2) - \log (x^2-3) - 3 \int \frac{1}{x^2-3} \, dx$$

$$= \log \left[\frac{(x-2)^2}{(x^2-3)} \right] + \frac{3}{2\sqrt{3}} \log \left(\frac{\sqrt{3}+x}{\sqrt{3}-x} \right) + D$$

This can be further simplified if necessary.

5. Let $4x + 3 = u$; then $du = 4\,dx$

$$\therefore \int (4x+3)^{10} \, dx = \int \frac{u^{10}}{4} \, du = \frac{u^{11}}{44} + C = \frac{(4x+3)^{11}}{44} + C.$$

6. Let $\log x = u$ and $dx = dv$

Then $\int \log x \, dx = (\log x)x - \int x \frac{1}{x} \, dx$

$$= x \log x - x + C$$

$$= x (\log x - 1) + C.$$

Exercises 7.4

1. (a) $\dfrac{\partial z}{\partial x} = y + 2xy + y^2$

$\dfrac{\partial z}{\partial y} = x + x^2 + 2xy$

(b) $\dfrac{\partial z}{\partial x} = 6xy + 4y^2 + 6y$

$\dfrac{\partial z}{\partial y} = 3x^2 + 8xy + 6x$

(c) $\dfrac{\partial z}{\partial x} = 2x \sin y - y^2 \sin x$

$\dfrac{\partial z}{\partial y} = x^2 \cos y + 2y \cos x$

(d) $\dfrac{\partial z}{\partial x} = (x + y)e^{x+y} + e^{x+y}$

$\dfrac{\partial z}{\partial y} = (x + y)e^{x+y} + e^{x+y}$

(e) $\dfrac{\partial z}{\partial x} = \dfrac{2x}{x^2 + y^2}$

$\dfrac{\partial z}{\partial y} = \dfrac{2y}{x^2 + y^2}$

(f) $\dfrac{\partial z}{\partial x} = \dfrac{(x - y) - (x + y)}{(x - y)^2} = \dfrac{-2y}{(x - y)^2}$

$\dfrac{\partial z}{\partial x} = \dfrac{(x - y) - (x + y)(-1)}{(x - y)^2} = \dfrac{2x}{(x - y)^2}$

(g) $\dfrac{\partial z}{\partial x} = \dfrac{(y + \sin x) - (x + \sin y)\cos x}{(y + \sin x)^2}$

$\dfrac{\partial z}{\partial y} = \dfrac{(y + \sin x)\cos y - (x + \sin y)}{(y + \sin x)^2}.$

2. Marginal product of capital is

$\dfrac{\partial q}{\partial K} = 4(K - 30) + 2(L - 20)$

$= 80 + 20 = 100$ when $K = 50, L = 30$

Marginal product of labour is

$\dfrac{\partial q}{\partial L} = 6(L - 20) + 2(K - 30) = 60 + 40 = 100$

when $K = 50$ and $L = 30$.

3. (a) Partial elasticity of demand for apples with respect to price of apples is

$\dfrac{\partial q}{\partial p_a}\dfrac{p_a}{q} = \dfrac{(-2p_a - p_o)\,p_a}{q} = -\dfrac{14(5)}{219} = -\dfrac{70}{219}$

when $p_a = 5$, $p_o = 4$ and $q = 219$.

(b) Partial elasticity of demand for apples with respect to price of oranges is

$$\frac{\partial q}{\partial p_0}\frac{p_0}{q} = \frac{(6 - p_a)p_0}{q} = \frac{4}{219}$$

when $p_a = 5$, $p_0 = 4$ and $q = 219$.

4. Since $Q = AL^\alpha K^\beta$ then $\dfrac{\partial Q}{\partial L} = \dfrac{\alpha Q}{L}$ and so $\dfrac{\partial Q L}{\partial L Q} = \alpha$.

5. Let $w = a_{11}x_1 + a_{12}x_2$ and $z = a_{21}x_1 + a_{22}x_2$ so that the first-order partial derivatives are

$$w_1 = a_{11}, w_2 = a_{12}, z_1 = a_{21}, z_2 = a_{22} \text{ and the Jacobian is}$$

$$|\mathbf{J}| = \begin{vmatrix} w_1 & w_2 \\ z_1 & z_2 \end{vmatrix} = \begin{vmatrix} a_{11} & a_{12} \\ a_{21} & a_{22} \end{vmatrix} = |\mathbf{A}| \text{ as required.}$$

Exercises 7.7

1. $dq = \dfrac{\partial q}{\partial L} dL + \dfrac{\partial q}{\partial C} dC$

   $= (3L^2 - 3 + 4C) dL + (4L - 2C) dC$

   $= 317(1) + 30(1) = 347$ when $L = 10$ and $C = 5$

   If $dC = 0$ and $dL = 1$, $dq = 317$.

2. $dq = \dfrac{\partial q}{\partial p_a} dp_a + \dfrac{\partial q}{\partial p_0} dp_0$

   $= (-2p_a - p_0) dp_a + (6 - p_a) dp_0$

   $= -14(1) + 1(1) = -13.$

3. $\dfrac{dz}{dt} = \dfrac{\partial z}{\partial x}\dfrac{dx}{dt} + \dfrac{\partial z}{\partial y}\dfrac{dy}{dt}$

   (a) $\dfrac{dx}{dt} = e^t(-\sin t) + e^t \cos t, \qquad \dfrac{dy}{dt} = e^t \cos t + e^t \sin t$

   $\dfrac{dz}{dt} = 2x(e^t)(\cos t - \sin t) + (-2y)(e^t)(\cos t + \sin t)$

(b) $\dfrac{dx}{dt} = 2te^t + t^2\,e^t, \quad \dfrac{dy}{dt} = \cos t$

$\dfrac{dz}{dt} = \left(\dfrac{1}{x+y}\right)(2t + t^2)e^t + \left(\dfrac{1}{x+y}\right)\cos t$

(c) $\dfrac{dx}{dt} = e^t, \quad \dfrac{dy}{dt} = \dfrac{1}{t}$

$\dfrac{dz}{dt} = (2xye^{x^2})e^t + (e^{x^2})\left(\dfrac{1}{t}\right)$

(d) $\dfrac{dx}{dt} = 1 - \dfrac{1}{t^2}, \quad \dfrac{dy}{dt} = 1 + \dfrac{1}{t^2}$

$\dfrac{dz}{dt} = (2x + y)\left(1 - \dfrac{1}{t^2}\right) + (x + 2y)\left(1 + \dfrac{1}{t^2}\right).$

4. (a) Since $Q_t = Ae^{rt}L_t^\alpha K_t^\beta$
   then $Q_L = \alpha Ae^{rt}L_t^{\alpha-1}K_t^\beta = \alpha Q/L$ and $Q_K = \beta Q/K$
   so $MRS = Q_L/Q_K = \alpha K/\beta L$ which is unaffected by r.

   (b) From above, $(K/L) = (\beta/\alpha)(MRS)$ and so the elasticity of substitution is

   $$\sigma = \frac{d(K/L)/(K/L)}{d(MRS)/MRS} = \frac{(\beta/\alpha)\,MRS}{K/L} = 1 \text{ as before}$$

   Therefore technical progress (of the form given in the question) does not affect the elasticity of substitution with the Cobb–Douglas production function.

5. Since $C = a + bY$, $C_Y = b$ and as $I = c - er$, $I_r = -e$
   (a) Taking the total derivative of $Y = C + I + G$,

   $$dY = C_Y\,dY + I_r\,dr + dG$$

   $$= b\,dY - e\,dr + dG \tag{1}$$

   Since G is fixed, $dG = 0$ and (1) can be written

   $$(1 - b)\,dY = -e\,dr \text{ and so } \frac{dY}{dr} = \frac{-e}{(1-b)} \text{ which is negative.}$$

   (b) Taking the total derivative of $M/P = fY - gr$ gives

$$\frac{dM}{P} - M\frac{dP}{P^2} = f\,dY - g\,dr \tag{2}$$

Now P and r are fixed and so $dr = 0 = dP$. Hence (1)

becomes $\qquad (1-b)\,dY = dG$ and (2) is $\dfrac{dM}{P} = f\,dY$

Therefore, $\qquad \dfrac{dM}{dG} = \dfrac{fP}{(1-b)}\dfrac{dY}{dY} = \dfrac{fP}{(1-b)}.$

Exercises 7.9

1. $\dfrac{dy}{dx} = -\dfrac{(\partial z/\partial x)}{(\partial z/\partial y)}$

(a) $\dfrac{dy}{dx} = -\dfrac{(-8xy^2 + 12y + 4)}{3y^2 - 8x^2y + 6x^2}$

(b) $\dfrac{dy}{dx} = -\dfrac{(2xy^2 + e^x)}{2x^2y - e^y}$

(c) $\dfrac{dy}{dx} = -\dfrac{[(3/(x+y)] + 8x}{[(3/(x+y)] - 2y}$

(d) $\dfrac{dy}{dx} = -\dfrac{[\cos(x+y) - 3y - 10y^2]}{\cos(x+y) - 3x - 20xy}.$

2. When $p = 2$, $q = 12.18$.
   By differentiating term by term

$$2q\frac{dq}{dp} = \frac{(2+p)(-2p) - (500 - p^2)}{(2+p)^2} + p\frac{dq}{dp} + q$$

Hence $\dfrac{dq}{dp} = \dfrac{1}{(2q-p)}\left[\dfrac{-4p - 500 - p^2}{(2+p)^2} + q\right]$

and elasticity when $p = 2$ and $q = 12.18$ is

$$\frac{p}{q}\frac{dq}{dp} = \frac{2}{12.18}\left(\frac{1}{22.36}\right)\left[-\frac{512}{16} + 12.18\right] = -0.15\,.$$

3. Differentiating $Q^2 = 8KL - 3Q + L^2 + 2K^2$ with respect to L: $2QQ_L = 8K - 3Q_L + 2L$

and $Q_L = (8K + 2L)/(2Q + 3)$

Similarly for K, $2QQ_K = 8L - 3Q_K + 4K$

and $Q_K = (8L + 4K)/(2Q + 3)$

The marginal rate of substitution is

$$MRS = \frac{Q_L}{Q_K} = \frac{(8K + 2L)}{(8L + 4K)} = \frac{4(K/L) + 1}{2(K/L) + 4} \tag{1}$$

For the elasticity of substitution (σ), we need $d(K/L)/d(MRS)$, and so rearranging (1),

$$[2(K/L) + 4] \, MRS = 4(K/L) + 1$$

Differentiating term by term, with respect to MRS,

$$[2(K/L) + 4] + MRS[2 \, d(K/L)/d(MRS)] = 4 \, d(K/L)/d(MRS)$$

and solving, $\dfrac{d(K/L)}{d(MRS)} = \dfrac{2(K/L) + 4}{4 - 2MRS}$

Therefore, $\sigma = \dfrac{d(K/L)}{d(MRS)} \dfrac{MRS}{K/L} = \dfrac{[(K/L) + 2] \, MRS}{[2 - MRS] \, K/L}$

$$= \frac{[(K/L) + 2][4(K/L) + 1]}{7(K/L)}.$$

Exercises 7.12

1. Replace x and y by tx and ty and show $f(tx, ty) = t^k f(x, y)$ where k is the degree of homogeneity: (a) $k = 3$; (b) $k = -1$; (c) $k = 3$.

2. Euler's theorem: $xz_x + yz_y = kz$
   For (a), $z_x = 3x^2 - 2xy$, $z_y = 9y^2 - x^2$
   hence $x(3x^2 - 2xy) + y(9y^2 - x^2) = 3f(x, y)$

   For (b), $z_x = \dfrac{(4x^3)y - 12x^2(xy + y^2)}{(4x^3)^2}$, $z_y = \dfrac{x + 2y}{4x^3}$

   hence $xz_x + yz_y = -f(x, y)$.

3. Own-price elasticity is

$$\frac{p}{q} \frac{\partial q}{\partial p} = \frac{p}{q}(4r - 4p)$$

Cross-elasticity is

$$\frac{r}{q}\frac{\partial q}{\partial r} = \frac{r}{q}(20r + 4p)$$

Sum is 2, the degree of homogeneity.

4. Replacing L and K by tL and tK, the first gives $t^{0.8}q$ and the second gives $t^{1.1}q$ as required.

5. The partial derivatives of $q = p_2 Y / p_1^2$ are

$$\frac{\partial q}{\partial p_2} = \frac{Y}{p_1^2}, \frac{\partial q}{\partial p_1} = \frac{-2p_2 Y}{p_1^3} \text{ and } \frac{\partial q}{\partial Y} = \frac{p_2}{p_1^2}$$

and so the elasticities are

$$\frac{p_2}{q}\frac{\partial q}{\partial p_2} = \frac{p_2 Y}{q p_1^2} = 1, \frac{p_1}{q}\frac{\partial q}{\partial p_1} = \frac{-p_1}{q}\frac{2p_2 Y}{p_1^3} = -2, \frac{Y}{q}\frac{\partial q}{\partial Y} = \frac{Y p_2}{q p_1^2} = 1$$

Notice that the sum of the elasticities is zero

The degree of homogeneity is found by multiplying each variable

by t to give $\dfrac{t p_2 t Y}{(t p_1)^2} = t^0 q$ and so is zero

Euler's theorem states that, for $q = f(p_1, p_2, Y)$

$$p_1 f_1 + p_2 f_2 + Y f_Y = kq$$

where k is the degree of homogeneity (zero here) and f_1 and f_2 indicate partial derivatives with respect to p_1 and p_2

Substituting in,

$$p_1 \frac{(-2p_2 Y)}{p_1^3} + p_2 \frac{(Y)}{p_1^2} + Y \frac{(p_2)}{p_1^2} = -2q + q + q = 0$$

as required.

6. For $q = AL^\alpha K^\beta$, $q_L = \alpha q / L$ and $q_K = \beta q / K$ so that the slope of an isoquant at (L^0, K^0) is $-q_L / q_K = -\alpha K^0 / \beta L^0$, and at (kL^0, kK^0) it is $-\alpha k K^0 / \beta k L^0 = -\alpha K^0 / L^0 \beta$. Therefore this Cobb–Douglas production function is homothetic.

Exercises 7.14

1. (a) $z = x^2 - xy + y^2$
   $$z_x = 2x - y \qquad z_y = -x + 2y$$

$$z_{xx} = 2 \qquad\qquad z_{yy} = 2$$
$$z_{xy} = -1$$

(b) $z = (x + y)^3$
$$z_x = 3(x + y)^2 \qquad z_y = 3(x + y)^2$$
$$z_{xx} = 6(x + y) \qquad z_{yy} = 6(x + y).$$
$$z_{xy} = 6(x + y)$$

2. Find when $z_x = 0 = z_y$ and check for
   Maximum: $z_{xx}z_{yy} > (z_{xy})^2$, $z_{xx} < 0$, $z_{yy} < 0$
   Minimum: $z_{xx}z_{yy} > (z_{xy})^2$, $z_{xx} > 0$, $z_{yy} > 0$
   Saddle point: $z_{xx}z_{yy} < (z_{xy})^2$, $z_{xx} = 0 = z_{yy}$

   (a) $z_x = 0 = z_y$ when $x = 0$, $y = 0$ and $x = \frac{2}{3}$, $y = -\frac{2}{3}$

   $$z_{xx} = 6x, \quad z_{yy} = 2, \quad z_{xy} = 2$$

   For $x = 0$, $y = 0$, tests fail – further investigation needed.

   For $x = \frac{2}{3}$, $y = -\frac{2}{3}$, minimum with $z = -\dfrac{4}{27}$.

   (b) $z_x = 0 = z_y$ when $x = \frac{1}{2}$ and $y = \frac{1}{4}$. Maximum value is $z = 1$.

3. Method is as in Problem 2 above.

   (a) $z_x = 0 = z_y$ when $x = 2$ and $y = 0.5$.
   $z_{xy} = 0$, $z_{xx} = 2$, $z_{yy} = 2$ so minimum value of $z = 45.75$.

   (b) $z_x = 0 = z_y$ when $x = \pm \sqrt{(0.67)}$, $y = 1.5$.
   Minimum is $z = 196.7$ when $x = \sqrt{(0.67)}$, $y = 1.5$

   (c) $z_x = 0 = z_y$ when $x = 3$, $y = 4$ and $x = -3$, $y = 4$.
   Minimum is $z = 180$ when $x = 3$, $y = 4$.

4. Here $q = 36K + 120L - 3K^2 - 4L^2 - 10 + 2KL$

   and so $q_L = 120 - 8L + 2K$

   $$q_K = 36 - 6K + 2L$$

   and setting these to zero and solving, $K = 12$, $L = 18$. Also, $q_{KK} = -6$, $q_{LL} = -8$, $q_{KL} = 2$ so that the conditions for a maximum are satisfied. The maximum is $q = 1286$.

5. Profits, π = total revenue − total costs
   $$= p_1q_1 + p_2q_2 - (q_1^2 + 3q_1q_2 + q_2^2)$$

   and substituting from the demand functions for p_1 and p_2,

   $$\pi = 40q_1 - q_1^2 + 60q_2 - 2q_2^2 - q_1^2 - 3q_1q_2 - q_2^2$$

$$= 40q_1 - 2q_1^2 + 60q_2 - 3q_2^2 - 3q_1q_2$$

The first-order derivatives are

$$\pi_1 = 40 - 4q_1 - 3q_2 \text{ and } \pi_2 = 60 - 6q_2 - 3q_1$$

Setting these to zero gives $q_1 = 4$, $q_2 = 8$ and the second-order derivatives are $\pi_{11} = -4$, $\pi_{22} = -6$, $\pi_{12} = -3$ and so we have a maximum value of profits of 320.

6. (a) With price discrimination, profits,

$$\pi = p_1q_1 + p_2q_2 - TC$$

$$= 105q_1 - 25q_1^2 + 50q_2 - 2q_2^2 - 10 - 5q_1 - 2q_2$$

$$= 100q_1 - 25q_1^2 + 48q_2 - 2q_2^2 - 10$$

The first-order derivatives are $\pi_1 = 100 - 50q_1$ and $\pi_2 = 48 - 4q_2$ and setting these to zero gives $q_1 = 2$, $q_2 = 12$
Here $\pi_{11} = -50$, $\pi_{22} = -4$, $\pi_{12} = 0$ and so we have a maximum for which $\pi = 378$. The prices are $p_1 = 55$ in the peak period and $p_2 = 26$ in the off-peak period.

(b) With no price discrimination the demand functions are $p = 105 - 25q_1$ and $p = 50 - 2q_2$. From these we obtain $q_1 = 4.2 - 0.04p$ and $q_2 = 25 - 0.5p$
Profits are given by $\pi = p(q_1 + q_2) - TC$ and so it is convenient to substitute for q_1 and q_2 in TC to get everything in terms of p:
$TC = 10 + 5(4.2 - 0.04p) + 2(25 - 0.5p)$ and so

$$\pi = p(29.2 - 0.54p) - 10 - 21 + 0.2p - 50 + p$$

$$= 30.4p - 0.54p^2 - 81$$

To maximise this we set $d\pi/dp = 0$, giving $0 = 30.4 - 1.08p$ or $p = 28.15$ and check that $d^2\pi/dp^2 = -1.08$ is negative as required. With $p = 28.15$ the profits are 346.9 and $q_1 = 3.074$, $q_2 = 10.93$. With price discrimination, $\pi = 378$, $q_1 = 2$, $p_1 = 55$, $q_2 = 12$ and $p_2 = 26$.

Exercises 7.18

1. Using a Lagrangian multiplier, as in section 7.16,
   $z' = 100x_1 - 2x_1^2 + 60x_2 - x_2^2 + \lambda(x_1 + x_2 - 41)$
   Setting $z_1' = 0$, $z_2' = 0$, $z_\lambda' = 0$ gives $x_1 = 20.33$
   $x_2 = 20.67$ which are the maximising values, $z = 2019.3$ (compare $z = 2000$ when the constraint was $x_1 + x_2, = 40$, and $\lambda = -20$).

Alternatively, substitute $x_1 = 41 - x_2$ in the expression for z and maximise as in section 5.7.

2. For the unrestricted case, $z_x = 0 = z_y$ gives $x = 2$, $y = 7$ and $z = 89$ as the minimum value (see section 7.13 for method).

   For the restricted case, using a Lagrangian multiplier, the first-order partial derivatives are zero when $x = 10/7$, $y = 43/7$, and the minimum value of z is 635/7.

3. (a) Using a Lagrangian multiplier shows that the first-order partial derivatives are zero when $x = 200/41$, $y = 160/41$, and the minimum value of z is 1600/41.

   (b) Using a Lagrangian multiplier shows that the first-order partial derivatives are zero when $x = 48/17$, $y = 36/17$, and the maximum value of z is 162/17.

   (c) Using a Lagrangian multiplier shows that the first-order partial derivatives set equal to zero give

   $$0 = 6xy^2 + \lambda, \quad 0 = 6x^2y + 3\lambda, \quad 0 = x + 3y - 18$$

   These are satisfied by three pairs of values:

   $$x = 0, \quad y = 6, \quad \text{minimum value } z = 0$$
   $$x = 18, \quad y = 0, \quad \text{minimum value } z = 0$$
   $$x = 9, \quad y = 3, \quad \text{maximum value } z = 2187$$

4. The constraint is $4l + 5c = 28$. Using a Lagrangian multiplier, we find that the first-order partial derivatives are zero when $l = 93/49$ and $c = 200/49$. The maximum value of x is 144.0.

5. The budget constraint is $100 = 5x + 10y$ and so let $z' = 6xy + \lambda(100 - 5x - 10y)$. The partial derivatives are $z'_x = 6y + 5\lambda$, $z'_y = 6x + 10\lambda$, $z'_\lambda = 100 - 5x - 10y$ and setting these to zero gives $x = 2y$ so that $x = 10$, $y = 5$ and $\lambda = 6$. To check whether this is a maximum, the Hessian is, with $z = 6xy$ and $u = -5x - 10y + 100$,

$$\begin{vmatrix} z_{xx} & z_{xy} & u_x \\ z_{yx} & z_{yy} & u_y \\ u_x & u_y & 0 \end{vmatrix} = \begin{vmatrix} 0 & 6 & -5 \\ 6 & 0 & -10 \\ -5 & -10 & 0 \end{vmatrix} = 600$$

and as this is positive, $x = 10$, $y = 5$ gives a maximum with $\mathbf{U} = 6xy = 300$.

Exercises 8.8

1. The situation shown in Table 8.4 can be expressed as

$$2x_1 + 3x_2 \leqslant 25$$

$$4x_1 + x_2 \leqslant 35$$

where x_1 and x_2 are the numbers of units of product A and B which are manufactured. x_1 and x_2 must both therefore be $\geqslant 0$

The contribution is given by $9x_1 + 7x_2$.

TABLE 8.4

	No. of hours required per unit of product		Total no. of hours available
	A	B	
Machine U	2	3	25
Machine V	4	1	35
Contribution per unit of product	9	7	

Slack variables x_3 and x_4 are inserted in the equations in order to obtain a first feasible solution. The Simplex tableau is as shown in Tableau 1.

Tableau 1

	x_1	x_2	x_3	x_4	p	p_i/a_{ij}	
x_3	2	3	1	0	25	25/2	
x_4	4	1	0	1	35	35/4	←departing variable
z	−9	−7	0	0	0		
	↑ entering variable						

Using the method of row and column operations successive tableaux are obtained as shown in Tableaux 2 and 3.

The final tableau shows that the maximum contribution is obtained by manufacturing 8 units of product A and 3 units of product B, in which case $z = 93$.

Check $8 \times 9 + 3 \times 7 = 72 + 21 = 93$.

Tableau 2

	x_1	x_2	x_3	x_4	p	p/a_{ij}	
x_3	0	$\frac{5}{2}$	1	$-\frac{1}{2}$	$\frac{15}{2}$	3	$\leftarrow$ departing variable
x_1	1	$\frac{1}{4}$	0	$\frac{1}{4}$	$\frac{35}{4}$	35	
z	0	$-\frac{19}{4}$	0	$\frac{9}{4}$	$\frac{315}{4}$		

$$\uparrow$$

entering variable

Tableau 3

	x_1	x_2	x_3	x_4	p
x_2	0	1	$\frac{2}{5}$	$-\frac{1}{5}$	3
x_1	1	0	$-\frac{1}{10}$	$\frac{3}{10}$	8
z	0	0	$\frac{19}{10}$	$\frac{26}{20}$	93

The value of increasing capacity on machine U by one unit is 19/10 and that on machine V is 26/20. These values are obtained directly from the bottom line of the final tableau. They can be verified by using the simplex method with the constraints adjusted by one unit of capacity for machines U and V separately.

2. (a) (i) $A = 9\frac{1}{2}$, $B = 2$. (ii) $A = 11$, $B = 1$
   (b) $A = 0$, $B = 8\frac{1}{3}$ (c) $A = 7\frac{1}{2}$, $B = 5$

3. (a) Let x_1 = number of units of A produced
   x_2 = number of units of B produced
   then the problem is to max $z = 3x_1 + 2x_2$

subject to $\quad x_1 + 4x_2 \leqslant 60 \quad$ (labour)

$$2x_1 + x_2 \leqslant 40 \quad \text{(land)}$$

$$3x_1 + x_2 \leqslant 50 \quad \text{(machines)}$$

and $x_1 \geqslant 0$, $x_2 \geqslant 0$, $x_3 \geqslant 0$

Introducing three slack variables, x_3, x_4 and x_5 the constraints are

$$x_1 + 4x_2 + x_3 = 60$$

$$2x_1 + x_2 + x_4 = 40$$

$$3x_1 + x_2 + x_5 = 50$$

The first tableau is:

	x_1	x_2	x_3	x_4	x_5	p
x_3	1	4	1	0	0	60
x_4	2	1	0	1	0	40
x_5	3	1	0	0	1	50
z	−3	−2	0	0	0	0

The first feasible solution is $x_3 = 60$, $x_4 = 40$, $x_5 = 50$
The entering variable is x_1 and the departing variable x_5
Dividing row 3 by 3 and combining the new row 3 with the other rows gives:

	x_1	x_2	x_3	x_4	x_5	p
x_3	0	11/3	1	0	−1/3	130/3
x_4	0	1/3	0	1	−2/3	20/3
x_1	1	1/3	0	0	1/3	50/3
z	0	−1	0	0	1	50

The entering variable is x_2 and the departing variable x_3
Dividing row 1 by 11/3 and combining the new row with the other rows gives:

	x_1	x_2	x_3	x_4	x_5	p
x_2	0	1	3/11	0	−1/11	130/11
x_4	0	0	−1/11	1	−21/33	90/33
x_1	1	0	−1/11	0	12/33	140/11
z	0	0	3/11	0	10/11	680/11

Since all the coefficients on the bottom row are positive the solution is $x_1 = 140/11$, $x_2 = 130/11$ and $z = 680/11$.

(b) The slack variable for land is x_4 and from the bottom row of the final tableau above the shadow price is zero. Therefore there is some land not being used (check the land constraint $2x_1 + x_2 \leqslant 40$). It is not worthwhile renting extra land since it would not increase profits.

4. (a) Let u_1, u_2 and u_3 be the number of units of bread, cheese and meat consumed. The total cost is $z = 2u_1 + 4u_2 + 5u_3$ which is to be minimised subject to the constraints:

$$\text{energy:} \quad u_1 + 2u_2 + \ u_3 \geqslant 10$$

$$\text{protein:} \quad u_1 + 3u_2 + 6u_3 \geqslant 15$$

and $u_1 \geqslant 0$, $u_2 \geqslant 0$, $u_3 \geqslant 0$.

(b) The dual problem is to max $z' = 10x_1 + 15x_2$ subject to

$$x_1 + \ x_2 \leqslant 2$$

$$2x_1 + 3x_2 \leqslant 4$$

$$x_1 + 6x_2 \leqslant 5$$

and $x_1 \geqslant 0$, $x_2 \geqslant 0$. Introducing three slack variables x_3, x_4 and x_5, the first tableau is

	x_1	x_2	x_3	x_4	x_5	p
x_3	1	1	1	0	0	2
x_4	2	3	0	1	0	4
x_5	1	6	0	0	1	5
z'	−10	−15	0	0	0	0

The first feasible solution is $x_3 = 2$, $x_4 = 4$, $x_5 = 5$
The entering variable is x_2 and the departing variable is x_5
Dividing row 3 by 6 and combining the new row with the other rows:

	x_1	x_2	x_3	x_4	x_5	p
x_3	5/6	0	1	0	−1/6	7/6
x_4	3/2	0	0	1	−1/2	3/2
x_2	1/6	1	0	0	1/6	5/6
z'	−15/2	0	0	0	5/2	12.5

The entering variable is x_1 and the departing variable is x_4
Dividing the second row by 3/2 and combining the new row with the other rows:

	x_1	x_2	x_3	x_4	x_5	p
x_3	0	0	1	−5/9	1/9	1/3
x_1	1	0	0	2/3	−1/3	1
x_2	0	1	0	−1/9	2/9	2/3
z'	0	0	0	5	0	20

The coefficients on the bottom row are positive and so the solution is $x_1 = 1$, $x_2 = 2/3$ and $z' = 20$. The only positive shadow price is for x_4, the slack variable associated with the price of cheese, and it would be advantageous to buy more cheese if the price is less than 5.

Exercises 9.4

1. Let $Y = VX$, so that $dY = V\,dX + X\,dV$
   Substituting and rearranging gives

   $$(a_1V - b_1 - a_2V^2 + b_2V)\,dX = (a_2VX - b_2X)\,dV$$

   or

   $$\frac{dX}{X} = \frac{(a_2V - b_2)\,dV}{a_1V - b_1 - a_2V^2 + b_2V}$$

   which can be integrated given the values of a_1, a_2, b_1 and b_2.

2. Let q be the quantity demanded at price p. Then we require

$$\frac{p}{q}\frac{dq}{dp} = -n$$

or

$$\frac{dq}{q} = -n\frac{dp}{p}$$

Integrating, we obtain

$$\log q = -n \log p + A' \quad \text{or} \quad q = Ap^{-n}$$

Thus $qp^n = A$ is the general form required.

3. Put $y = vx$, $dy = v\,dx + x\,dv$; then, on re-arranging,

$$\frac{v\,dv}{24 - 2v^2} = \frac{dx}{x}$$

or

$$-\tfrac{1}{4}\log(24 - 2v^2) = \log x + A'$$

This becomes

$$A = (24 - 2v^2)x^4 = 24x^4 - 2x^2\,y^2$$

and using $x = 2$, $y = 4$ gives $A = 256$.

4. Separable variables: $c = ax - \dfrac{bx^2}{2} + A$.

5. (a) $\dfrac{dp}{dt} = kap$ or $\dfrac{dp}{p} = ka\,dt$

Integrating gives

$$\log p = kat + A' \quad \text{or} \quad p = Ae^{kat}$$

(b) $\dfrac{dp}{dt} = k(ap + bp^2)$ or $\dfrac{dp}{p(a + bp)} = k\,dt$

Using partial fractions (see Section 6.14), we have

$$\frac{dp}{ap} - \frac{b\,dp}{a(a + bp)} = k\,dt$$

or

$$\frac{1}{a}\log p - \frac{1}{a}\log(a + bp) = kt + A'$$

$$\frac{p}{(a + bp)} = Ae^{akt}.$$

6. Substituting gives

$$b\frac{dY}{dt} = aY \quad \text{or} \quad b\frac{dY}{Y} = a\ dt$$

Hence, $\quad b \log Y = at + A' \quad$ or $\quad Y^b = Ae^{at}$.

7. (a) $x\ dy = (x - y)\ dx$

Put $y = vx$, $dy = v\ dx + x\ dv$, then $(-1 + 2v)\ dx = -x\ dv$

or $\qquad\qquad \dfrac{dx}{x} + \dfrac{dv}{2v - 1} = 0$

so $\quad \log x + \frac{1}{2} \log (2v - 1) = A' \quad$ or $\quad 2xy - x^2 = A$.

(b) Let $X = y + 1$ and $Y = 2y + x$; then

$$\frac{dY}{dX} = \frac{dY/dx}{dX/dx} = \frac{2(dy/dx) + 1}{dy/dx} = \frac{2(X/Y) + 1}{X/Y}$$

and $\qquad X\ dY = (2X + Y)\ dX$

Let $Y = VX$, $dY = V\ dX + X\ dV$

Hence $2\ dX = X\ dV$ and $2 \log X = V + A$

Thus $\qquad 2 \log (y + 1) = \left(\dfrac{2y + x}{y + 1}\right) + A \cdot$

(c) Rewriting as

$$\frac{dy}{dx} + 2y = x(1 + x)$$

we have e^{2x} as an integrating factor and the integral of the left hand side is ye^{2x}. The right hand side is

$$\int e^{2x} (x + x^2)\ dx = \frac{(x + x^2)e^{2x}}{2} - \int \frac{e^{2x}}{2}(1 + 2x)\ dx$$

$$= \frac{(x + x^2)e^{2x}}{2} - \frac{(1 + 2x)e^{2x}}{4} + \int \frac{e^{2x}}{4}(2)\ dx$$

$$= \left(\frac{2x^2 - 1}{4}\right)e^{2x} + \frac{e^{2x}}{4} + A = \frac{x^2 e^{2x}}{2} + A$$

Hence the solution is

$$ye^{2x} = \frac{x^2 e^{2x}}{2} + A .$$

(d)   If

$$\frac{dy}{dx} + 4y = x^3$$

then e^{4x} is an integrating factor and integrating gives

$$ye^{4x} = \int x^3 e^{4x} \, dx = e^{4x} \left[\frac{x^3}{4} - \frac{3x^2}{16} + \frac{3x}{32} - \frac{3}{128} \right] + A .$$

8. Since the market clears, $q^s = q^d$ and so

$$-a + bp = e - fp + \frac{dp}{dt} \quad \text{or} \quad \frac{dp}{dt} = (b + f)p - (a + e)$$

or

$$\frac{dp}{(b + f)p - (a + e)} = dt$$

For convenience, multiply each side by $(b + f)$ and integrating gives

$$\log \, [(b + f)p - (a + e)] = (b + f)t + \log A$$

or

$$(b + f)p - (a + e) = Ae^{(b+f)t}$$

and so

$$p = [(a + e) + Ae^{(b+f)t}]/(b + f)$$

Exercises 9.6

In each case put $y = e^{mx}$ etc. and form the reduced equation $am^2 + bm + c$. The roots are m_1 and m_2

1. $m_1 = 4,$ $m_2 = -3,$ $y = k_1 e^{4k} + k_2 e^{-3x}$

2. $m_1 = 1,$ $m_2 = 2,$ $y = k_1 e^{2x} + k_2 e^{x}$

3. $m_1 = 3,$ $m_2 = 3,$ $y = (k_1 + k_2 x)e^{3x}$

4. $m_1 = 1,$ $m_2 = 1,$ $y = (k_1 + k_2 x)e^{x}$

5. $m_1 = 4,$ $m_2 = -4,$ $y = k_1 e^{4x} + k_2 e^{-4x}$

6. $m_1 = 4i,$ $m_2 = -4i,$ $y = k_3 \cos 4x + k_4 \sin 4x$
                                                   $= A \cos (4x - \epsilon)$

7. $m_1 = -1 + 3i$, $m_2 = -1 - 3i$, $\quad y = e^{-x}(k_3 \cos 3x + k_4 \sin 3x)$
$$= Ae^{-x} \cos(3x - \epsilon)$$

Exercises 9.8

1. (a) Complementary function: $m_1 = 8$, $m_2 = 2$
$$y = k_1 e^{8x} + k_2 e^{2x}$$

Particular integral: try $y = K_1 + K_2 x$
then $y = 5/64 + x/8$
Solution is
$$y = k_1 e^{8x} + k_2 e^{2x} + \frac{5}{64} + \frac{x}{8}$$

The initial conditions give $k_1 = 0.1979$, $k_2 = -0.3542$

(b) Complementary function [from Exercise 9.5, Question 5],
$$y = k_1 e^{4x} + k_2 e^{-4x}$$

Particular integral: try $y = ae^x$, then $y = -e^x/15$.
Solution is
$$y = -\frac{e^x}{15} + k_1 e^{4x} + k_2 e^{-4x}$$

The initial conditions give $k_1 = \frac{3}{4}$, $k_2 = \frac{1}{4}$.

(c) Complementary function [from Exercise 9.5, Question 4],
$$y = (k_1 + k_2 x)e^x$$

Particular integral: try $y = ae^{2x}$; then $y = e^{2x}$
Solution is $y = (k_1 + k_2 x)e^x + e^{2x}$
and the initial conditions give $k_1 = 1$, $k_2 = -1$.

(d) Complementary function: $m_1 = -1 + 2i$, $m_2 = -1 - 2i$:
$$y = Ae^{-x} \cos(2x - \epsilon)$$

Particular integral: try $y = ax + b$; then $y = \frac{x}{5} - \frac{2}{25}$

Solution is $y = \frac{x}{5} - \frac{2}{25} + Ae^{-x} \cos(2x - \epsilon)$

The initial conditions give $A \cos(-\epsilon) = 0$
$$1 = -2A \sin(-\epsilon)$$

and hence $\epsilon = -\pi/2$ and $A = -\frac{1}{2}$.

(e) Complementary function: $m_1 = 0$, $m_2 = 1$

$$y = A + Be^x$$

Particular integral: try $y = a \cos x + b \sin x$; then

$$y = -\cos x - \sin x.$$

Solution is $y = A + Be^x - \cos x - \sin x$
Initial conditions give $A = -1$, $B = 3$.

2. Solve homogeneous part by putting $y = e^{mx}$ etc; then $m_1 = 2$, $m_2 = -3$ and

$$y = Ae^{2x} + Be^{-3x}$$

Particular integral: try $y = a + bx + cx^2$; then
$a = -7/6$, $b = -1$, $c = -3$.
Solution is $y = Ae^{2x} + Be^{-3x} - (\frac{7}{6}) - x - 3x^2$
Initial conditions give $A = \frac{11}{10}$, $B = \frac{1}{15}$.

Exercises 9.12

1. Differentiating (1),

$$\frac{dp}{dt} = -b\frac{dU}{dt} + f\frac{dp^e}{dt} \tag{4}$$

and using (2) and (3),

$$\frac{dp}{dt} = be(m - p) + fk(p - p^e)$$

Differentiating:

$$\frac{d^2p}{dt^2} = -be\frac{dp}{dt} + fk\frac{dp}{dt} - fk\frac{dp^e}{dt}$$

and using (4) to eliminate dp^e/dt,

$$\frac{d^2p}{dt^2} = -be\frac{dp}{dt} + fk\frac{dp}{dt} - k\left\{\frac{dp}{dt} + b\frac{dU}{dt}\right\}$$

Using (3) to remove dU/dt and simplifying,

$$\frac{d^2p}{dt^2} + (be + k[1 - f])\frac{dp}{dt} + bekp = bekm$$

which reduces to (6) in the text when $f = 1$.

2. (a) $q^s = q^d$ gives $f\dfrac{d^2p}{dt^2} + c\dfrac{dp}{dt} + (a + h)p = b + g$

For the homogeneous equation, putting $p = e^{mt}$ gives $fm^2 + cm + (a + h) = 0$ and the roots are complex, giving a cyclical solution, if $c^2 - 4(a + h)f < 0$.

(b) Putting the values into the equation,

$$\frac{d^2p}{dt^2} + 8\frac{dp}{dt} + 15p = 1800$$

and the auxiliary equation is $m^2 + 8m + 15 = 0$ which has the roots -3 and -5. The particular integral is from putting $p = K$ and $K = 1800/15 = 120$ and so the solution is $p = Ae^{-3t} + Be^{-5t} + 120$. The condition $t = 0$, $p = 130$ gives $130 = A + B + 120$ or $A + B = 10$. The condition $t = 0$, $dp/dt = -2$ gives $-2 = -3A - 5B$. Solving, $A = 24$, $B = -14$ and $p = 24e^{-3t} - 14e^{-5t} + 120$. As t increases the value of p quickly approaches 120.

Exercises 10.7

1. The general solution of $Y_{t+1} = \lambda Y_t + K$ is

$$Y_t = A\lambda^t + \frac{K}{1 - \lambda}$$

where A is a constant

(a) $\lambda = 1.5$, $K = 3$, $Y_t = A(1.5^t) - 6$
Initial condition gives $A = 8$

t	0	1	2	3	4	5
Y_t	2	6	12	21	34.5	54.75

(b) $\lambda = 0.9$, $K = 2$, $Y_t = A(0.9^t) + 20$
Initial condition gives $A = -17$

t	0	1	2	3	4	5
Y_t	3	4.7	6.23	7.61	8.85	9.96

(c) $\lambda = 1$, $K = 0$, $Y_t = A$.
Initial condition gives $A = 6$

t	0	1	2	3	4	5
Y_t	6	6	6	6	6	6

(d) $\lambda = 1.1$, $K = 4$, $Y_t = A(1.1^t) - 40$
Initial condition gives $A = 41$

t	0	1	2	3	4	5
Y_t	1	5.1	9.61	14.57	20.03	26.03

2. Let Y_t be the value of the savings after t years; then $Y_0 = 15$, $Y_1 = 1.05(15) + 15$ and in general $Y_{t+1} = 1.05 \, Y_t + 15$. The solution is $Y_t = A(1.05)^t - 300$ and $Y_0 = 15$ gives $A = 315$. $Y_{21} = 535.8$.

3. The difference equation is $p_t = -1.5p_{t-1} + 3$ and the solution is $p_t = -\frac{1}{5}(-1.5)^t + 1.2$, which is 'exploding' since $1.5 > 1$.

4. The difference equation is $p_t = 0.5p_{t-1} + 3$ with the solution $p_t = -(-0.5)^t + 2$, which is stable since $0.5 < 1$. The equilibrium price is 2.

5. The difference equation is $p_t = -p_{t-1} + 8/3$ with the solution $p_t = [2(-1)^t + 4]/3$ which oscillates since $|-1| = 1$.

6. The difference equation is $0.9Y_t = 0.3Y_t - 0.3Y_{t-1}$ or $Y_t = -0.5Y_{t-1}$, and the solution is $Y_t = 100(-0.5)^t$.

7. Substituting into the pricing equation:

$$P_{t+1} = p_t - 0.3(0.4p_t - 50 - 150 + 0.6p_t)$$
$$= 0.7p_t + 60$$

The solution is $p_t = A(0.7^t) + 60/(1 - 0.7) = A(0.7^t) + 200$
As $t \to \infty$, p converges on the value 200.

8. Substituting into the first equation:

$$Y_t = bY_{t-1} + c + a(Y_t - Y_{t-1})$$

or
$$Y_t = \frac{(b - a)}{(1 - a)} Y_{t-1} + \frac{c}{(1 - a)}$$

For stability, the coefficient on Y_{t-1} has to lie between 1 and -1 or $-1 < (b - a)/(1 - a) < 1$.

Exercises 10.12

1. Difference equation is $Y_t = (c + b) Y_{t-1} - b Y_{t-2}$ and putting $Y_t = \lambda^t$ gives the characteristic equation $\lambda^2 = (c + b) \lambda - b$

   (a) Roots are $\lambda_1 = 0.7 + 0.1i$, $\quad \lambda_2 = 0.7 - 0.1i$
   (b) Roots are $\lambda_1 = 0.8$, $\qquad\qquad \lambda_2 = 0.8$
   (c) Roots are $\lambda_1 = 0.845$, $\qquad\quad \lambda_2 = 0.355$.

2. Solve homogeneous part by putting $Y_t = \lambda_t$ and obtaining the roots λ_1 and λ_2 of the characteristic equation. The equilibrium or particular solution is found by trying expressions of the same form as $f(t)$.

   (a) $\lambda_1 = 2$, $\lambda_2 = -1$. Try $Y_t = Z$ as particular solution then $Y_t = -1.5$. Solution is $Y_t = A(2^t) + B(-1)^t - 1.5$ and initial values give $A = 7/3$, $B = 7/6$.

   (b) $\lambda_1 = 2$, $\lambda_2 = 1$. Try $Y_t = Z$ as particular solution – fails. Try $Y_t = Zt$; $Y_{t+1} = Zt + Z$, $Y_{t+2} = Zt + 2Z$, then $Z = -4$. Solution is $Y_t = A(2^t) + B - 4t$ and initial conditions give $A = 5$, $B = -4$.

   (c) $\lambda_1 = 2$, $\lambda_2 = 2$. Try $Y_t = Z(3^t)$; then $Z = 1$. Solution is $Y_t = (k_1 + k_2 t)(2^t) + 3^t$ and initial conditions give $k_1 = 2$, $k_2 = -1.5$.

   (d) $\lambda_1 = 2$, $\lambda_2 = 1$. Try $Y_t = Zt(2^t)$; then $Z = +1.5$. Solution is $Y_t = (A(2^t) + B + 1.5t(2^t)$ and initial conditions give $A = 1$, $B = 0$.

   (e) $\lambda_1 = -0.5$, $\lambda_2 = -0.5$. Try $Y_t = Z_0 + Z_1 t$; then $Z_0 = -\frac{2}{9}$, $Z_1 = \frac{1}{3}$. Solution is $Y_t = (k_1 + k_2 t)(-0.5)^t - \frac{2}{9} + (t/3)$ and initial conditions give $k_1 = -88/9$, $k_2 = 8$.

   (f) $\lambda_1 = -2 + i$, $\lambda_2 = -2 - i$ hence $a = -2$, $b = 1$, $r = \sqrt{5}$ and $\cos \theta = -2/\sqrt{5}$ so $\theta = -26° \, 34' = -0.4636$ radians. Try $Y_t = Z$; then $Z = 0.4$ and the solution is

   $$Y_t = A(\sqrt{5})^t \cos(-0.4636t - \epsilon) + 0.4$$

(g) $\lambda_1 = 2 + 2i$, $\lambda_2 = 2 - 2i$, hence $a = 2$, $b = -2$, $r = \sqrt{8}$ and $\cos \theta = 2/\sqrt{8} = 0.7071$, so $\theta = \pi/4$. Try $Y_t = k_1 + k_2 t$ and $k_1 = 2/25$, $k_2 = \frac{1}{5}$. The solution is $Y_t = A(\sqrt{8})^t \cos [\pi(t/4) - \epsilon] + (2/25) + t/5$).

3. Equation is $(1 - \alpha) Y_t = (\beta + \gamma) Y_{t-1} - \gamma Y_{t-2} + \delta$

(a) $\lambda_1 = 0.625 + 0.599i$, $\lambda_2 = 0.625 - 0.599i$
   hence $a = 0.625$, $b = 0.599$, $r = 0.8657$, $\theta = 0.7642$ radians.
   Try $Y_t = Z$ and $Z = 5$.
   Solution is $Y_t = A(0.8657)^t \cos (0.7642t - \epsilon) + 5$.

(b) $\lambda_1 = 0.4 + 0.49i$, $\lambda_2 = 0.4 - 0.49i$; hence $a = 0.4$, $b = 0.49$,
   $r = 0.6324$, $\theta = 0.886$ radians. Try $Y_t = Z$ and $Z = 3.33$. The
   solution is $Y_t = A(0.6324)^t \cos (0.886t - \epsilon) + 3.33$.

4. The method is to use all three equations and to eliminate U and p^e.
   Differencing (1),

$$p_t - p_{t-1} = -b(U_t - U_{t-1}) + f(p_t^e - p_{t-1}^e)$$

Substituting from (3),

$$p_t - p_{t-1} - be(m - p_t) = f(p_t^e - p_{t-1}^e)$$

Using (2), $p_t - p_{t-1} - be(m - p_t) = fk(p_{t-1} - p_{t-1}^e)$
and since, from (1), $fp_t^e = p_t - a + bU_t + c$

$$p_t - p_{t-1} - be(m - p_t) = fkp_{t-1} - k(p_{t-1} - a + bU_{t-1} + c)$$

Simplifying: $(1 + be)p_t + (k - kf - 1)p_{t-1} - bem - ak + ck$

$$= -bkU_{t-1}$$

Differencing again to eliminate U,

$$(1 + be)(p_t - p_{t-1}) + (k - kf - 1)(p_{t-1} - p_{t-2}) = -bk(U_{t-1} - U_{t-2})$$

Using (3) and simplifying,

$$(1 + be)p_t - (kf - k + 2 + be - bke)p_{t-1} + (1 - k + kf)p_{t-2}$$

$$= bkem$$

Comparing this with the result in the text:

$$(1 + be)p_{t+1} - (2 + be[1 - k])p_t + p_{t-1} = bkem$$

it can be seen that they are the same if $f = 1$, but if this is not the case then the dynamics are affected.

5. Substituting in the national income identity,

$$Y_t = aY_{t-1} + b(Y_t - Y_{t-1}) + cY_{t-1} + dG_{t-1}$$

or $$(1 - b)Y_t = (a - b + c)Y_{t-1} + dG_{t-1}$$

From the national income identity, $G_t = Y_t - C_t - I_t$ and so $G_{t-1} = Y_{t-1} - C_{t-1} - I_{t-1} = Y_{t-1} - aY_{t-2} - bY_{t-1} + bY_{t-2}$ and so $(1 - b)Y_t = (a - b + c)Y_{t-1} + d(1 - b)Y_{t-1} + d(b - a)Y_{t-2}$ or $(1 - b)Y_t = (a - b + c + d - bd)Y_{t-1} - (b - a)dY_{t-2}$ which is the required equation.

(a) If $d = 0$ the equation is $(1 - b)Y_t = (a - b + c)Y_{t-1}$ as required.

(b) With $a = 0.6$, $b = 2$, $c = 0.1$ and $d = 0.5$ the equation is $-Y_t = -1.8Y_{t-1} - 0.7Y_{t-2}$ or $Y_t - 1.8Y_{t-1} - 0.7Y_{t-2} = 0$. The roots of the characteristic equation are -0.33 and 2.13 and so the general solution is $Y_t = A(-0.33)^t + B(2.13)^t$. The initial conditions $Y_0 = 1$ and $Y_1 = 1$ give $1 = A + B$ and $1 = -0.33A + 2.13B$. The solution is $A = 0.45$ and $B = 0.55$ so that $Y_t = 0.45(-0.33)^t + 0.55(2.13)^t$. As t increases the first term will tend to zero and the second term approach infinity.

Further Reading

This book is intended as an introduction to the use of mathematics in economics and business. As a supplement to the material covered here, and for those who like to use a work-book with many worked problems to accompany a text, the following excellent book is recommended:

A. J. Mabbett, *Work Out Mathematics for Economists* (Basingstoke: Macmillan, 1986).

Once the basics have been mastered, the student should be able to benefit from the many more advanced books which are available for those wishing to study the subject seriously. Two recommended mathematics books for economists are:

A. C. Chiang, *Fundamental Methods of Mathematical Economics* (New York: McGraw-Hill, 1984) 3rd edn.

R. C. Read, *A Mathematical Background for Economists and Social Scientists* (Englewood Cliffs, N.J.: Prentice-Hall, 1972).

There are a large number of books which use mathematics as a means of developing economic theory. The following are suitable for intermediate level students:

A. Koutsoyiannis, *Modern Microeconomics* (Basingstoke: Macmillan, 1984).

J. M. Henderson and R. E. Quandt, *Microeconomic Theory* (New York: McGraw-Hill, 1980) 3rd edn.

R. Dornbusch and S. Fischer, *Macroeconomics* (New York: McGraw-Hill, 1987) 4th edn.

W. M. Scarth, *Macroeconomics* (Toronto: Harcourt Brace Jovanovich, 1988).

One of the major applications of mathematics in economics is in the area of economic modelling and forecasting. Introductions to these topics are given in:

K. Holden, D. A. Peel and J. L. Thompson, *Economic Forecasting: an introduction* (Cambridge: Cambridge University Press, 1990).

R. S. Pindyck and D. L. Rubinfeld, *Econometric Models and Economic Forecasts* (New York: McGraw-Hill, 1990) 3rd edn.

Applications of quantitative methods in marketing and marketing research are the subject of many books, and two recent ones are:

G. Oliver, *Marketing Today* (New York: Prentice-Hall, 1990) 3rd edn.

C. McDaniel and R. Gates, *Contemporary Marketing Research* (St Paul, Minnesota: West Publishing, 1991).

For details of the formulae for the solution of cubic equations, see:

J. Parry Lewis, *An Introduction to Mathematics for Students of Economics* (Basingstoke: Macmillan, 1969) 2nd edn.

For both cubic and quartic equations, see:

F. Gerrish, *Pure Mathematics*, vol. 2 (Cambridge: Cambridge University Press, 1960).

Further applications of mathematics to finance, and to the use of mortality tables, are contained in:

F. Ayres, *Theory and Problems of Finance* (New York: Schaum, 1963).

For a detailed treatment of the properties of matrices, see:

G. Hadley, *Linear Algebra* (Reading, Mass: Addison-Wesley, 1961).

For the application of matrices in econometrics, see:

J. Johnston, *Econometric Methods* (New York: McGraw-Hill, 1984) 3rd edn.

Introductions to the uses of input–output analysis in regional economics are provided by:

H. Armstrong and J. Taylor, *Regional Economics and Policy* (Deddington: Philip Allan, 1985).

P. N. Balchin and G. H. Bull, *Regional and Urban Economics* (London: Harper & Row, 1987).

A more general treatment of input–output analysis is given by:

R. O'Connor and E. W. Henry, *Input–Output Analysis and its Applications* (London: Griffen, 1975).

The inventory models discussed in Chapter 5 assume a known demand. The extension of the model to probabilistic demand is covered in books on operations research, for example:

D. R. Anderson, D. J. Sweeney and T. A. Williams, *An Introduction to Management Science* (St Paul, Minnesota: West Publishing, 1991).

This also discusses linear programming in detail as well as other applications of mathematics to management problems. A general treatment of difference and differential equations is provided by:

W. J. Baumol, *Economic Dynamics* (New York: Macmillan, 1959) 2nd edn.

Finally, since any list of books inevitably dates, the student is advised to look at the current journals in economics and business to see the latest applications. A start could be made with *The American Economic Review*, *The Economic Journal*, *Management Science*, *The Journal of Marketing* and *The Journal of Marketing Research*.

Index